AF712408

Practical Biomedical Signal Analysis Using MATLAB®

Series in Medical Physics and Biomedical Engineering

Series Editors: John G Webster, Slavik Tabakov, Kwan-Hoong Ng

Other recent books in the series:

Physics for Diagnostic Radiology, Third Edition
P P Dendy and B Heaton (Eds)

Nuclear Medicine Physics
J J Pedroso de Lima (Ed)

Handbook of Photonics for Biomedical Science
Valery V Tuchin (Ed)

Handbook of Anatomical Models for Radiation Dosimetry
Xie George Xu and Keith F Eckerman (Eds)

Fundamentals of MRI: An Interactive Learning Approach
Elizabeth Berry and Andrew J Bulpitt

Handbook of Optical Sensing of Glucose in Biological Fluids and Tissues
Valery V Tuchin (Ed)

Intelligent and Adaptive Systems in Medicine
Oliver C L Haas and Keith J Burnham

A Introduction to Radiation Protection in Medicine
Jamie V Trapp and Tomas Kron (Eds)

A Practical Approach to Medical Image Processing
Elizabeth Berry

Biomolecular Action of Ionizing Radiation
Shirley Lehnert

An Introduction to Rehabilitation Engineering
R A Cooper, H Ohnabe, and D A Hobson

The Physics of Modern Brachytherapy for Oncology
D Baltas, N Zamboglou, and L Sakelliou

Electrical Impedance Tomography
D Holder (Ed)

Contemporary IMRT
S Webb

Series in Medical Physics and Biomedical Engineering

Practical Biomedical Signal Analysis Using MATLAB®

K J Blinowska
University of Warsaw, Poland

J Żygierewicz
University of Warsaw, Poland

CRC Press
Taylor & Francis Group
Boca Raton London New York

CRC Press is an imprint of the
Taylor & Francis Group, an **informa** business
A TAYLOR & FRANCIS BOOK

MATLAB® and Simulink® are trademarks of The MathWorks, Inc. and are used with permission. The MathWorks does not warrant the accuracy of the text or exercises in this book. This book's use or discussion of MATLAB® and Simulink® software or related products does not constitute endorsement or sponsorship by The MathWorks of a particular pedagogical approach or particular use of the MATLAB® and Simulink® software.

CRC Press
Taylor & Francis Group
6000 Broken Sound Parkway NW, Suite 300
Boca Raton, FL 33487-2742

© 2012 by Taylor & Francis Group, LLC
CRC Press is an imprint of Taylor & Francis Group, an Informa business

No claim to original U.S. Government works

Printed in the United States of America on acid-free paper
Version Date: 20110805

International Standard Book Number: 978-1-4398-1202-0 (Hardback)

This book contains information obtained from authentic and highly regarded sources. Reasonable efforts have been made to publish reliable data and information, but the author and publisher cannot assume responsibility for the validity of all materials or the consequences of their use. The authors and publishers have attempted to trace the copyright holders of all material reproduced in this publication and apologize to copyright holders if permission to publish in this form has not been obtained. If any copyright material has not been acknowledged please write and let us know so we may rectify in any future reprint.

Except as permitted under U.S. Copyright Law, no part of this book may be reprinted, reproduced, transmitted, or utilized in any form by any electronic, mechanical, or other means, now known or hereafter invented, including photocopying, microfilming, and recording, or in any information storage or retrieval system, without written permission from the publishers.

For permission to photocopy or use material electronically from this work, please access www.copyright.com (http://www.copyright.com/) or contact the Copyright Clearance Center, Inc. (CCC), 222 Rosewood Drive, Danvers, MA 01923, 978-750-8400. CCC is a not-for-profit organization that provides licenses and registration for a variety of users. For organizations that have been granted a photocopy license by the CCC, a separate system of payment has been arranged.

Trademark Notice: Product or corporate names may be trademarks or registered trademarks, and are used only for identification and explanation without intent to infringe.

Visit the Taylor & Francis Web site at
http://www.taylorandfrancis.com

and the CRC Press Web site at
http://www.crcpress.com

Contents

About the Series

The *Series in Medical Physics and Biomedical Engineering* describes the applications of physical sciences, engineering, and mathematics in medicine and clinical research.

The series seeks (but is not restricted to) publications in the following topics:

- Artificial organs
- Assistive technology
- Bioinformatics
- Bioinstrumentation
- Biomaterials
- Biomechanics
- Biomedical engineering
- Clinical engineering
- Imaging
- Implants
- Medical computing and mathematics
- Medical/surgical devices
- Patient monitoring
- Physiological measurement
- Prosthetics
- Radiation protection, health physics, and dosimetry
- Regulatory issues
- Rehabilitation engineering
- Sports medicine
- Systems physiology
- Telemedicine
- Tissue engineering
- Treatment

The *Series in Medical Physics and Biomedical Engineering* is an international series that meets the need for up-to-date texts in this rapidly developing field. Books in the series range in level from introductory graduate textbooks and practical handbooks to more advanced expositions of current research.

The *Series in Medical Physics and Biomedical Engineering* is the official book series of the International Organization for Medical Physics.

The International Organization for Medical Physics

The International Organization for Medical Physics (IOMP), foundedin 1963, is a scientific, educational, and professional organization of 76 national adhering organizations, more than 16,500 individual members, several corporate members, and four international regional organizations.

IOMP is administered by a council, which includes delegates from each of the adhering national organizations. Regular meetings of the council are held electronically as well as every three years at the World Congress on Medical Physics and Biomedical Engineering. The president and other officers form the executive committee, and there are also committees covering the main areas of activity, including education and training, scientific, professional relations, and publications.

Objectives

- To contribute to the advancement of medical physics in all its aspects
- To organize international cooperation in medical physics, especially in developing countries
- To encourage and advise on the formation of national organizations of medical physics in those countries which lack such organizations

Activities

Official journals of the IOMP are *Physics in Medicine and Biology* and *Medical Physics and Physiological Measurement*. The IOMP publishes a bulletin, *Medical Physics World*, twice a year, which is distributed to all members.

A World Congress on Medical Physics and Biomedical Engineering is held every three years in cooperation with IFMBE through the International Union for Physics and Engineering Sciences in Medicine (IUPESM). A regionally based international conference on medical physics is held between world congresses. IOMP also sponsors international conferences, workshops, and courses. IOMP representatives contribute to various international committees and working groups.

The IOMP has several programs to assist medical physicists in developing countries. The joint IOMP Library Programme supports 69 active libraries in 42 developing countries, and the Used Equipment Programme coordinates equipment donations. The Travel Assistance Programme provides a limited number of grants to enable physicists to attend the world congresses. The IOMP Web site is being developed to include a scientific database of international standards in medical physics and a virtual education and resource center.

Information on the activities of the IOMP can be found on its Web site at www.iomp.org.

Preface

This book is intended to provide guidance for all those working in the field of biomedical signal analysis and application, in particular graduate students, researchers at the early stages of their careers, industrial researchers, and people interested in the development of the methods of signal processing. The book is different from other monographs, which are usually collections of papers written by several authors. We tried to present a coherent view on different methods of signal processing in the context of their application. Not only do we wish to present the current techniques of biomedical signal processing, we also wish to provide a guidance for which methods are appropriate for the given task and given kind of data.

One of the motivations for writing this book was our longstanding experience in reviewing manuscripts submitted to journals and to conference proceedings, which showed how often methods of signal processing are misused. Quite often, methods, which are sophisticated but at the same time non-robust and prone to systematic errors, are applied to tasks where simpler methods would work better. In this book we aim to show the advantages and disadvantages of different methods in the context of their applications.

In the first part of the book we describe the methods of signal analysis, including the most advanced and newest methods, in an easy and accessible way. We omitted proofs of the theorems, sending the reader to the more specialized mathematical literature when necessary. In order to make the book a practical tool we refer to MATLAB® routines when available and to software freely available on the Internet.

In the second part of the book we describe the application of the methods presented in the first part of the book to the different biomedical signals: electroencephalogram (EEG), electrocorticogram (ECoG), event-related potential (ERP), electrocardiogram (ECG), heart rate variability signal (HRV), electromyograms (EMG), electroenterograms (EEnG), and electrogastrograms (EGG). The magnetic fields connected with the activity of brain (MEG) and heart (MCG) are considered as well. Methods for acoustic signals—phonocardiograms (PCG) and otoacoustic emissions (OAE) analysis—are also described. Different approaches to solving particular problems are presented with indication to which methods seem to be most appropriate for the given application. Possible pitfalls, which may be encountered in cases of application of the concrete method, are pointed out.

We hope that this book will be a practical help to students and researchers in

---

MATLAB and Simulink are registered trademarks of the MathWorks, Inc. For product information, please contact: The MathWorks, Inc. 3 Apple Hill Drive Natick, MA 01760-2098 USA Tel: 508-647-7000 Fax: 508-647-7001 E-mail: info@mathworks.com Web: www.mathworks.com.

choosing appropriate methods, designing their own methods, and adding new value to the growing field of open biomedical research.

Acknowledgments

We would like to thank Maria Piotrkiewicz, Jan Żebrowski, and Roman Maniewski for consultations and supplying us with valuable materials.

Authors

K. J. Blinowska obtained her Ph.D. in 1969 and her Sc.D. (Doctor Habilitatus) in 1979 from the Faculty of Physics, Warsaw University. She has been with Warsaw University since 1962 where she was employed as an assistant, assistant professor, and associate professor, and in 1994 she became a full professor. In 1974, Dr. Blinowska created at the Faculty of Physics the didactic specialty of biomedical physics. In 1976, she established the Laboratory of Medical Physics (which then became the Department of Biomedical Physics), focusing its research on biomedical signal processing and modeling. From 1984–2009, she was Director of Graduate Studies in Biomedical Physics and Head of the Department of Biomedical Physics. She promoted over 40 M.Sc. and 10 Ph.D. graduates. Prof. Blinowska is author of over 180 scientific publications.

Prof. Blinowska was a visiting professor at the University of Southern California from 1981–1982, University of Alberta from 1990–1991, and University of Lisbon in 1993. She gave invited lectures at several universities, including: Oxford University, Heidelberg University, University of Amsterdam, Politechnico di Milano, University of North Carolina, and Johns Hopkins Medical University.

Prof. Blinowska's research is focused on biomedical time-series processing and modeling, with emphasis on the signals connected with the activity of the nervous system: electroencephalograms (EEG), magnetoencephalograms (MEG), event-related potentials (ERP), local field potentials (LFP), and otoacoustic emissions (OAE). She has also been involved in statistical analysis of medical data, computer-aided diagnosis, neuroinformatics, and application of Internet databases for neuroscience. She has extensive experience in the development of new methods of advanced time-series analysis for research and clinical applications. In particular, she introduced to the repertoire of signal processing methods—Directed Transfer Function (DTF)—now a widely used method for determination of directed cortical connectivity. Dr. Blinowska was the first to apply advanced methods of time-frequency analysis (wavelets and matching pursuit) to biomedical signal analysis. She also applied the signal processing methods to the quantification of genetic information. Her studies of OAE signal—noninvasive, objective test of hearing impairments—contributed to the understanding of the mechanisms of sound perception and diagnosis of hearing impairments.

She has been a principal investigator in numerous grants from the Polish Ministry of Higher Education and Science, coordinator of the Foggarty grant from the NIH, subcontractor of Concerted Action on Otoacoustic Emissions sponsored by the European Union, and a Polish coordinator for the DECIDE project operating in the framework of the 7th EU program. Prof. Blinowska currently serves as a reviewer for leading biomedical engineering journals. She has been acting as an expert for the European Commission, and an expert evaluator for the Swedish Research Council, Spanish Ministry of Health, Marsden Fund of New Zealand, and the Qatar National Research Fund.

J. Żygierewicz was born in Warsaw, Poland in 1971. He received his M.Sc. (1995) and Ph.D. (2000) degrees at the Faculty of Physics of Warsaw University. His Ph.D. thesis concerned the analysis of sleep EEG by means of adaptive approximations (matching pursuit) and modeling of the sleep spindles generation. His research interests concern time-frequency analysis of EEG and MEG signals, especially event-related power changes of these signals. He developed methodology for statistical analysis of event-related synchronization and desynchronization in EEG and MEG. His research also encompasses realistic neuronal network models that provide insight into the mechanisms underlying the effects observed in EEG and MEG signals.

Dr. Żygierewicz is an author of 19 scientific papers in peer-reviewed journals and has made numerous contributions to international conferences. He currently acts as a reviewer for *Clinical Neurophysiology, Journal of Neuroscience Methods,* and *Medical & Biological Engineering & Computing*.

Presently, he is an assistant professor in the Department of Biomedical Physics, Faculty of Physics, Warsaw University, and has promoted 15 M.Sc. students. He has been involved in the creation and implementation of the syllabus for neuroinformatics studies at the University of Warsaw. Dr. Żygierewicz lectures on: signal processing, statistics, mathematical modeling in biology, and artificial neural networks, which are all accompanied by MATLAB practices.

List of Abbreviations

AIC	Akaike information criterion
ANN	artificial neural networks
ApEn	approximate entropy
AR	autoregressive model
ARMA	autoregressive moving average model
BAEP	brain stem auditory evoked potentials
BSPM	body surface potential mapping
BSR	burst suppression ratio
BSS	blind source separation
CAP	cyclic alternating pattern
CSD	current source density
CWT	continuous wavelet transform
DFA	detrended fluctuation analysis
DFT	discrete Fourier transform
DWT	discrete wavelet transform
ECG	electrocardiogram
ECoG	electrocorticogram
EEG	electroencephalogram
EEnG	electroenterogram
EF	evoked fields
EGG	electrogastrogram
EIG	electrointestinogram

EMD	empirical mode decomposition
EMG	electromyogram
EOG	electrooculogram
EP	evoked potentials
ERC	event-related causality
ERD	event-related desynchronization
ERP	event-related potential
ERS	event-related synchronization
FA	factor analysis
FAD	frequency amplitude damping (method)
FDR	false discovery rate
fECG	fetocardiogram
FFT	fast Fourier transform
FIR	finite impulse response
fMCG	feto magnetocardiogram
FT	Fourier transform
FWE	family wise error
FWER	family wise error rate
GAD	Gabor atom density
GFP	global field power
GS	generalized synchronization
HHT	Hilbert-Huang transform
HMM	hidden Markov model
HSD	honesty significant difference test
IC	independent component
ICA	independent component analysis
IDFT	inverse discrete Fourier transform

iEEG	intracranial electroencephalogram
IIR	infinite impulse response
IQ	information quantity
KL	Kullback Leibler (entropy)
LDA	linear discriminant analysis
LDS	linear dynamic system
LFP	local field potentials
LTI	linear time invariant
MCG	magnetocardiogram
MCP	multiple comparison problem
MDL	minimum description length (criterium)
mECG	maternal electrocardiogram
MEG	magnetoelectroencephalogram
MEnG	magnetoenterogram
MGG	magnetogastrogram
MI	mutual information
MLP	multilayer perceptron
MMSE	minimum mean square error
MP	matching pursuit
MPC	multiple comparison problem
MPF	median power frequency
MTM	multi taper method
MU	motor unit
MUAP	motor unit action potential
OAE	otoacoustic emissions
PCA	principal component analysis
PCG	phonocardiogram

PCI	phase clustering index
PLV	phase locking value
PM	Poincaré map
PS	phase synchronization
PSD	power spectral density
PSP	post-synaptic potential
REM	rapid eye movement
RP	recurrence plot
SaEn	sample entropy
sEMG	surface electromyogram
SIQ	subband information quantity
SnPM	statistical non-parametric mapping
SOAE	spontaneous otoacoustic emissions
SOBI	second order blind inference
SPM	statistical parametric mapping
SSOAE	synchronized spontaneous otoacoustic emissions
STFT	short time Fourier transform
SVD	singular value decomposition
SWA	slow wave activity
TE	transfer entropy
TWA	T-wave alternans
WP	wavelet packets
WT	wavelet transform
WVD	Wigner-Ville distribution

1

Introductory concepts

1.1 Stochastic and deterministic signals, concepts of stationarity and ergodicity

A signal is a physical quantity that can carry information. Physical and biological signals may be classified as deterministic or stochastic. The stochastic signal contrary to the deterministic one cannot be described by a mathematical function. An example of a deterministic signal may be the time course of a discharging capacity or the position of a pendulum. Typical random process may be the number of particles emitted by the radioactive source or the output of a noise generator. Physiological signals can be qualified as stochastic signals, but they usually consist of a deterministic and a random component. In some signals, the random component is more pronounced while in others the deterministic influences prevail. An example of the stochastic signal, where random component is important may be EEG. Another class of signals can be represented by an ECG which has a quite pronounced deterministic component related to propagation of the electrical activity in the heart structures, although some random component coming from biological noise is also present.

A process may be observed in time. A set of observations of quantity x in function of time t forms the time series $x(t)$. In many cases the biophysical time series can be considered as a realization of a process, in particular the stochastic process.

If K will be the assembly of k events ($k \in K$) and to each of these events we assign function $x_k(t)$ called realization of the process $\xi(t)$, the stochastic process can be defined as a set of functions:

$$\xi(t) = \{x_1(t), x_2(t), \ldots, x_N(t)\} \tag{1.1}$$

where $x_k(t)$ are the random functions of variable t.

In the framework of the theory of stochastic processes the physical or biophysical process can be described by means of the expected values of the estimators found by the ensemble averaging over realizations. The expected value of stochastic process is an average over all realizations of the process weighted by the probabilities of their occurrence. The mean value $\mu_x(t_1)$ of the stochastic process $\xi(t)$ in the time t_1 can be found by summation of the actual values of each realization in time t_1 weighted by the probability of the occurrence of the given realization $p(x_k,t_1)$:

$$\mu_x(t_1) = E\left[\xi(t_1)\right] = \lim_{N\to\infty} \sum_{k=1}^{N} x_k(t_1)p(x_k,t_1) \tag{1.2}$$

$E[.]$ denotes expected value. In general the expected value of the given function $f(\xi)$ may be expressed by:

$$E\left[f\left(\xi\left(t_1\right)\right)\right] = \lim_{N\to\infty} \sum_{k=1}^{N} f\left(x_k(t_1)\right)p(x_k,t_1) \tag{1.3}$$

If the probability of occurrence of each realization is the same, which frequently is the case, the equation (1.3) is simplified:

$$E\left[f\left(\xi\left(t_1\right)\right)\right] = \lim_{N\to\infty} \frac{1}{N} \sum_{k=1}^{N} f\left(x_k(t_1)\right) \tag{1.4}$$

In particular, function $f(\xi)$ can represent moments or joint moments of the processes $\xi(t)$. Moment of order n is : $f(\xi) = \xi^n$. In these terms mean value (1.2) is a first order moment and mean square value ψ^2 is the second order moment of the process:

$$\psi^2(t_1) = E\left[\xi^2\left(t_1\right)\right] = \lim_{N\to\infty} \sum_{k=1}^{N} x_k^2(t_1)p(x_k,t_1) \tag{1.5}$$

Central moments m_n about the mean are calculated in respect to the mean value μ_x. The first central moment is zero. The second order central moment is variance:

$$m_2 = \sigma_x^2 = E\left[\left(\xi - \mu_x\right)^2\right] \tag{1.6}$$

where σ is the standard deviation. The third order central moment in an analogous way is defined as:

$$m_3 = E\left[\left(\xi - \mu_x\right)^3\right] \tag{1.7}$$

Parameter β_1 related to m_3:

$$\beta_1 = \frac{m_3}{m_2^{3/2}} = \frac{m_3}{\sigma^3} \tag{1.8}$$

is called skewness, since it is equal to 0 for symmetric probability distributions of $p(x_k,t_1)$.

Kurtosis:

$$\beta_2 = \frac{m_4}{m_2^2} \tag{1.9}$$

is a measure of flatness of the distribution. For normal distribution kurtosis is equal to 3. A high kurtosis distribution has a sharper peak and longer, fatter tails, in contrast to a low kurtosis distribution which has a more rounded peak and shorter thinner tails.

Often instead of kurtosis parameter e - excess of curtosis: $e = \beta_2 - 3$ is used. The subtraction of 3 at the end of this formula is often explained as a correction to make the kurtosis of the normal distribution equal to zero. For the normally distributed variables (variables whose distribution is described by Gaussian), central odd moments are equal to zero and central even moments take values:

$$m_{2n+1} = 0 \qquad m_{2n} = (2n-1)m_2^{2n} \tag{1.10}$$

Calculation of skewness and kurtosis can be used to assess if the distribution is roughly normal. These measures can be computed using functions from the MATLAB statistics toolbox: `skewness` and `kurtosis`.

The relation of two processes $\xi(t) = \{x_1(t), \dots, x_N(t)\}$ and $\eta(t) = \{y_1(t), \dots, y_N(t)\}$ can be characterized by joint moments. Joint moment of the first order $R_{xy}(t)$ and joined central moment $C_{xy}(t)$ of process $\xi(t)$ are called, respectively, cross-correlation and cross-covariance:

$$R_{xy}(t_1, \tau) = E\left[\xi(t_1)\eta(t_1 + \tau)\right] \tag{1.11}$$

$$C_{xy}(t_1, \tau) = E\left[(\xi(t_1) - \mu_x(t_1))(\eta(t_1 + \tau) - \mu_y(t_1))\right] \tag{1.12}$$

where τ is the time shift between signals x and y.

A special case of the joint moments occurs when they are applied to the same process, that is $\xi(t) = \eta(t)$. Then the first order joint moment $R_x(t)$ is called autocorrelation and joined central moment $C_x(t)$ of process $\xi(t)$ is called autocovariance.

Now we can define:

Stationarity: For the stochastic process $\xi(t)$ the infinite number of moments and joint moments can be calculated. If all moments and joint moments do not depend on time, the process is called stationary in the strict sense. In a case when mean value μ_x and autocorrelation $R_x(\tau)$ do not depend on time the process is called stationary in the broader sense, or weakly stationary. Usually weak stationarity implies stationarity in the strict sense, and for testing stationarity usually only mean value and autocorrelation are calculated.

Ergodicity: The process is called ergodic when its mean value calculated in time (for the infinite time) is equal to the mean value calculated by ensemble averaging (according to equation 1.2). Ergodicity means that one realization is representative for the whole process, namely that it contains the whole information about the process. Stationarity of a process implies its ergodicity. For ergodic processes we can describe the properties of the process by averaging one realization over time, instead of ensemble averaging.

Under the assumption of ergodicity moment of order n is expressed by:

$$m_n = \lim_{T \to \infty} \int_0^T x^n(t) p(x)\, dt \tag{1.13}$$

1.2 Discrete signals

In nature, most of the signals of interest are some physical values changing in time or space. The biomedical signals are continuous in time and in space.

On the other hand we use computers to store and analyze the data. To adapt the natural continuous data to the digital computer systems we need to digitize them. That is, we have to sample the physical values in certain moments in time or places in space and assign them a numeric value—with a finite precision. This leads to the notion of two processes: sampling (selecting discrete moments in time) and quantization (assigning a value of finite precision to an amplitude).

1.2.1 The sampling theorem

Let's first consider sampling. The most crucial question is how often the signal $f(t)$ must be sampled? The intuitive answer is that, if $f(t)$ contains no frequencies[1] higher than F_N, $f(t)$ cannot change to a substantially new value in a time less than one-half cycle of the highest frequency; that is, $\frac{1}{2F_N}$. This intuition is indeed true. The Nyquist-Shannon sampling theorem [Shannon, 1949] states that:

> If a function $f(t)$ contains no frequencies higher than F_N cps, it is completely determined by giving its ordinates at a series of points spaced $\frac{1}{2F_N}$ seconds apart.

The frequency F_N is called the Nyquist frequency and $2F_N$ is the minimal sampling frequency. The "completely determined" phrase means here that we can restore the unmeasured values of the original signal, given the discrete representation sampled according to the Nyquist-Shannon theorem (Figure 1.1).

A reconstruction can be derived via sinc function $f(x) = \frac{\sin \pi x}{\pi x}$. Each sample value is multiplied by the sinc function scaled so that the zero-crossings of the sinc function occur at the sampling instants and that the sinc function's central point is shifted to the time of that sample, nT, where T is the sampling period (Figure 1.1 b). All of these shifted and scaled functions are then added together to recover the original signal (Figure 1.1 c). The scaled and time-shifted sinc functions are continuous, so the sum is also continuous, which makes the result of this operation a continuous signal. This procedure is represented by the Whittaker-Shannon interpolation formula. Let $x[n] := x(nT)$ for $n \in \mathbb{Z}$ be the nth sample. We assume that the highest frequency present in the sampled signal is F_N and that it is smaller than half of the sampling frequency $F_N < \frac{1}{2}F_s$. Then the function $f(t)$ is represented by:

$$f(t) = \sum_{n=-\infty}^{\infty} x[n] \operatorname{sinc}\left(\frac{t-nT}{T}\right) = \sum_{n=-\infty}^{\infty} x[n] \frac{\sin \pi(2F_s t - n)}{\pi(2F_s t - n)} \tag{1.14}$$

[1]Frequencies are measured in cycles per second—cps, or in Hz—[Hz]=$\frac{1}{[s]}$.

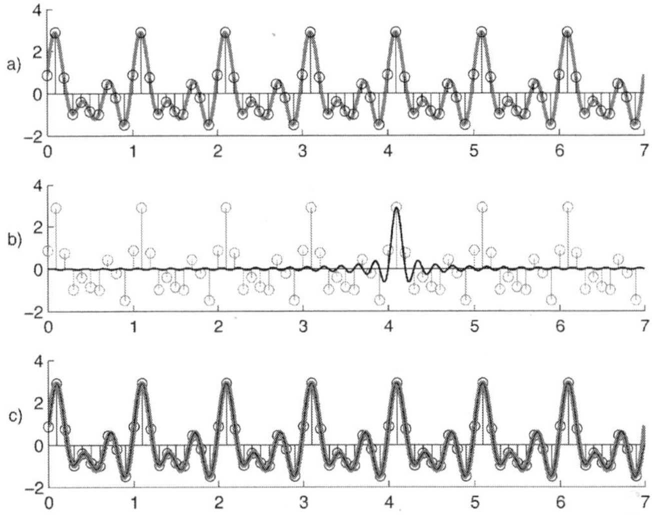

FIGURE 1.1: Illustration of sampling and interpolating of signals. a) The continuous signal (gray) is sampled at points indicated by circles. b) The impulse response of the Whittaker-Shannon interpolation formula for a selected point. c) Reconstruction (black) of signal computed according to (1.14).

1.2.1.1 Aliasing

What happens if the assumption of the sampling theorem is not fulfilled and the original signal contains frequencies higher than the Nyquist frequency? In such cases we observe an effect called *aliasing*—different signal components become indistinguishable (aliased). If the signal of frequency $f_0 \in \left(\frac{1}{2}F_s, F_s\right)$ is sampled with frequency F_s then it has the same samples as the signal with frequency $f_1 = F_s - f_0$. Note that $|f_1| < \frac{1}{2}F_s$. The sampled signal contains additional low frequency components that were not present in the original signal. An illustration of that effect is shown in Figure 1.2.

1.2.2 Quantization error

When we measure signal values, we usually want to convert them to numbers for further processing. The numbers in digital systems are represented with a finite precision. The analog to digital converter (ADC) is characterized by the number of bits N it uses to represent numbers. The full range R of measurement values is divided into 2^N levels. The quantization error can be estimated as being less than $\frac{R}{2^N}$ (Figure 1.3).

This error sometimes has to be taken into consideration, especially when the amplitudes of measured signals span across orders of magnitude. An example here can be EEG measurement. Let's assume that we have adjusted the amplification of signal

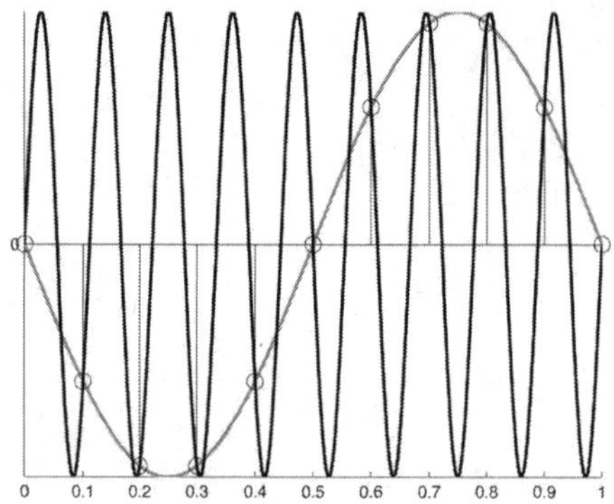

FIGURE 1.2: Illustration of the aliasing effect. Samples taken every 0.1 s from a 9 Hz sinusoid (black) are exactly the same as these taken from the 1 Hz sinusoid (gray).

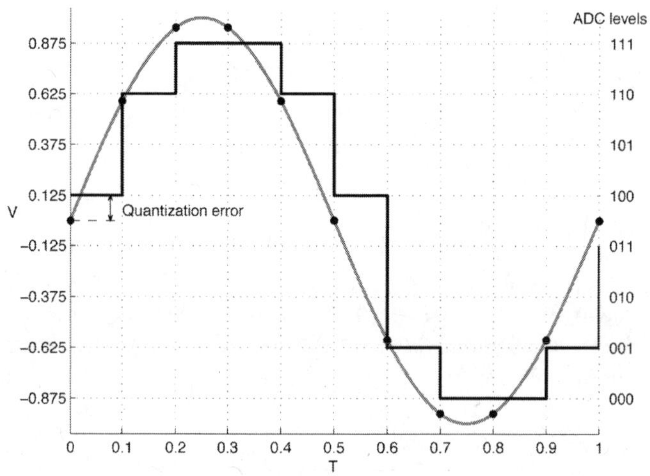

FIGURE 1.3: Illustration of the quantization error for a 3 bit ADC. The digitized representation (black line) of the continuous sinusoid (gray line). The range of 2 [V] is divided into 8 levels. The sampled signal values (black dots) are rounded to the nearest ADC level.

so that $\pm 200 \mu V$ covers the full range of a 12 bit ADC. This range is divided into bins of $400/2^{12} = 400/4096 \approx 0.1 \; \mu V$. It means that we measure the amplitude of the signal with precession $\pm 0.05 \; \mu V$.

1.3 Linear time invariant systems

In signal processing there is an important class of systems called linear time invariant systems—LTI (in case of sampled signals this is sometimes named linear shift invariant). We can think of the system as a box which modifies the input with a linear operator L to produce the output:

$$input \longrightarrow \boxed{L\{input\}} \longrightarrow output$$

The basic properties of such a system are:

1. Linearity: superposition of inputs produces superposition of outputs, formally: if an input $x_1(t)$ produces output $y_1(t)$:
$$L\{x_1(t)\} = y_1(t)$$
and input $x_2(t)$ produces output $y_2(t)$
$$L\{x_2(t)\} = y_2(t)$$
then the output of the superposition will be:
$$L\{a_1x_1(t) + a_2x_2(t)\} = a_1L\{x_1(t)\} + a_2L\{x_2(t)\} = a_1y_1(t) + a_2y_2(t)$$

2. Time invariance: for a given input the system produces identical outputs no matter when we apply the input. More formally:
if $L\{x(t)\} = y(t)$ then $L\{x(t+T)\} = y(t+T)$

The important property of LTI systems is that they are completely characterized by their impulse response function. The impulse response function can be understood as the output of the system due to the single impulse[2] at the input. It is so, because we can think of the input signal as consisting of such impulses. In case of a discrete signal it is very easy to imagine. In case of continuous signal we can imagine it as a series of infinitely close, infinitely narrow impulses. For each input impulse, the system reacts in the same way. It generates a response which is proportional (weighted by the amplitude of the impulse) to impulse response function. The responses to consecutive impulses are summed up with the response due to the former inputs (Figure 1.4). Such operation is called *convolution*. For convolution we shall use symbol: $*$. The process is illustrated in Figure 1.4. Formally the operation of the LTI system can be expressed in the following way. Let's denote the impulse response function as $h[n]$. Next, let us recall the definition of the Kronecker delta:

$$\delta[n] = \begin{cases} 1 & \text{if } n = 0 \\ 0 & \text{if } n \neq 0 \end{cases}, \qquad \text{and } n \in \mathbb{Z} \tag{1.15}$$

[2]In case of continuous time system the impulse is the Dirac's delta—an infinitely sharp peak bounding unit area; in case of discrete systems it is a Kronecker delta—a sample of value 1 at the given moment in time.

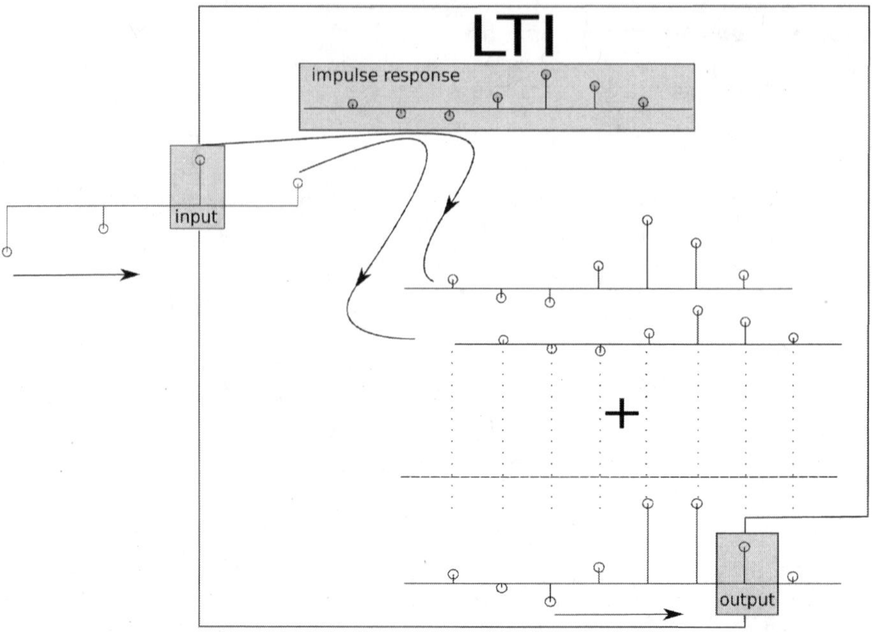

FIGURE 1.4: Idea of output production by an LTI system—convolution of input spikes with the impulse response function. The impulse response of the LTI system is multiplied by the input impulse. This is the response to the current input. The current response is added to the time-shifted responses to the previous inputs.

Using this function any discrete sequence $x[n]$ can be expressed as:

$$x[n] = \sum_k x[k]\delta[n-k] \tag{1.16}$$

and the output of the LTI system at time n due to single impulse at moment k as[3]:

$$h_k[n] = L\{\delta[n-k]\} = h[n-k] \tag{1.17}$$

The output of the LTI system

$$y[n] = L\{x[n]\} \tag{1.18}$$

can be computed by substituting (1.16) into (1.18):

$$y[n] = L\{\sum_k x[k]\delta[n-k]\} \tag{1.19}$$

Due to the linearity of L and property (1.17) we have:

$$y[n] = \sum_k x[k]L\{\delta[n-k]\} = \sum_k x[k]h[n-k] = (x*h)[n] \tag{1.20}$$

[3]In this formula we also use the time invariance property.

For any linear operator there is a class of functions, called *eigenfunctions*, that are not distorted when the operator is applied to them. The only result of application of the operator to its eigenfunction is multiplication by a number (in general it can be a complex number). Complex exponentials are the eigenfunctions for the LTI system. For an LTI system with signals in the real domain[4] such eigenfunctions are the sinusoids. It follows from the linearity of LTI system and the Euler formulas:

$$\begin{aligned}\cos x &= \frac{1}{2}\left(e^{ix}+e^{-ix}\right)\\ \sin x &= \frac{1}{2i}\left(e^{ix}-e^{-ix}\right)\end{aligned} \tag{1.21}$$

Please note, that when expressing a real signal in the language of complex exponentials we must introduce a negative valued frequency since any oscillation in a real signal described by f cycles per second is represented by a pair of complex exponentials with frequencies f and $-f$. Since complex exponentials are eigenfunctions of LTI system, from equation (1.21) it follows that when a sinusoid is passed through the LTI system it can change the amplitude and phase but cannot change the frequency. This property is very useful when dealing with the LTI systems, since sinusoids form a basis in the space of real functions. Hence we can express exactly any real function as a (possibly infinite) sum (or integral) of sinusoids with specific amplitudes, frequencies, and phases. This property is the reason Fourier analysis is used so extensively in signal processing.

1.4 Duality of time and frequency domain

In the previous section we noticed that a tool which allows to translate input signal to a sum or integral of sinusoids would be very handy when dealing with LTI systems. Such a tool is the Fourier transform. In fact, there is always a pair of transforms: one, from time domain to the frequency domain, we shall denote as $\mathcal{F}\{x(t)\}$ and the inverse transform, from frequency to time domain, is denoted as $\mathcal{F}^{-1}\{X(\omega)\}$. The operation of the transformation is illustrated in the scheme below:

$$x(t) \underset{\mathcal{F}^{-1}}{\overset{\mathcal{F}}{\rightleftarrows}} X(\omega)$$

In time domain we think of a signal $x(t)$ as a series of values at certain moments in time; in the frequency domain the same signal $X(\omega)$ is thought of as a specific set of frequencies, and each frequency has its own amplitude and phase. Those two

[4]It means that the signal values are real numbers.

representations of the signal are equivalent. That is we can transform without any loss of information signals from time to frequency representation and vice versa.

The frequency can be expressed in radians per second—in this case we shall denote it as ω or in cycles per second—in this case we shall denote it as f. Both quantities are related: $f = 2\pi\omega$.

Depending on the signal, different kinds of the Fourier transform are used. They will be described below.

1.4.1 Continuous periodic signal

Let us first consider the simplest case: the signal $x(t)$ is periodic with period T. Such a signal can be expressed as a series:

$$x(t) = \sum_{n=-\infty}^{\infty} c_n e^{-i\frac{2\pi t}{T}n} \tag{1.22}$$

where:

$$c_n = \frac{1}{T}\int_0^T x(t)e^{i\frac{2\pi t}{T}n}\,dt \tag{1.23}$$

This fact can be easily checked by substitution:

$$\begin{aligned}
\int_0^T x(t)e^{i\frac{2\pi t}{T}k}\,dt = & \\
&= \int_0^T \sum_{n=-\infty}^{\infty} c_n e^{-i\frac{2\pi t}{T}n} e^{i\frac{2\pi t}{T}k}\,dt \\
&= \sum_{n=-\infty}^{\infty} \int_0^T c_n e^{-i\frac{2\pi t}{T}(k-n)}\,dt \\
&= \sum_{n=k} \int_0^T c_n\,dt + \sum_{n\neq k} \underbrace{\int_0^T c_n e^{-i\frac{2\pi t}{T}(k-n)}\,dt}_{=0} \\
&= T c_n
\end{aligned} \tag{1.24}$$

We can think of c_n in expression (1.23) as the contribution of the frequency $f_n = \frac{n}{T}$ to the signal $x(t)$

$$c_n = X\left(\frac{n}{T}\right) \tag{1.25}$$

Hence a periodic signal can be expressed by the linear combination of complex exponentials with a discrete set of frequencies.

1.4.2 Infinite continuous signal

We can extend the formula (1.23) for aperiodic signals. The trick is that we consider the whole infinite aperiodic signal domain as a single period of an infinite periodic signal. In the limit $T \to \infty$ we obtain:

$$x(t) = \int_{-\infty}^{\infty} X(f)e^{-i2\pi ft}\,df \tag{1.26}$$

$$X(f) = \int_{-\infty}^{\infty} x(t)e^{i2\pi ft}\,dt \tag{1.27}$$

1.4.3 Finite discrete signal

In practice we deal with discrete signals of finite duration. The Fourier transform that operates on this kind of signal is called discrete Fourier transform (DFT) and the algorithms that implement it are FFT (fast Fourier transform).

The DFT formula can be derived from (1.23). The signal to be transformed is N samples long $x = \{x[0],\ldots,x[n],\ldots x[N-1]\}$ and the samples are taken every T_s seconds. It is assumed that the finite signal x is just one period of the infinite periodic sequence with period $T = N \cdot T_s$. The process of sampling can be written as $x[n] = x(nT_s) = x(t)\delta(t - nT_s)$. Substituting this into (1.23) gives:

$$X[k] = \frac{1}{T}\int_0^T x(t)\delta(t-nT_s)e^{i\frac{2\pi t}{T}k}\,dt = \frac{1}{NT_s}\sum_{n=0}^{N-1} x[n]e^{i\frac{2\pi knT_s}{T}}T_s = \frac{1}{N}\sum_{n=0}^{N-1} x[n]e^{i\frac{2\pi}{N}kn} \tag{1.28}$$

From the above formula it follows that k in the range $k = 0,\ldots,N-1$ produces different components in the sum. From the Euler formulas (1.21) it follows that for a real signal a pair of conjunct complex exponentials is needed to represent one frequency. Thus for real signals there are only $N/2$ distinct frequency components. The inverse discrete Fourier transform (IDFT) is given by

$$x[n] = \sum_{k=0}^{N-1} X[k]e^{-i\frac{2\pi}{N}kn} \qquad n = 0,\ldots,N-1. \tag{1.29}$$

Note, that the signs of the exponents and the normalization factors by which the DFT and IDFT are multiplied (here $1/N$ and 1) are conventions, and may be written differently by other authors. The only requirements of these conventions are that the DFT and IDFT have opposite sign exponents and that the product of their normalization factors be $1/N$.

1.4.4 Basic properties of Fourier transform

Given signals $x(t)$, $y(t)$, and $z(t)$ we denote their Fourier transforms by $X(f)$, $Y(f)$, and $Z(f)$, respectively. The Fourier transform has the following basic properties: [Pinsky, 2002].

Linearity: For any complex numbers a and b:

$$z(t) = ax(t) + by(t) \quad \Rightarrow \quad Z(f) = a \cdot X(f) + b \cdot Y(f)$$

Translation: For any real number t_0:

$$z(t) = x(t - t_0) \quad \Rightarrow \quad Z(f) = e^{-2\pi i t_0 f} X(f)$$

Modulation: For any real number f_0:

$$z(t) = e^{2\pi f_0} x(t) \quad \Rightarrow \quad Z(f) = X(f - f_0)$$

Scaling: For all non-zero real numbers a:

$$z(t) = x(at) \quad \Rightarrow \quad Z(f) = \frac{1}{|a|} X\left(\frac{f}{a}\right)$$

The case $a = -1$ leads to the time-reversal property, which states:

$$z(t) = x(-t) \quad \Rightarrow \quad Z(f) = X(-f)$$

Conjugation:

$$z(t) = x(t)^* \quad \Rightarrow \quad Z(f) = X(-f)^*$$

The * symbol throughout the book denotes the complex conjugation.

Convolution theorem:

$$y(t) = (x \ast z)(t) \quad \Leftrightarrow \quad Y(f) = X(f) \cdot Z(f) \tag{1.30}$$

This theorem works also in the opposite direction:

$$Y(f) = (X \ast H)(f) \quad \Leftrightarrow \quad y(t) = x(t) \cdot z(t) \tag{1.31}$$

This theorem has many applications. It allows to change the convolution operation in one of the dual (time or frequency) spaces into the multiplication in the other space. Combined with the FFT algorithm the convolution theorem allows for fast computations of convolution. It also provides insight into the consequences of windowing the signals, or applications of filters.

1.4.5 Power spectrum: the Plancherel theorem and Parseval's theorem

Let's consider the $x[n]$ as samples of a voltage across a resistor with the unit resistance $R = 1$. Then the $P = x[n]^2/R$ is the power dissipated by that resistor. By analogy in the signal processing language a square absolute value of a sample is called *instantaneous signal power*.

If $X[k]$ and $Y[k]$ are the DFTs of $x[n]$ and $y[n]$, respectively, then the Plancherel theorem states:

$$\sum_{n=0}^{N-1} x[n] y^*[n] = \frac{1}{N} \sum_{k=0}^{N-1} X[k] Y^*[k] \tag{1.32}$$

where the star denotes complex conjugation. Parseval's theorem is a special case of the Plancherel theorem and states:

$$\sum_{n=0}^{N-1} |x[n]|^2 = \frac{1}{N} \sum_{k=0}^{N-1} |X[k]|^2. \tag{1.33}$$

Due to the Parseval's theorem (1.33) we can think of $\frac{1}{N}|X[k]|^2$ as the portion of signal power carried by the complex exponential component of frequency indexed by k. If we process real signals, then the complex exponential in Fourier series come in conjugate pairs indexed by k and $N-k$ for $k \in 1, \dots, N/2$. Each of the components of the pair carries half of the power related to the oscillation with frequency $f_k = \frac{k}{N} F_s$ (F_s is the sampling frequency). To recover the total power of oscillations at frequency f_k we need to sum the two parts. That is:

$$P(f_k) = \frac{1}{N}\left(|X[k]|^2 + |X[N-k]|^2\right) \tag{1.34}$$

For real signals the power spectrum is often displayed by plotting only the power (1.34) for the positive frequencies.

1.4.6 Z-transform

A more general case of the discrete signal transformation is the Z transform. It is especially useful when considering parametric models or filters.

In a more general context, the Z transform is a discrete version of the Laplace transform. For a discrete signal $x[n]$ the Z transform is given by:

$$X(z) = Z\{x[n]\} = \sum_{n=0}^{\infty} x[n] z^{-n} \tag{1.35}$$

where $z = Ae^{i\phi}$ is a complex number. The discrete Fourier transform is a special case of the Z transform.

Properties of the Z transform:

Linearity:

$$Z\{a_1 x_1[n] + a_2 x_2[n]\} = a_1 X_1(z) + a_2 X_2(z)$$

Time translation:

$$Z\{x[n-k]\} = z^{-k} X(z)$$

Transform of an impulse:

$$Z\{\delta[n]\} = 1$$

The transform of an impulse together with the time translation yields:

$$Z\{\delta[n-n_0]\} = z^{-n_0}$$

Taking into account the linearity of Z we can compute the transform of a linear combination of the p signal samples:

$$\begin{aligned} Z\{x[n] + a_1 x[n-1] + \cdots + a_p x[n-p]\} &= (1 + a_1 z^{-1} + \cdots + a_p z^{-p}) X(z) \\ &= A(z) X(z) \end{aligned} \tag{1.36}$$

This result will be very useful for discussion of AR model and filtering.

1.4.7 Uncertainty principle

In the previous sections we considered signals represented in either time or frequency domain. For stationary signals it is enough to know one of the representations. Real-life signals are often a mixture of stationary and non-stationary elements, e.g., alpha spindles, sleep spindles, K-complexes, or other graphoelements together with ongoing background EEG activity. We are often interested in characterizing the non-stationary transients. It is natural to think about their frequency f_0 and localization in time t_0. It is also obvious that such a transient has certain duration in time—that is, it is not localized in a single time point but it has a span in time. The time span can be characterized by σ_t. The localization of the transient in the frequency domain also has a finite resolution characterized by frequency span σ_f. Those two spans are bounded by the uncertainty principle. Before we continue with the formal notation, let's try to understand the principle heuristically. Let's think about a fragment of a sinusoid observed over time $T = (t - \sigma_t, t + \sigma_t)$, as shown in the Figure1.5. We can calculate its frequency dividing the number of periods observed during time T by the length of T. As we shrink the time T the localization in time of the fragment of the sinusoid becomes more precise, but less and less precise is estimation of the number of cycles, and hence the frequency.

Now let's put it formally. We treat the signal energy representation in time $|x(t)|^2$ and representaion in frequency $|X(f)|^2$ as probability distributions with normalizing factor $E_x = \int_{-\infty}^{\infty} |x(t)|^2 dt$. Then we can define mean time:

$$t_0 = \frac{1}{E_x} \int_{-\infty}^{\infty} t\,|x(t)|^2\, dt \tag{1.37}$$

Mean frequency:

$$f_0 = \frac{1}{E_x} \int_{-\infty}^{\infty} f\,|X(f)|^2\, df \tag{1.38}$$

Time span:

$$\sigma_t^2 = \frac{1}{E_x} \int_{-\infty}^{\infty} (t - t_0)^2 |x(t)|^2\, dt \tag{1.39}$$

Frequency span:

$$\sigma_f^2 = \frac{1}{E_x} \int_{-\infty}^{\infty} (f - f_0)^2 |X(f)|^2\, df \tag{1.40}$$

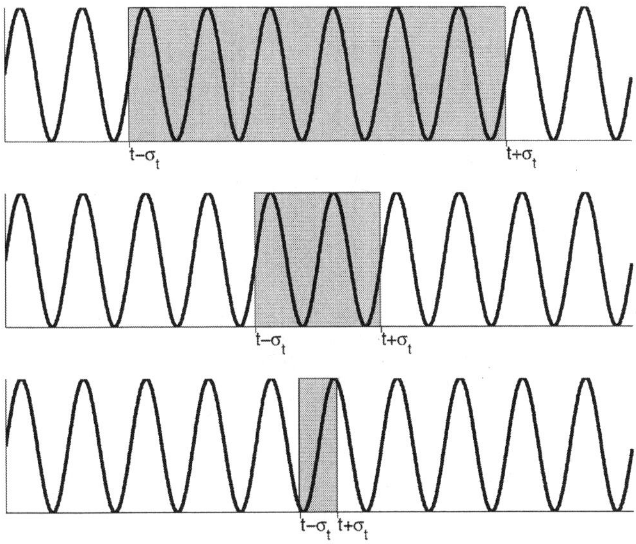

FIGURE 1.5: Illustration of the uncertainty principle. As the observation time shrinks, the time localization of the observation improves, but the estimate of the frequency deteriorates.

It turns out that the product of the time and frequency span is bounded [Folland and Sitaram, 1997]:

$$\sigma_t^2 \sigma_f^2 \geq \frac{1}{16\pi^2} \tag{1.41}$$

The equality is reached by Gabor functions (Gaussian envelope modulated by cosine). It is important to realize this property, especially when working with time-frequency representations of a signal. Many different methods of time-frequency representations make various trade-offs between time and frequency span but in each of them the inequality (1.41) holds.

1.5 Hypotheses testing

1.5.1 The null and alternative hypothesis

The key to success in statistical hypothesis testing is the correct formulation of the problem. We need to specify it as two options: the null hypothesis H_0 and the alternative one H_1. The two hypotheses must be disjoint and complementary. We usually try to put the option which we would like to reject as the null hypothesis, since we can control the probability of erroneous rejection of the null hypothesis.

To actually perform a test, a function called *statistic* $S(x)$ is needed. We assume that the data are random variables and represent a sample taken from some population characterized by properties which are in accordance with the null hypothesis. Statistic is a function of a random variable. The main thing that must be known about the statistic is the probability with which it takes different values for the random variables conforming to the null hypothesis.

If for the given data the computed statistics is $S_x = S(x)$ then the p value returned by the test is the probability of observing statistics with values equal or more extreme than S_x. If the p value is high, then we assume that the data conform to the null hypothesis. But if the probability of observing such a value of the statistic is low then we can doubt that the data agree with the null hypothesis. Consequently we reject that hypothesis and accept the alternative one. The critical level of probability used to make the decision is called the significance level α. It expresses how low the p value must be to doubt the null hypothesis.

1.5.2 Types of tests

To select the type of test we need to answer the question: do we know the probability distribution from which the data were sampled?

Yes, we know or can assume, or can transform, e.g., by Box-Cox transform [Box and Cox, 1964], the data to one of the known probability distributions. In this case we select appropriate classical parametric tests based on the normal, t, F, χ^2 or some other known statistics. In MATLAB Statistics Toolbox many such tests are available e.g., `ttest`, `ttest2`, `anova`, `chi2gof`. To test the normality assumption we can use Lilliefors' test [Lilliefors, 1967], implemented as `lillietest`, or use a qualitative graphical test implemented as `normplot`.

No, we do not know the probability distribution. In this case we have two possibilities:

- Use a classial non-parametric test e.g.:

  Wilcoxon rank sum test — tests, if two independent samples come from identical continuous distributions with equal medians, against the alternative that they do not have equal medians. In MATLAB Statistics Toolbox it is implemented as `ranksum`.

  Wilcoxon signed rank test — One-sample or paired-sample Wilcoxon signed rank test. It tests, if a sample comes from a continuous distribution symmetric about a specified median, against the alternative that it does not have that median. In MATLAB Statistics Toolbox it is implemented as `signrank`.

  Sign test — One-sample or paired-sample sign test. It tests, if a sample comes from an arbitrary continuous distribution with a specified median, against the alternative that it does not have that median. In MATLAB Statistics Toolbox it is implemented as `signtest`.

- Use a resampling (bootstrap or permutation) test [Efron and Tibshirani, 1993]. In this type of test in principle any function of the data can be used as a statistic. We need to formulate a statistical model of the process that generates the data. The model is often very simple and relies on appropriate resampling of the original dataset, e.g., we draw a random sample from the original dataset with replacement. It is crucial that the model (the resampling process) conforms to the null hypothesis. The model is simulated many times and for each realization the statistic is computed. In this way the empirical distribution of the statistic is formed. The empirical distribution of the statistic is used then to evaluate the probability of observing the value of statistic equal or more extreme than the value for the original dataset.

1.5.3 Multiple comparison problem

Let us assume that we perform a t test with the significance level α. This means that under the true null hypothesis H_0 we can observe with the probability α value greater than the critical:

$$\begin{aligned} P(t \geq t_\alpha) &= \alpha \\ P(t < t_\alpha) &= 1 - \alpha \end{aligned} \tag{1.42}$$

For n independent tests, the probability that none of them will not give t value greater than the critical is $(1-\alpha)^n$. Thus the probability of observing at least one of the n values exceeding the critical one (probability of the family wise error FWER) is:

$$P(t \geq t_\alpha; \text{in n tests}) = P^{FWER} = 1 - (1-\alpha)^n \tag{1.43}$$

Let us consider tests performed at significance level $\alpha = 0.05$, The probability computed from (1.43) gives the chance of observing extreme values of statistic for data conforming to the null hypothesis just due to fluctuations. For a single test the above formula gives $P(t \geq t_\alpha; \text{in 1 test}) = 0.05$, for $n = 10$, $P(t \geq t_\alpha; \text{in 10 test}) \approx 0.4$, for $n = 100$, $P(t \geq t_\alpha; \text{in 100 test}) \approx 0.994$.

In case of some dependence (e.g., correlation) among the tests, the above described problem is less severe but also present. There is clearly a need to control the error of false rejections of null hypothesis due to multiple comparison problem (MCP).

Table 1.1 summarizes the possible outcomes of m hypothesis tests. As a result of the application of a test we obtain one of two true statements: true null hypothesis is accepted (number of such cases is denoted U) or false null hypothesis is rejected (number of such cases is denoted S). There is a possibility to commit one of two types of errors. Type I error is when a true null hypothesis is rejected (the number of such cases is denoted as V); Type II error—the alternative hypothesis is true, but the null hypothesis is accepted (denoted as T). The total number of rejected null hypotheses is denoted by R.

The total number of tested hypotheses m is known. The number of true null hypotheses m_0 and the number of false null hypotheses $m_1 = m - m_0$ are unknown.

The number of cases V, T, U, S, and R are treated as random variables, but only R can be observed. The family wise error rate $FWER$ is the probability of falsely rejecting one or more true null hypotheses among all the hypotheses when performing multiple tests:

$$FWER = P(V \geq 1) = 1 - P(V = 0) \tag{1.44}$$

In studies where one specifies a finite number of a priori inferences, families of hypotheses are defined for which conclusions need to be jointly accurate or by which hypotheses are similar in content or purpose. If these inferences are unrelated in terms of their content or intended use (although they may be statistically dependent), then they should be treated separately and not jointly [Hochberg and Tamhane, 1987].

Sometimes it is not necessary to control the $FWER$ and it is sufficient to control the number of falsely rejected null hypotheses—the false discoveries. The false discovery rate (FDR) is defined as expected proportion of incorrectly rejected null hypotheses:

$$FDR = E\left[\frac{V}{V+S}\right] = E\left[\frac{V}{R}\right] \tag{1.45}$$

TABLE 1.1: Outcomes of m hypothesis tests.

	# declared non-significant (H_0 accepted)	# declared significant (H_0 rejected)	Total
# true null hypotheses	U	V	m_0
# false null hypotheses	T	S	$m - m_0$
Total	$m - R$	R	m

Below we briefly describe three approaches to MCP: correction of the significance level α, statistical maps, and false discovery rate (FDR).

1.5.3.1 Correcting the significance level

The most straightforward approach is known as Bonferroni correction, which states that if one performs n hypotheses tests on a set of data, then the statistical significance level that should be used for each hypothesis separately should be reduced n times in respect to the value that would be used if only one hypothesis were tested. For example, when testing two hypotheses, instead of an α value of 0.05, one should use α value of 0.025. The Bonferroni correction is a safeguard against multiple tests of statistical significance on the same data. On the other hand this correction is conservative in case of correlated tests, which means that the significance level gets lower than necessary to protect against rejections of null hypothesis due to fluctuations.

In special, but very common cases of comparison of a set of mean values it is suggested to consider Tukey's HSD (honestly significant difference) test. Tukey's method considers the pairwise differences. Scheffé's method applies to the set of estimates of all possible contrasts among the factor level means. An arbitrary contrast is a linear combination of two or more means of factor levels whose coefficients add up to zero. If only pairwise comparisons are to be made, the Tukey method will result in a narrower confidence limit, which is preferable. In the general case when many or all contrasts might be of interest, the Scheffé method tends to give narrower confidence limits and is therefore the preferred method [Armitage et al., 2002].

1.5.3.2 Parametric and nonparametric statistical maps

The multiple comparison problem is critical for comparison of images or volume data collected under different experimental conditions. Usually such images or volumes consist of a huge number of elements: pixels, voxels, or resels (resolution elements). For each element statistical models (parametric or non-parametric) are assumed. Hypotheses expressed in terms of the model parameters are assessed with univariate statistics.

In case of the parametric approach (statistical parametric mapping) the general linear models are applied to describe the variability in the data in terms of experimental and confounding effects, and residual variability. In order to control the $FWER$ adjustments are made, based on the number of resels in the image and the theory of continuous random fields in order to set a new criterion for statistical significance that adjusts for the problem of multiple comparisons [Friston et al., 2007]. This methodology, with application to neuroimaging and MEG/EEG data, is implemented in SPM—a MATLAB software package, written by members and collaborators of the Wellcome Trust Centre for Neuroimaging. SPM is free but copyright software, distributed under the terms of the GNU General Public Licence. SPM homepage: `http://www.fil.ion.ucl.ac.uk/spm/`.

In case of the non-parametric approach (statistical non-parametric mapping, SnPM) the idea is simple: if the different experimental conditions do not make any different effect on the measured quantity, then the label assignment to the conditions is arbitrary. Any reallocation of the labels to the data would lead to an equally plausible statistic image. So, considering the statistic images associated with all possible re-labelings of the data, we can derive the distribution of statistic images possible for this data. Then we can test the null hypothesis of no experimental effect by comparing the statistic for the actual labeling of the experiment with this relabeled distribution. If, out of N possible relabelings the actual labeling gives the rth most extreme statistic, then the probability of observing that value under the null hypothesis is r/N. The details are worked out in [Holmes et al., 1996, and Nichols and Holmes, 2002]. SnPM is implemented as a free MATLAB toolbox: `http://www.sph.umich.edu/ni-stat/SnPM/#dox`.

1.5.3.3 False discovery rate

In case of independent tests, Benjamini and Hochberg [Benjamini and Hochberg, 1995] showed that the Simes' procedure [Simes, 1986], described below, ensures that its expected fraction of false rejection of null hypothesis, is less than a given q. This procedure is valid when the m tests are independent. Let $H_1 \dots H_m$ be the null hypotheses and $P_1 \dots P_m$ their corresponding p-values. Order these values in increasing order and denote them by $P_{(1)} \dots P_{(m)}$. For a given q, find the largest k such that:

$$P_{(k)} \leq \frac{k}{m} q \tag{1.46}$$

Then reject all $H_{(i)}$ for $i = 1, \dots, k$.

In case of dependent tests Benjamini and Yekutieli [Benjamini and Yekutieli, 2001] proposed correction of the threshold (1.46) such that:

$$P_{(k)} \leq \frac{k}{m \cdot c(m)} q \tag{1.47}$$

where:

- $c(m) = 1$ if the tests are independent, or are positively correlated
- $c(m) = \sum_{i=1}^{m} \frac{1}{i}$ if the tests are negatively correlated

1.6 Surrogate data techniques

The surrogate data concept was introduced by [Theiler et al., 1992] and further developed by [Efron and Tibshirani, 1993, Andrzejak et al., 2003a] primarily to distinguish non-linear from stochastic time series. However, the method could be used as well to test for the consistent dependencies between multivariate time series.

The method of surrogate data is basically an application of the *bootstrap* method to infer the properties of the time series. In the surrogate data test the null hypothesis concerns the statistical model that generates the signal. Any function of data can be used as statistics. In general the test consists of the following steps:

1. select a function of data to be the statistics;
2. compute the statistics for the original signal;
3. generate new signal according to the model assumed in the null hypothesis which shares with the original signal as many properties as possible (e.g., mean, variance, and Fourier spectrum);
4. compute the statistics for the new signal;

5. repeat steps 3 and 4 many times to construct the distribution of the statistics;

6. use that distribution to compute the probability of observing the original or more extreme value of statistics for processes conforming to the null hypothesis.

Two approaches to surrogate data construction will be described below. The simplest question one would like to answer is whether there is evidence for any dynamics at all in the time series, i.e., if there are any relations between the consecutive samples. The null hypothesis in this case is that the observed data is fully described by a series in which each sample is an independent random variable taken from identical distribution. The surrogate data can be generated by shuffling the time-order of the original time series. The surrogate data will have the same amplitude distribution as the original data, but any temporal correlations that may have been in the original data are destroyed.

The test for linearity corresponds to the null hypothesis that all the structure in the time series is given by the autocorrelation function, or equivalently, by the Fourier power spectrum. The test may be performed by fitting autoregressive model to the series and examination of the residuals, or by randomizing the phases of the Fourier transform. The second approach is recommended, since it is more stable numerically. The main steps of the procedure are the following:

1. compute the Fourier transform $X(f)$ of the original data $x(t)$;

2. generate a randomized version of the signal $Y(f) = X(f)e^{i\phi(f)}$ by multiplying each complex amplitude $X(f)$ by $e^{i\phi(f)}$, where $\phi(f)$ is a random value, independently drawn for each frequency f, from the interval $[0, 2\pi]$;

3. in order to get real components from the inverse Fourier transform symmetrize the phases, so that $\phi(f) = -\phi(-f)$;

4. perform the inverse transform.

In this way we obtain surrogate data $y(t)$ characterized by the same power spectrum as the original time series $x(t)$, but with the correlation structure destroyed. Then the statistical test is performed comparing the estimators (i.e., statistics) obtained from original and the distribution of that estimator obtained from surrogate data.

In case of multivariate time series the surrogates with randomized phases may be used to test for the presence of the phase dependencies between the signals. The rejection of the null hypothesis assuming independence between the time series does not mean the presence of the non-linearities. In biological time series analysis the Occham's razor approach should be used: we should seek the simplest model consistent with the data and the surrogate data tests are helpful in this respect.

2

Single channel (univariate) signal

2.1 Filters

Signals may contain many frequency components. Some of them may be of interest like frequency bands of EEG rhythms; others like power line artifacts may be undesired. To remove the unwanted part of spectral contents of the signal one usually uses filters. In order to be implementable the filter must be *causal*. That is, the filter response must only depend on the current and past inputs. If the filter has to depend also on the past outputs it still can be causal by introducing the feedback with appropriate delays.

We shall discuss here only the digital filters implemented in software (not the hardware electronic devices used, e.g., as anti-aliasing filters before sampling). Most of the digital filters perform the following action: they compute a linear combination of a number n_b of past inputs x and a number of n_a of past outputs y to evaluate current output $y[n]$:

$$\begin{aligned} y[n] = b(1) * x[n] + b(2) * x[n-1] + \ldots + b(n_b + 1) * x[n - n_b] \\ - a(2) * y[n-1] - \ldots - a(n_a + 1) * y[n - n_a] \end{aligned} \tag{2.1}$$

The maximum delay, in samples, used in creating each output sample is called the *order* of the filter. In the difference-equation representation (2.1), the order is the larger of numbers n_b and n_a. There are three possibilities:

- If $n_a = 0$ and $n_b \neq 0$ then the filter is called finite impulse response (FIR), which means that once the filter is fed with one non-zero pulse the response will have finite duration. After n_b time intervals the output sets back to zero. This filter sometimes is called moving average, MA.

- If $n_b = 0$ and $n_a \neq 0$ the filter has an infinite impulse response (IIR). After a single non-zero sample the filter could produce some output forever. It is also called recursive or autoregressive (AR) filter.

- If $n_b \neq 0$ and $n_a \neq 0$ this is the most general filter type. It is also called IIR or ARMA—autoregressive moving average.

The IIR filters usually have lower orders than FIR filters with the same attenuation properties.

The operation of the filter is much easier to understand in the frequency domain. First let's reorder the equation (2.1):

$$y[n]+a(2)*y[n-1]+...+a(n_a+1)*y[n-n_a] \\ =b(1)*x[n]+b(2)*x[n-1]+...+b(n_b+1)*x[n-n_b] \quad (2.2)$$

Application of Z transform (Sect. 1.4.6) to both sides of (2.2) yields:

$$A(z)Y(z)=B(z)X(z) \quad (2.3)$$

From this we get:

$$Y(z)=A(z)^{-1}B(z)X(z)=H(z)X(z) \quad (2.4)$$

Function H in (2.4) is called the frequency response function, and it has the form:

$$H(z)=\frac{b(1)+b(2)z^{-1}+\cdots+b(n_b+1)z^{-n_b}}{a(1)+a(2)z^{-1}+\cdots+a(n_a+1)z^{-n_a}} \quad (2.5)$$

This function is a ratio of two polynomials. We can factor the numerator and denominator to obtain:

$$H(z)=g\frac{(1-q_1z^{-1})(1-q_2z^{-1})\dots(1-q_{n_b}z^{-1})}{(1-p_1z^{-1})(1-p_2z^{-1})\dots(1-p_{n_a}z^{-1})} \quad (2.6)$$

The numbers $\{q_1,q_2,\dots,q_{n_b}\}$ are zeros of the numerator and are called zeros of the transfer function. The numbers $\{p_1,p_2,\dots,p_{n_a}\}$ are zeros of the denominator and are called poles of the transfer function. The filter order equals the number of poles or zeros, whichever is greater. We can obtain frequency dependent transfer function $H(f)$ substituting $z=e^{i2\pi f}$. The function assigns to each frequency f a complex number with magnitude M and phase ϕ: $H(f)=M(f)e^{i\phi(f)}$. From equation (2.4) we see that the operation of a filter is multiplication of each of the Fourier components of the signal by the complex number $H(f)$; that is, filter changes the magnitude and phase of the component. A very useful measure can be derived from the phase of the frequency response function—the group delay. The group delay, defined as:

$$\tau_g(f)=-\frac{d\phi(f)}{df} \quad (2.7)$$

is a measure of the time delay introduced by the filter to the signal component of frequency f. From the formula (2.7) we see that if the phase ϕ depends linearly on the frequency then $\tau_g(f)=const$. This means that all frequency components are equally delayed. The phase structure of the signal is not distorted, which is essential, if the filtering is used as a preprocessing step for more advanced methods of signal processing that are phase sensitive. Such a linear phase delay is the property of the FIR filters. In case of off-line applications the delays introduced by the filter, in any form (not only the linear ones), can be corrected. The technique relies on applying the same filter twice. After filtering the signal in the forward direction, the filtered sequence is reversed in time and run again through the filter. Due to the time reversal property of the Fourier transform (see Sect. 1.4.4) the component $X(f)$ of the original signal is multiplied effectively by

$$H_{eff}(f) = A(f)e^{i\phi(f)} \cdot A(f)e^{-i\phi(f)} = A(f)^2 \tag{2.8}$$

yielding the zero-phase double order filter. In MATLAB this technique is implemented as `filtfilt` function. The standard filtering corresponding to equation 2.1 is implemented as a `filter` function.

2.1.1 Designing filters

Practical applications of filters require that they meet certain requirements. Often the characteristics of a filter is expressed as properties of the magnitude responses (the absolute value of the transfer function) such as: the cut-off frequency or frequency band to be attenuated or passed, the amount of attenuation of the unwanted spectral components, steepness of the filter, or the order of the transfer function.

In specifying filter characteristics a unit called decibel [dB] is commonly used. Two levels of signal P and P_0 differ by n decibels, if

$$n = 10\log_{10}\frac{P}{P_0}$$

Terms used in specifying filter characteristics are:

Wp - pass band, frequency band that is to be passed through filter without alternation

Ws - stop band, frequency band which has to be attenuated

Rp - allowed pass band ripples, expressed in [dB]

Rs - required minimum attenuation in the stop band, expressed in [dB]

The requirements concerning filter characteristics are usually contradictory, e.g., for a given filter order the increase of steepness of the filter results in bigger ripples. Therefore some optimization process has to be applied and the resulting filter is a compromise between the set of the conditions. A decision must be made which of the possible filter designs is best for the planned use.

Below we describe briefly the functions from the MATLAB Signal Processing Toolbox that facilitate the design and the validation of the filters. In the filter design function in MATLAB the frequencies are expressed as a fraction of the Nyquist frequency F_N (namely are scaled so that $F_N = 1$). FIR filters can be designed by means of the following functions:

fir1 - allows for design of classical lowpass, bandpass, highpass, bandstop filters. It implements an algorithm based on Fourier transform. If the ideal transfer function is $H_{id}(f_n)$, its inverse Fourier transform is $h_{id}(n)$ and $w(n)$ is a window (by default a Hamming window) then the filter coefficients are $b(n) = w(n)h_{id}(n)$ for $n \in [1,N]$, N is the filter order. This filter has group delay $\tau_g = N/2$.

fir2 - this function allows to specify arbitrary piecewise linear magnitude response. The algorithm first interpolates the desired magnitude response onto a dense evenly spaced grid of points, computes the inverse Fourier transform of the result, and multiplies it by the specified window.

firls - allows to specify arbitrary piecewise characteristics of magnitude response and minimizes the sum of squared errors between the desired and the actual magnitude response based on algorithm given in [Park and Burrus, 1987, pp. 54–83].

firpm - function implements the Parks-McClellan algorithm [Park and Burrus, 1987, p.83]. It designs filters that minimize the maximum error between the desired frequency response and the actual frequency response. Filters designed in this way exhibit an equiripple behavior in their frequency response. The Parks-McClellan FIR filter design algorithm is perhaps the most popular and widely used FIR filter design methodology.

Functions for designing IIR filters:

butter - Butterworth filter gives smooth and monotonic magnitude response function.

cheby1 - designs Chebyshev Type I filter. These filters are equiripple in the passband and monotonic in the stopband.

cheby2 - designs Chebyshev Type II filter. Filters are monotonic in the passband and equiripple in the stopband. Type II filters do not roll off as fast as type I filters, but are free of passband ripple.

ellip - designs elliptic filter. Elliptic filters offer steeper rolloff characteristics than Butterworth or Chebyshev filters, but are equiripple in both the pass- and stopbands. In general, elliptic filters, compared to other filters, meet given performance specifications with the lowest order.

Once the filter coefficients are computed it is necessary to analyze the properties of the resulting filter. The transfer function for given filter coefficients $\{a_n\}$ and $\{b_n\}$ can be computed according to equation (2.5); in MATLAB it can be evaluated with `freqz` function. The absolute value of this function gives the frequency magnitude response. A plot of overlaid desired and actual magnitude response allows to check whether the actual magnitude response function is close enough to the desired one (Figure 2.1 a). One should pay attention to the steepness of the magnitude response edges or in other words to the width of the transition between the stop-, and passband. Increasing the filter order should make the transition steeper. One should also consider the magnitude of ripples. They can be made smaller by decreasing the filter order or broadening the transition width.

If the filter is to be used in standard, one direction mode, it is also advisable to observe the group delay function (2.7) for delays of different frequency components. This function can be evaluated by means of `grpdelay` from Signal Processing Toolbox (Figure 2.1 b).

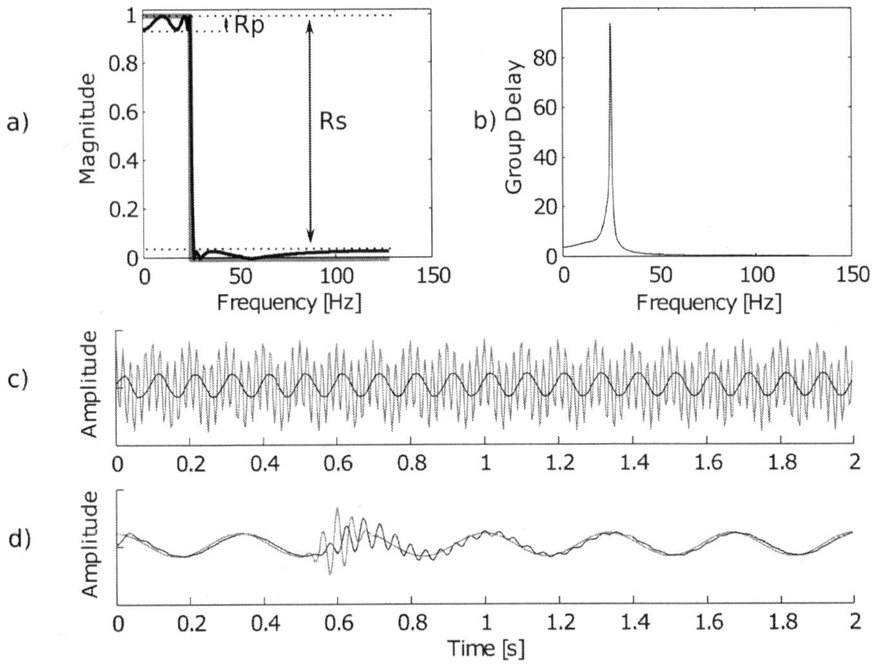

FIGURE 2.1: Illustration of filter properties. a) Magnitude of frequency response function. Gray rectangular outline—the ideal response, black—the magnitude of the designed elliptic 5th order filter. Arrows indicate allowed passband ripples (Rp) and required minimum attenuation in the stop band (Rs). b) Group delay function. c) Application of the designed filter to a signal composed of a 10 Hz and 50 Hz sinusoid: input—gray, output—black line. d) Illustration of the delay and the edge effects of the filter: the input signal (gray) is a 3 Hz sinusoid with a 25 Hz transient at 0.6 s; the output black line shows (i) in the first 0.1 s an edge distortion of the filter (ii) delays: the output 3 Hz sinusoid is delayed slightly compared to the input; the 25 Hz transient is more delayed and spread, which follows from the characteristics of the group delay function (b).

2.1.2 Changing the sampling frequency

In previous sections we described the main field of filter applications—the selection of relevant frequency bands from the signals and suppression of unwanted frequency bands. Here, we would like to mention one more application where the filters are indispensable: the process of resampling the signal at another sampling frequency. Let's imagine that we need to reduce the sampling frequency (*downsample*) of the signal by half. The simplest idea could be skipping every other sample in the original signal. However in most cases this would spoil the signal due to the aliasing (see Sect. 1.2.1.1). In order to do it properly, one needs to take care that the assumption of sampling theorem (Sect. 1.2.1) is fullfiled; that is, the signal contains only frequencies below the Nyquist frequency of the resultant downsampled signal. This is usually achieved by lowpass filtering the original signal before downsampling. If the original signal s_0 is sampled at frequency F_{s_0} and the desired frequency is $F_{s_1} = F_{s_0}/q$ for an integer q, then the frequency band of the signal s_0 needs to be reduced to the range $(0, \frac{1}{2q}F_{s_0})$. In practice, one needs to set the cut-off frequency of the lowpass filter smaller then $\frac{1}{2q}F_{s_0})$, because the filter roll-off is finite. In MATLAB Signal Processing Toolbox the downsampling with anti-aliasing filtering is implemented in function `decimate`.

The opposite process of increasing the sampling rate (*interpolation*) can be accomplished in many ways, without the explicit use of filters. However, the interpolation can also be carried out in the following way. If the sampling frequency has to be increased by integer factor p then between each two samples of the original signal s_0 one needs to insert $p-1$ zeros, and then filter the resulting signal with a lowpass filter. The cut-off frequency of the filter should be set below the Nyquist frequency of the original signal, to remove all the high frequency components introduced to the signal by the discontinuities at the inserted samples. In MATLAB Signal Processing Toolbox the filter interpolation is implemented as function `interp`.

Finally, using a combination of the above described procedures it is possible to change sampling rate by rational factor $\frac{p}{q}$. First the signal is interpolated to p times higher sampling frequency and then downsampled by q. In MATLAB Signal Processing Toolbox the algorithm is implemented as function `resample`.

2.1.3 Matched filters

In many applications one would like to detect in the signal some specific structures, matching the template s and treating the other signal components as noise. Formally, the signal may be described as composed of the template—s and additive noise v:

$$x = s + v \tag{2.9}$$

The matched filter is the linear filter that maximizes the output signal-to-noise ratio. The solution is the filter with transfer function h which is complex conjugate time reversal of $\hat{h}$ given by [Turin, 1960]:

$$\hat{h} = Ks \tag{2.10}$$

where $K = \frac{1}{\sqrt{s^H R_v^{-1} s}} R_v^{-1}$, R_v is covariance matrix of the noise, H denotes Hermitian (conjugate) transpose. In MATLAB matched filtering with template s of signal x and noise v can be realized with the following code:

```
R= cov(v);
K= R^(-1)./sqrt(s'*R^(-1)*s);
h_hat=K*s;
h=fliplr(h_hat');
y=filter(h,1,x);
```

where y is the output of the filter.

2.1.4 Wiener filter

Typical filters described in the beginning of Sect. 2.1 are designed to have a specific frequency response. It is not the case for a Wiener filter, introduced by Norbert Wiener [Wiener, 1949].

Let's assume that a certain system generates signal u. We record the signal with an apparatus which has a known impulse response r. We assume that the measurement is corrupted by the additive Gaussian white noise v. This can be expressed as:

$$x[n] = y[n] + v[n] = \sum_i r[n-i]u[i] + v[n] \tag{2.11}$$

In such a system u cannot be accessed directly. However, minimizing the square error we can find $\hat{u}$ which estimates u :

$$\hat{u}[n] = \arg\min E\left[\sum_n |u[n] - \hat{u}[n]|^2\right] \tag{2.12}$$

From the Parseval theorem it follows, that the above relation will also hold for the Fourier transforms of the respective signals:

$$\hat{U}[f] = \arg\min E\left[\sum_f |U[f] - \hat{U}[f]|^2\right] \tag{2.13}$$

Taking advantage of the convolution theorem (Sect. 1.4.4) for the signal $y[n] = \sum_i r[n-i]u[i]$ we obtain:

$$Y[f] = U[f]R[f] \tag{2.14}$$

Assuming that the estimator has the form:

$$\hat{U}[f] = \frac{X[f]\Phi[f]}{R[f]} \tag{2.15}$$

It can be shown that condition (2.13) is satisfied for:

$$\Phi[f] = \frac{|Y[f]|^2}{|Y[f]|^2 + \langle |V[f]|^2 \rangle} \tag{2.16}$$

$\Phi[f]$ is called a Wiener filter. Thus:

$$H[f] = \frac{\Phi[f]}{R[f]} = \frac{1}{R[f]} \frac{|Y[f]|^2}{|Y[f]|^2 + \langle |V[f]|^2 \rangle} \tag{2.17}$$

gives the frequency domain representation of the transfer function of the optimal filter.

In practice neither $Y[f]$ nor $V[f]$ is known. The empirical estimate of $\Phi[f]$ may be computed as:

$$\hat{\Phi}[f] = \begin{cases} \frac{S_x[f] - \hat{S}_v[f]}{S_x[f]} & for\ S_x[f] > \hat{S}_v[f] \\ 0 & for\ S_x[f] \leq \hat{S}_v[f] \end{cases} \tag{2.18}$$

where $S_x[f]$ is the power spectrum of x and $\hat{S}_v[f]$ is the approximation of the noise power spectrum.

2.2 Probabilistic models

2.2.1 Hidden Markov model

An excellent introduction to the Hidden Markov models (HMM) concept, estimation and illustrative applications can be found in [Rabiner, 1989]. Here we review the basic facts only.

Let us consider a system which can be in a number of *states*. In discrete time moments the state of the system may change. The transitions between states are probabilistic. The probability distribution of the next state depends only on the current state and it doesn't depend on the way the system reached the current state. If the states of the system can be observed directly then it can be mathematically described as a regular Markov model. The Markov model is specified by a set of states and the probabilities of transitions between each pair of states. An extension of the Markov model allowing for much wider applications is a Hidden Markov model (HMM). It is hidden in the sense that the states of the system are not observed directly, but the observations of the model depend in the probabilistic way on the states of the model. Although the states are not observed directly, often some physical sense can be attached to them. As the system evolves in time the state transitions occur and the system passes through a sequence of states $S(t)$. This sequence is reflected in a sequence of observations $Y(t)$. An important property of HMM is that the conditional probability distribution of the hidden state $S(t)$ at time t, depends only on the hidden state $S(t-1)$. Similarly, the observation $Y(t)$ depends only on the hidden state $S(t)$—both occurring at the same time t. A simple example of such a model is shown in Figure 2.2.

The HMM is specified by the pair of parameters: number of states (N) and number of possible observations per state (M), and additionally by three probability distributions governing: state transitions (A), observations (B), initial conditions (π). The HMMs can be used to generate possible sequences of observations. But more interesting applications of HMMs rely on:

- Estimating the model parameters (N, M, A, B, π) given the sequence of observations. This can be done with maximum likelihood parameter estimation using expectation-maximization Baum-Welch algorithm [Baum et al., 1970].

- Given a sequence of observations one can evaluate probability that the sequence was generated by a specific model (N, M, A, B, π).

Imagine that we have a set of observations known to be generated by a number of different models. We can estimate parameters of each of the models. Later, when a new sequence of observations is obtained, we could identify which model most probably generated that sequence. In this sense HMMs can be used as classifiers with supervised learning.

Another problem that can be addressed with HMM is the question of the most probable sequence of states given the model and the sequence of observations. The

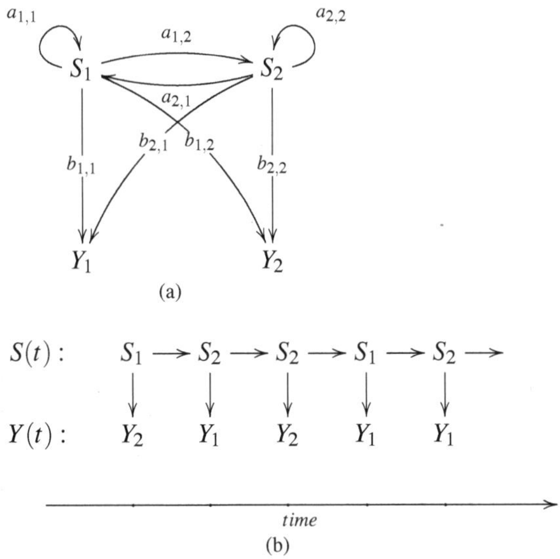

FIGURE 2.2: (a) An example of two state Hidden Markov model. S_1 and S_2 are states of the model, $a_{i,j}$ are probabilities of transition from state i to j. Y_1 and Y_2 are observations, $b_{i,j}$ are probabilities of observation Y_j if the system is in state S_i. (b) A possible sequence of states and observations. Arrows indicate the dependence.

solution to this problem is usually obtained with the Viterbi algorithm [Viterbi, 1967].

Hidden Markov Model (HMM) Toolbox for MATLAB written by Kevin Murphy supports maximum likelihood parameter estimation, sequence classification and computation of the most probable sequence. The toolbox can be downloaded from `http://www.cs.ubc.ca/~murphyk/Software/HMM/hmm.html`.

2.2.2 Kalman filters

A Kalman filter is a method for estimation of the state of a linear dynamic system (LDS) discretized in time domain. At time k the system is in a *hidden* state x_k. It is hidden since it can't be directly observed. The observer's knowledge about the state of the system comes from measurements z_k which are distorted by noise w_k. Formally it can be expressed as:

$$z_k = \mathbf{M}_k x_k + w_k \tag{2.19}$$

where $\mathbf{M}_k$ is the matrix describing the linear operation of taking the observation. The measurement noise w_k is assumed to come from a zero mean normal distribution with covariance matrix $\mathbf{R}_k$.

The current state of the system x_k is assumed to depend only on the previous state x_{k-1}, on the current value of a control vector u_k, and current value of a random perturbation v_k. Expressing it as an equation we get:

$$x_k = \mathbf{A}_k x_{k-1} + \mathbf{B}_k u_k + v_k \tag{2.20}$$

where $\mathbf{A}_k$ is the state transition matrix, $\mathbf{B}_k$ is the matrix transforming the control input, v_k is the process noise coming from the zero mean normal distribution with covariance matrix $\mathbf{Q}_k$. The matrixes $\mathbf{M}_k$, $\mathbf{R}_k$, $\mathbf{A}_k$, $\mathbf{B}_k$, and $\mathbf{Q}_k$ may vary in time. The initial state, and the noise vectors at each step $\{x_0, w_1, \ldots, w_k, v_1, \ldots, v_k\}$ are all assumed to be mutually independent.

In the estimation of the current state of the system all the above mentioned matrixes are assumed to be known. Two types of error can be defined: *a priori* estimate error:

$$e_k^- = x_k - \hat{x}_k^- \tag{2.21}$$

where $\hat{x}_k^-$ is the estimate of state x_k based only on the knowledge of previous state of the system, with covariance matrix given by:

$$\mathbf{P}_k^- = E[e_k^- e_k^{-T}] \tag{2.22}$$

and *a posteriori* estimate error, with the result z_k of observation known:

$$e_k = x_k - \hat{x}_k \tag{2.23}$$

with covariance matrix given by:

$$\mathbf{P}_k = E[e_k e_k^T] \tag{2.24}$$

Posterior estimate of the system state in current step can be obtained as a linear mixture of the *a priori* estimate of the system state and the error between the actual and estimated result of the observation:

$$\hat{x}_k = \hat{x}_k^- + \mathbf{K}(z_k - \mathbf{M}\hat{x}_k^-) \tag{2.25}$$

The matrix **K** is chosen such that the Frobenius norm of the *a posteriori* covariance matrix is minimized. This minimization can be accomplished by first substituting (2.25) into the above definition for e_k (2.23), substituting that into (2.24), performing the indicated expectations, taking the derivative of the trace of the result with respect to **K**, setting that result equal to zero, and then solving for **K**. The result can be written as:

$$\mathbf{K}_k = \mathbf{P}_k^- \mathbf{M}^T \left(\mathbf{M}\mathbf{P}_k^- \mathbf{M}^T + \mathbf{R}\right)^{-1} \tag{2.26}$$

The Kalman filter estimates the state of LDS by using a form of feedback control: the filter estimates the process state at some time and then obtains feedback in the form of (noisy) measurements. Two phases of the computation may be distinguished: predict and update. The predict phase uses the state estimate from the previous time step to produce an estimate of the current time step. In the update phase the current *a priori* prediction is combined with the current observation to refine the *a posteriori* estimate. This algorithm is presented in Figure 2.3. Excellent introduction to the

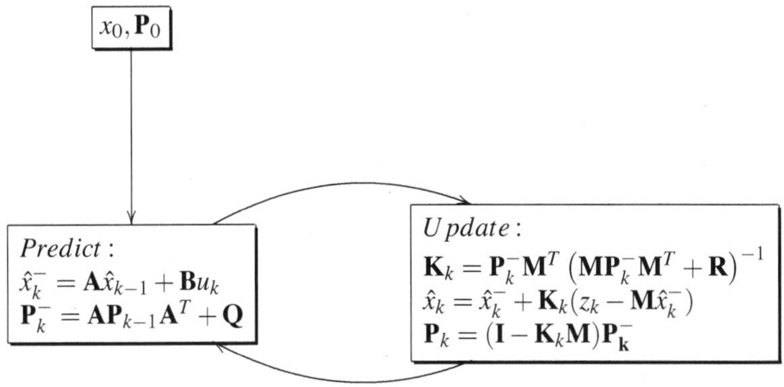

FIGURE 2.3: The computation of successive system states with Kalman filter.

Kalman filter may be found in [Welch et al., 1960].

MATLAB toolbox by Kevin Murphy avaliable at `http://people.cs.ubc.ca/~murphyk/Software/Kalman/kalman.html` provides Kalman filter formalism for filtering, smoothing, and parameter estimation for linear dynamical systems.

Many real dynamical systems do not exactly fit the model assumed by Kalman filter design. The not-modeled dynamics contributes to the component w_k in equation 2.19. In practice we cannot distinguish between the uncertainty of the measurement and the not modeled dynamics. We have to assume a certain value of $\mathbf{R}$. Too big $\mathbf{R}$ results in the slow adaptation of the filter and too small $\mathbf{R}$ leads to instability of the results. $\mathbf{R}$ controls the speed of adaptation and it has to be chosen in an optimal way.

2.3 Stationary signals

2.3.1 Analytic tools in the time domain

2.3.1.1 Mean value, amplitude distributions

In Sect. 1.1 formulas for calculation of moments such as mean, variance, standard deviation, correlation, etc., were based on ensemble averaging paradigm. Quite often we have only one realization of a process. Under the assumption of ergodicity one can calculate the estimators by averaging over time instead of averaging over an ensemble.

We have to bear in mind that in case of application of ergodicity assumption, the strict value of the estimator is obtained for an infinitely long time. In practice the shorter the time epochs, the higher the error of estimate will be.

2.3.1.2 Entropy and information measure

The concept of information measure, in statistics called entropy, was introduced by Shannon [Shannon, 1948]. It is connected with the probability of occurrence of a given effect. Let us assume that an event has M possibilities, and i^{th} possibility occurs with probability p_i, then the information connected with the occurrence of ith possibility will be:

$$I_i = -\log p_i \tag{2.27}$$

The expected value of the information is entropy:

$$En = -\sum_{i=1}^{M} p_i \log p_i \tag{2.28}$$

Entropy is a measure of uncertainty associated with the outcome of an event. The higher the entropy, the higher the uncertainty as to which possibility will occur. The highest value of entropy is obtained when all possibilities are equally likely.

In practice for time series the entropy is calculated from the amplitude distributions. The amplitude range A of a raw sampled signal is divided into K disjointed intervals I_i, for $i = 1,\ldots,K$. The probability distribution can be obtained from the ratio of the frequency of the samples N_i falling into each bin I_i and the total sample number N:

$$p_i = N_i/N \tag{2.29}$$

The distribution $\{p_i\}$ of the sampled signal amplitude is then used to calculate entropy measure according to (2.28).

2.3.1.3 Autocorrelation function

Correlation function was introduced in Sect. 1.1 by equation (1.11) in terms of ensemble averaging. Under the assumption of ergodicity, autocorrelation $R_x(\tau)$ and autocovariance functions $C_x(\tau)$ are defined by:

$$R_x(\tau) = \int_{-\infty}^{+\infty} x(t)x(t+\tau)dt \tag{2.30}$$

and

$$C_x(\tau) = \int_{-\infty}^{+\infty} (x(t)-\mu_x)(x(t+\tau)-\mu_x)dt \tag{2.31}$$

where τ is the delay, μ_x is mean of the signal (equation 1.2). Autocorrelation function $R_x(\tau)$ is always real and symmetric: $R_x(\tau) = R_x(-\tau)$. It takes maximal value for $\tau = 0$. It follows from equations (1.5) and (1.11) that $R_x(\tau = 0) = \Psi^2$ (the mean square value of the signal). Variance σ_x^2 is equal to the autocovariance function for time 0: $C_x(\tau = 0) = \sigma_x^2$. Autocorrelation of a periodic function is also periodic. Autocorrelation of noise decreases rapidly with the delay τ (Figure 2.4). We can see from Figure 2.4 c that autocorrelation can help in the extraction of the periodic signal from noise, even when the noise amplitude is higher than the signal.

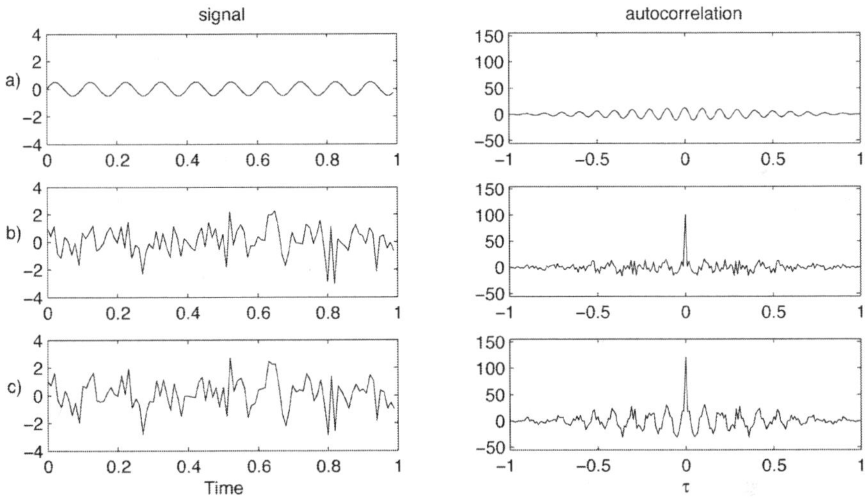

FIGURE 2.4: Illustration of autocorrelation function properties. Left column signals, right column autocorrelation functions of the corresponding signals. a) Periodic signal, b) zero mean Gaussian noise, c) sum of a) and b).

2.3.2 Analytic tools in the frequency domain

2.3.2.1 Estimators of spectral power density based on Fourier transform

In physiological signals quite often important information is connected with the occurrence of specific rhythms; therefore the analysis in frequency domain is of particular interest. A commonly used method of estimation of the power spectrum $S(f)$ is based on the Fourier transform:

$$S(f) = X(f)X^*(f) \tag{2.32}$$

$X(f)$ is the Fourier transform of the signal $s(t)$ defined by equation (1.27), where Fourier transform is defined for times $\in (-\infty, +\infty)$. Fourier transform for the sampled signal and for finite time of observation is expressed by formula (1.28), derived under the assumption that the signal given in the time window, outside it, is periodically repeated to infinity (Figure 2.5). In practice, taking the finite segment of the signal for $t \in (-T/2, T/2)$ is equivalent to multiplication of the signal by a function w:

$$w_{T/2}(t) = \begin{cases} 0 \text{ for } & t < -T/2 \\ w \text{ for } & -T/2 \leq t \leq T/2 \\ 0 \text{ for } & t > T/2 \end{cases} \tag{2.33}$$

where $w = 1$. This procedure introduces discontinuities at the edges of the window which disturb the spectrum. The discontinuities contribute to the spectrum at many frequencies, causing an effect known as spectral leakage. Multiplication in the time

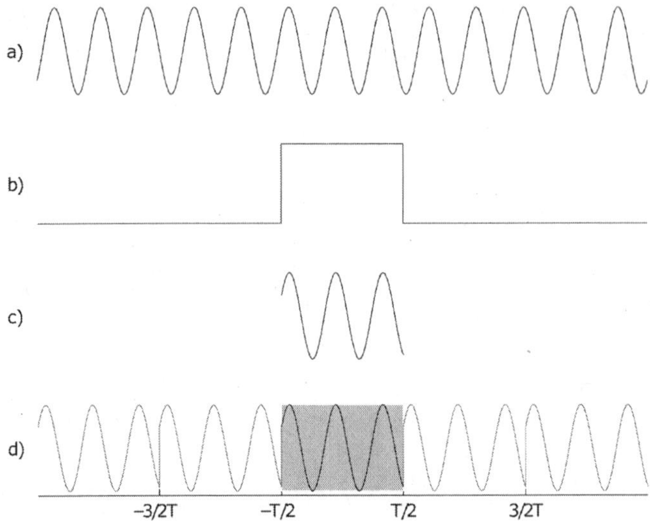

FIGURE 2.5: The illustration of assumptions of DFT. a) The assumed signal. b) Observation window (equation (2.33)) c) The observed fragment of signal. d) The signal from observation window (marked with gray rectangle) is periodically repeated to infinity. The discontinuities of the signal at the window edge introduce additional components to the frequency spectrum.

domain corresponds to convolution in the frequency domain (see equation 1.31). Therefore the transform we obtain can be considered as the convolution of the transform of idealized infinite signal $X(f)$ with the transform of the window function $W_{T/2}(f)$. This property deteriorates the spectral estimate; especially disturbing are the side lobes of function $W(f)$.

2.3.2.1.1 Choice of windowing function A commonly used technique to reduce the spectral leakage is the choice of the proper function $w_{T/2}(t)$. Usually one wants the window w to smoothly approach zero at both edges. In order to improve estimation of spectral power, windows of different shapes were introduced. Good windows are characterized by a highly concentrated central lobe with very low or quickly diminishing side lobes of their transforms. In MATLAB Signal Processing Toolbox there is a very convenient tool for studying windows properties: `wintool`. It displays the time and frequency representation of the selected window and evaluates its important characteristics:

- Leakage factor—ratio of power in the side lobes to the total window power
- Relative side lobe attenuation—difference in height from the main lobe peak to the highest side lobe peak
- Width of the main lobe at 3 dB below the main lobe peak

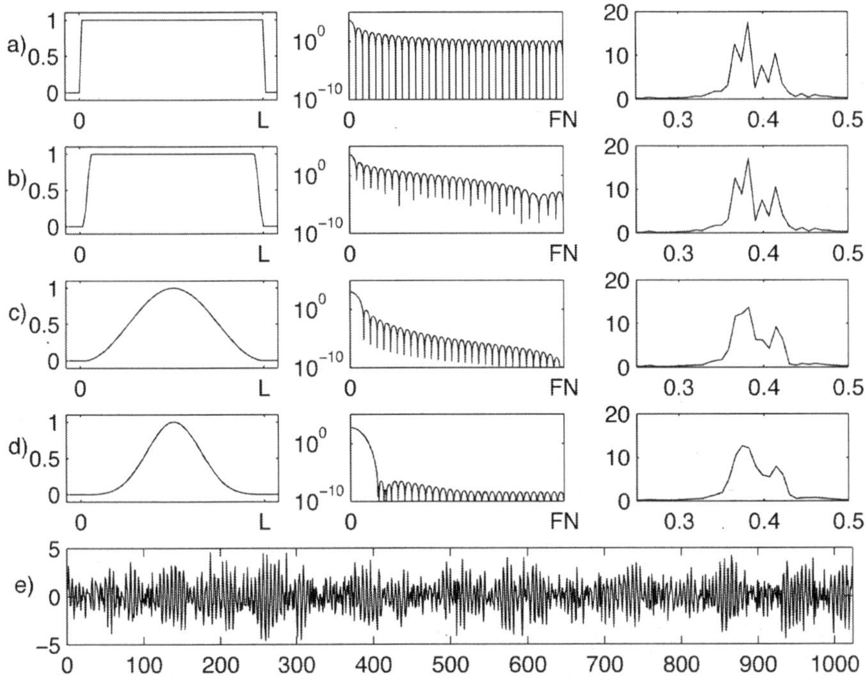

FIGURE 2.6: First column—window in time domain, second column—window in frequency domain, third column—spectra of signal e) obtained by application of respective window function and Fourier transform. e) signal composed of autoregressive time series with spectral peak around 0.375 and sinusoid with frequency 0.42. The windows are: a) rectangular b) Tukey with 10% round off c) Hann window d) Blackmann-Harris. Note the spectrum leakage effects in form of spurious peaks in case of rectangular and Tukey windows.

The avaliable windows are: `barthannwin`, `bartlett`, `blackman`, `blackmanharris`, `bohmanwin`, `chebwin`, `flattopwin`, `gausswin`, `hamming`, `hann`, `kaiser`, `nuttallwin`, `parzenwin`, `rectwin`, `triang`, `tukeywin`.

For a signal consisting of multiple frequencies the applied window has considerable influence on the detectability of individual spectral components [Harris, 1978]. Examples of windows and their properties are shown in Figure 2.6.

2.3.2.1.2 Errors of Fourier spectral estimate Fourier transform $X(f)$ is a complex number whose real $X_R(f)$ and imaginary $X_I(f)$ parts can be considered as uncorrelated random values of zero mean and equal variance. Since Fourier transform is a linear operation, components $X_R(f)$ and $X_I(f)$ have a normal distribution, if $x(t)$ has normal distribution. Therefore value:

$$|X(f)|^2 = X_R^2(f) + X_I^2(f) \tag{2.34}$$

is a sum of squares of two independent variables of normal distribution; hence each frequency component of estimator $\hat{S}(f)$ has a distribution given by:

$$\frac{\hat{S}(f)}{S(f)} = \frac{\chi_2^2}{2} \tag{2.35}$$

where χ_2^2 is a chi-square statistics of two degrees of freedom (corresponding to the real and imaginary parts of Fourier transform) [Bendat and Piersol, 1971]. Please note that the above expression is independent of the observation time T, which means that increasing of epoch T does not change the distribution function describing the error of the estimator. The increase of the time of observation T increases only the number of the frequency components of the spectrum. It means that the estimator of the spectral density obtained by means of Fourier transform is biased; also the error of the estimator is high. Namely for a given frequency f_1 the relative error of $S(f_1)$ is: $\varepsilon_r = \sigma_{S_{f_1}}/\mu_{S_{f_1}}$. For the distribution χ_n^2: $\sigma^2 = 2n$ and $\mu = n$, where n is a number of degrees of freedom. For $n = 2$ we get $\varepsilon_r = 1$, which means that each single frequency component of $S(f)$ has a relative error 100%. In consequence spectral power calculated by means of Fourier transform is strongly fluctuating.

In order to improve statistical properties of the estimate two techniques were introduced. The first one relies on averaging over neighboring frequency estimates. Namely we calculate smoothed estimator $\hat{S}_k$ of the form:

$$\hat{S}_k = \frac{1}{l}[S_k + S_{k+1} + \cdots + S_{k+l-1}] \tag{2.36}$$

Assuming that the frequency components S_i are independent, estimator S_k is characterized by the χ^2 distribution of number degrees of freedom equal $n = 2l$. The relative standard error of frequency estimate will be therefore: $\varepsilon_r = \sqrt{\frac{1}{l}}$.

Another way of improving power spectral estimate is averaging the estimates for successive time epochs:

$$\hat{S}_k = \frac{1}{q}[S_{k,1} + S_{k,2} + \cdots + S_{k,j} + \cdots + S_{k,q}] \tag{2.37}$$

where $S_{k,j}$ is the estimate of the frequency component k based on the time interval j. The number of degrees of freedom in this case equals q; therefore the relative error of single frequency estimate will be: $\varepsilon_r = \sqrt{\frac{1}{q}}$. This approach is known as Welch's method and is implemented in MATLAB Signal Processing Toolbox as `pwelch`.

Both of these methods require stationarity of the signal in the epoch long enough in order to: (i) perform averaging over frequency estimates, without excessive decreasing of frequency resolution or (ii) divide the epoch into segments long enough to provide necessary frequency resolution.

An alternative method, which can be useful in case of relatively short time epochs, is the multitaper method (MTM) described in [Thomson, 1982]. This method uses a sequence of windows that are orthogonal to each other (discrete prolate spheroidal sequences). Each window is used to compute the windowed periodogram of the signal. Subsequently the periodograms are averaged. This method is implemented in MATLAB Signal Processing Toolbox as `pmtm`.

2.3.2.1.3 Relation of spectral density and the autocorrelation function According to the Wiener-Chinchyn formula spectral density function $S(f)$ is a Fourier transform of the autocorrelation function $R(\tau)$:

$$S(f) = \int_{-\infty}^{\infty} R(\tau) e^{i2\pi f\tau} \, d\tau \tag{2.38}$$

Assuming that $R(\tau)$ exists and

$$\int_{-\infty}^{\infty} |R(\tau)| \, d\tau < \infty \tag{2.39}$$

$R(\tau)$ is connected to $S(f)$ by inverse Fourier transform.

$$R(\tau) = \int_{-\infty}^{\infty} S(f) e^{-i2\pi f\tau} \, df \tag{2.40}$$

From the above formula it follows that the integral of the spectral density is equal to $R(0)$. Usually, one-sided spectral density estimator $S(f)$ is calculated for $f \in (0, \infty)$, since for real signals $S(f)$ is a symmetric function.

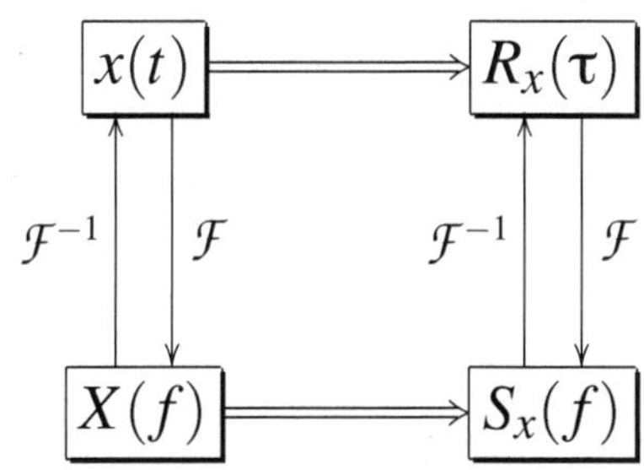

FIGURE 2.7: Illustration of relations between signal $x(t)$, its Fourier transform $X(f)$, its autocorrelation function $R_x(\tau)$, and its spectrum $S_x(f)$.

2.3.2.1.4 Bispectrum and bicoherence Polyspectra or higher order spectra provide supplementary information to the power spectrum. Third order polyspectrum is called bispectrum. The prefix bi refers not to two time series, but to the two frequencies of a single signal. Power spectrum according to equation 2.38 is a Fourier transform of autocorrelation function, which is also called second order cumulant.

Bispectrum is a Fourier transform of third order cumulant $R_3(\tau_1, \tau_2)$. For discrete case (sampled signal) correlation is expressed by:

$$R_2(\tau) = \sum_{t=-\infty}^{\infty} x(t)x(t+\tau) \tag{2.41}$$

and third order cumulant by:

$$R_3(\tau_1, \tau_2) = \sum_{t=-\infty}^{\infty} x(t)x(t+\tau_1)x(t+\tau_2) \tag{2.42}$$

Bispectrum (BS) is defined as:

$$BS(f_1, f_2) = X(f_1)X(f_2)X^*(f_1+f_2) \tag{2.43}$$

where X denotes Fourier transform of signal x. Bispectrum quantifies the relationship between the sinusoids at two primary frequencies f_1 and f_2 and the modulation component at frequency $f_1 + f_2$. Bispectrum is a function of the triplet of frequencies $(f_1, f_2, f_1 + f_2)$ incorporating both power and phase information. Bicoherence, which takes values from the range $[0-1]$, is defined as a squared normalized version of the bispectrum:

$$B^2(f_1, f_2) = \frac{|BS(f_1, f_2)|}{S(f_1)S(f_2)S(f_1+f_2)} \tag{2.44}$$

Bicoherence is a function which gives the information on non-linear interactions. It is a measure for quantifying the extent of phase coupling between different frequencies in the signal. Namely bicoherence measures the proportion of the signal energy at any bifrequency that is quadratically phase coupled.

In practical applications, the computation of bicoherence is limited by the quality of the data. The statistical error of spectral estimate is high as was mentioned already in Sect. 2.3.2.1.2. For bicoherence these errors cumulate due to the multiplication of spectral terms. Therefore reliable estimation of bicoherence is possible only for a high signal to noise ratio and requires long enough data allowing to compute an average estimate based on several equivalent segments of the signal.

2.3.2.2 Parametric models: AR, ARMA

A time series model that approximates many discrete stochastic and deterministic processes encountered in practice is represented by the ARMA filter difference equation:

$$x_t = \sum_{i=1}^{p} a_i x_{t-i} - \sum_{k=0}^{r} b_k y_{n-k} \tag{2.45}$$

Where x_t is the output sequence of a causal filter and y_t is an input driving sequence. In the ARMA model it is assumed that the driving sequence is white noise process of zero mean.

AR model is defined by the equation:

$$x_t = \sum_{i=1}^{p} a_i x_{t-i} + \varepsilon_t \tag{2.46}$$

where ε_t is a white noise. It has been shown [Marple, 1987] that the AR model of sufficiently high model order can approximate the ARMA model well. The determination of the model order and the coefficients for the AR model is much simpler than for the ARMA model which requires non-linear algorithms for the estimation of parameters. Many efficient algorithms of AR model fitting exist; therefore in the following we shall concentrate on AR model. More details on ARMA may be found in [Marple, 1987]. One more reason for application of the AR model in practice is the fact that in transfer function $H(z)$ (equation 2.6), which describes the spectral properties of the system, AR model coefficients determine the poles of transfer function, which correspond to the spectral peaks. MA coefficients determine the zeros of $H(z)$, which correspond to the dips of spectrum. In the practice of biomedical signals processing, we are usually more interested in the spectral peaks, since they correspond to the rhythms present in the time series.

2.3.2.2.1 AR model parameter estimation There is a number of algorithms for estimation of the AR model parameters. Here we present the Yule-Walker algorithm. First, equation (2.46) is multiplied by the sample at time $t-m$:

$$x_t x_{t-m} = \sum_{i=1}^{p} a_i x_{t-i} x_{t-m} + \varepsilon_t x_{t-m} \tag{2.47}$$

then we calculate expected values of the left and right sides of the equation taking advantage of the linearity of the expected value operator:

$$R(m) = E\{x_t x_{t-m}\} = \sum_{i=1}^{p} E\{a_i x_{t-i} x_{t-m}\} + E\{\varepsilon_t x_{t-m}\} \tag{2.48}$$

It is easy to see that the expected value $E\{x_t x_{t-m}\}$ is the autocorrelation function $R(m)$, and $E\{\varepsilon_t x_{t-m}\}$ is nonzero only for $t = m$ so:

$$R(m) = \sum_{i=1}^{p} a_i R(m-i) + \sigma_\varepsilon^2 \delta(m) \tag{2.49}$$

where $m = 0, \ldots, p$.

For $m > 0$ we can write a set of equations:

$$\begin{bmatrix} R(0) & R(-1) & \ldots & R(1-p) \\ R(1) & R(0) & R(-1) & \\ \vdots & & & \vdots \\ R(p-1) & \ldots & \ldots & R(0) \end{bmatrix} \begin{bmatrix} a_1 \\ a_2 \\ \vdots \\ a_p \end{bmatrix} = \begin{bmatrix} R(1) \\ R(2) \\ \vdots \\ R(p) \end{bmatrix} \tag{2.50}$$

and compute the coefficients a.

For $m = 0$ we have:

$$R(0) = \sum_{i=1}^{p} a_i R(-i) + \sigma_\varepsilon^2 \tag{2.51}$$

which allows us to compute σ_ε^2

Methods for estimation of AR model parameters implemented in MATLAB Signal Processing Toolbox [Orfanidis, 1988, Kay, 1988] are:

- `arburg` Estimate AR model parameters using Burg method
- `arcov` Estimate AR model parameters using covariance method
- `armcov` Estimate AR model parameters using modified covariance method
- `aryule` Estimate AR model parameters using Yule-Walker method

2.3.2.2.2 Choice of the AR model order When fitting the AR model to the signal we have to make an assumption that the signal can be described by the autoregressive process. The correct order of the process can be assessed by finding the minimum of the Akaike information criterion (AIC), which is a function of the model order p:

$$AIC(p) = \frac{2p}{N} + \log V \tag{2.52}$$

where p is the model order (the number of free parameters in the model), N - number of signal samples used for model parameters estimation, V - residual noise variance. The higher the model order, the less variance remains unaccounted; however an excessive number of model parameters increases the statistical uncertainty of their estimates. The AIC is a sort of cost function that we seek to minimize. The first element of that function expresses punishment for using high order model and the second element expresses a reward for reducing the unexplained variance.

AIC is the one mostly applied in practice, but other criteria may be commonly found in the literature, e.g., minimum description length *MDL* [Marple, 1987]:

$$MDL(p) = \frac{p \log N}{N} + \log V \tag{2.53}$$

MDL is said to be statistically consistent, since $p \log N$ increases with N faster than with p. The criteria for model order determination work well and give similar results when the data are reasonably well modeled by AR process. An example of an *AIC* function is shown in Figure 2.8.

2.3.2.2.3 AR model power spectrum Once the model is properly fitted to the signal all the signal's properties are contained in the model coefficients. Especially, we can derive an analytical formula for AR model power spectrum. First we rewrite the model equation (2.46):

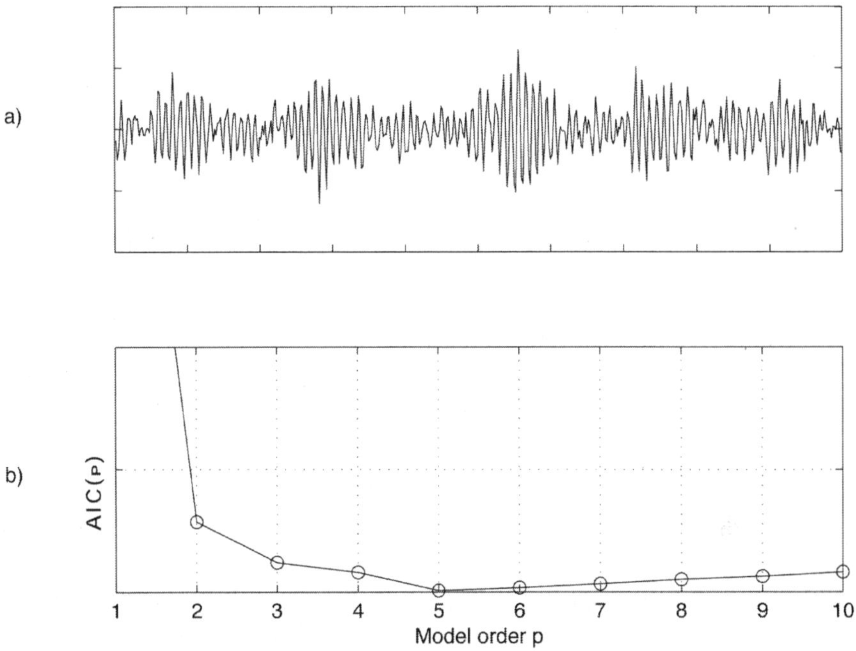

FIGURE 2.8: Example of *AIC* function. a) Signal generated with AR model order 5 ($a_1 = 0.2, a_2 = -0.5, a_3 = -0.3, a_4 = 0.1, a_5 = 0.2$), b) *AIC* for models of order p estimated for that signal. Note a minimum at $p = 5$.

$$\sum_{i=0}^{p} a_i x_{t-i} = \varepsilon_t \tag{2.54}$$

Applying Z transform to both sides of the above equation and taking into account its properties (Sect. 1.4.6) we get:

$$A(f)X(f) = E(f) \tag{2.55}$$

multiplying equation (2.55) by A^{-1} leads to:

$$X(f) = A^{-1}(f)E(f) = H(f)E(f) \tag{2.56}$$

where we defined $H(f) = A^{-1}(f)$. From the definition of power spectrum we obtain:

$$S(f) = X(f)X^*(f) = H(f)VH^*(f) \tag{2.57}$$

Expressing H(s) in terms of model coefficients:

$$H(z) = \frac{1}{\sum_{k=0}^{p} a_k z^{-k}} \tag{2.58}$$

we get:

$$S(f) = \frac{V}{2\pi} \frac{1}{\left|\sum_{k=0}^{p} a_k z^{-k}\right|^2}, \qquad z = e^{i2\pi f} \tag{2.59}$$

The AR spectral estimate is maximum entropy method. The entropy of information (strictly speaking the entropy rate for infinite process) is connected with power spectral density function by the formula [Smylie et al., 1973]:

$$En = \frac{1}{4F_N} \int_{-F_N}^{+F_N} \log S(f) df \tag{2.60}$$

It was pointed out by [Ulrych and Bishop, 1975] that AR spectral estimate is equivalent to the estimate fulfilling maximum entropy condition, which means that the AR estimate expresses maximum uncertainty with respect to the unknown information but is consistent with the known information.

An illustrative comparison of parametric and non-parametric spectra estimates for EEG signal is shown in Figure 2.9. Estimates obtained with the AR model and

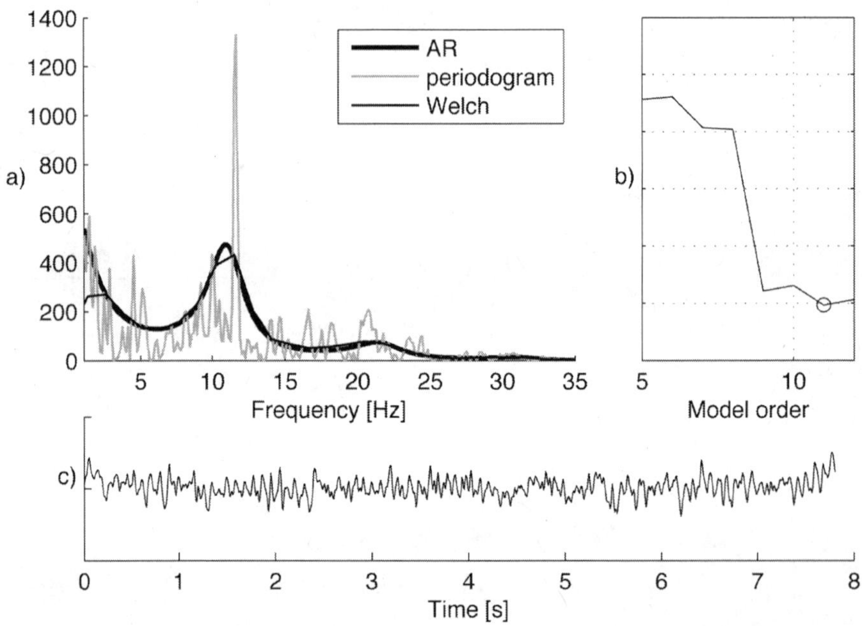

FIGURE 2.9: Comparison of AR and Fourier spectra. a) Spectra: thick line—spectrum of AR model with the lowest AIC, thin black line—power spectral density estimate via Welch's method window length is 1/10 signal length and the overlap is half the window length, gray line–modified periodogram with Blackmann-Harris window. b) *AIC* for different model orders, estimated for the signal c) (minimum is marked with a circle). c) An epoch of EEG signal; sampling frequency 128Hz.

with Welch's method (averaged periodogram) show significant reduction of variation compared with modified periodogram. AR power spectrum is unbiased, free of windowing effects, and it has better statistical properties than Fourier spectrum, since the number of the degrees of freedom in case of AR is greater; namely it is equal to N/p, where N is the number of points of the data window and p is a model order. Correct estimation of the model order is important, but small deviations of p do not drastically change the power spectrum. In case of AR model fitting, the signal should not be oversampled. Sampling should be just slightly above the Nyquist frequency. Oversampling does not bring any new information and it causes redundancy. In case of the autoregressive model in order to account for all frequencies, especially low frequencies, we have to increase the number of steps backward in autoregressive process, which means that we have to increase the model order. Estimating more parameters (from the same amount of data) increases the uncertainty of the estimates.

2.3.2.2.4 Parametric description of the rhythms by AR model, FAD method

The transfer function $H(z)$ has maxima for z values corresponding to the zeroes of the denominator of the expression (2.58). These values of z are the poles z_j lying inside the unit circle $|z| = 1$ in the complex plane. The frequency axis $0 < f < f_N$ in polar coordinates (Figure2.10) corresponds to the counterclockwise traverse of the upper half of this unit circle. The angular position of z_j lying closest to the unit circle determines the frequency of the corresponding peak, and the radial distance depends on the damping of the relevant frequency component. From the coefficients of the AR model we can derive parameters which characterize oscillatory properties of the underlying process. Namely the transfer function $H(z)$ equation (2.58) for the corresponding continuous system takes a form:

$$H(z) = \sum_{j=1}^{p} C_j \frac{z}{z - z_j} \tag{2.61}$$

It can be interpreted as a system of parallel filters each of them corresponding to a characteristic frequency (Figure 2.11). For $z = z_j$ the resonance for filter j occurs. We can introduce parameters characterizing these filters [Franaszczuk and Blinowska, 1985]:

$$\alpha_j = \frac{1}{\Delta t} \log z_j, \quad \beta_j = -Re(\alpha_j), \quad \omega_j = Im(\alpha_j), \quad \phi_j = Arg(C_j), \quad B_j = 2|C_j| \tag{2.62}$$

where Δt is the sampling interval. The imaginary part of α_j corresponds to the resonance frequency ω_j and the real part to the damping factor β_j of the oscillator generating this frequency. By rewriting $H(z)$ in terms of parameters from (2.62) and performing inverse Laplace transform we find the impulse response function in the form of damped sinusoids (Figure 2.11):

$$h(t) = \sum_{j=1}^{p} B_j e^{-\beta_j t} cos(\omega_j t + \phi_j) \tag{2.63}$$

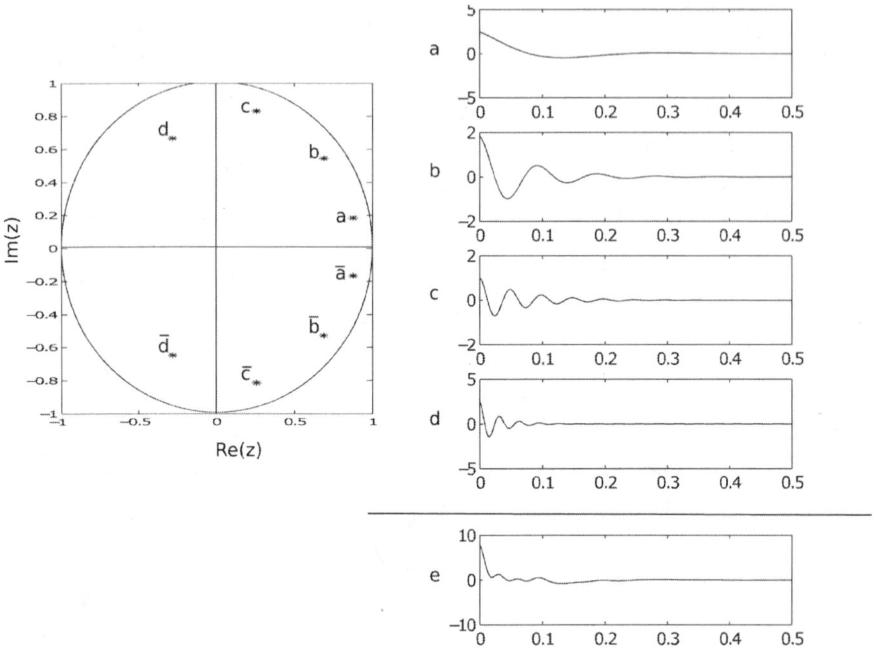

FIGURE 2.10: Left: poles of illustrative AR transfer function in the complex plane. For real signals the transfer function has pairs of complex conjugate poles (here: a and $\bar{a}$, b and $\bar{b}$, c and $\bar{c}$, d and $\bar{d}$). These poles z_j lie inside the unit circle $|z| = 1$ in the complex plane. The frequency f related to the pole $(0 < f < f_N)$ corresponds to the angular position of the pole. Right: traces a-d are impulse response functions corresponding to the poles a-d; each of them is in the form of an exponentially damped sinusoid. Trace e—total impulse response of the AR model is the sum of the components a, b, c, and d.

In this way we can find directly (not from the spectral peak positions) the frequencies of the rhythms present in the signal, which can be quite useful, especially for weaker components. We can also find the amplitudes and damping factors of the oscillations. The method of description of time series by means of parameters: frequency (ω_j), amplitude (B_j), damping (β_j) was named FAD. The number of identified oscillatory components depends on the model order; for even model order it is $p/2$. Model of odd order contains the non-oscillatory component $e^{-\beta}$, which accounts for the usually observed form of the power spectrum decaying with frequency. The correct estimation of the model order is important for description of signals in terms of FAD parameters. For a too low model order, not all components will be accounted for. For a too high model order, too many poles in the complex plane will appear. They will lie close to the main frequency components and will give the effect of "sharpening" the spectral peaks.

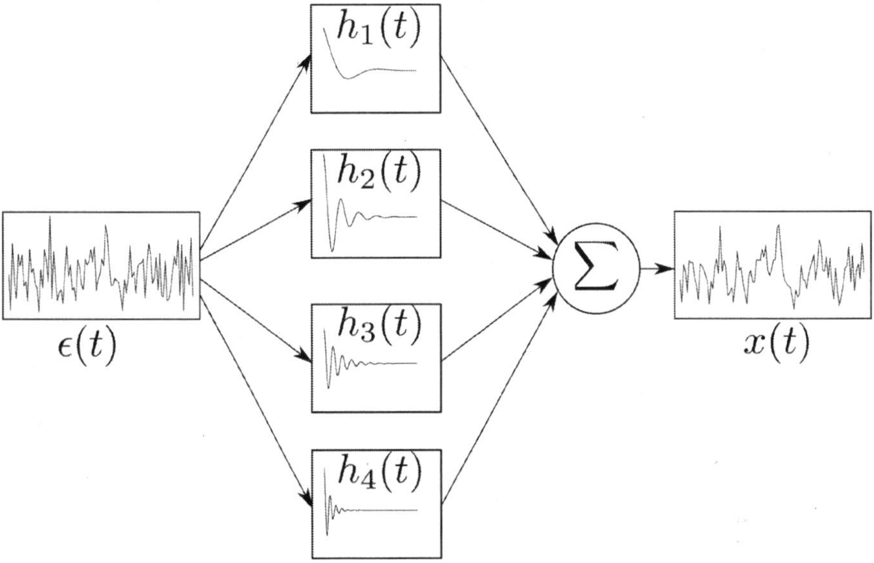

FIGURE 2.11: Transfer function of the AR model with four rhythms as a system of parallel filters each of them corresponding to a characteristic frequency (see equation 2.63).

2.4 Non-stationary signals

2.4.1 Instantaneous amplitude and instantaneous frequency

In case of stationary signals the concept of amplitude and frequency spectra is intuitive. For a non-stationary process it is more difficult to determine signal amplitude or frequency, since these quantities can vary in time, so it is useful to introduce instantaneous amplitude and frequency. To define the instantaneous amplitude and instantaneous frequency, first we need to introduce the concept of analytic signal.

Analytic signal is a complex signal $x_a(t)$ related to a real signal $x(t)$ by formula:

$$x_a(t) = x(t) + ix_h(t) \tag{2.64}$$

where $x_h(t)$ is the Hilbert transform of $x(t)$. The Hilbert transform is a linear operator that can be thought of as the convolution of $x(t)$ with the function $h(t) = \frac{1}{\pi t}$. The spectral power for $x_a(t)$ is nonzero only for positive f.

Thus approximation of the analytic signal can be obtained from the following algorithm:

1. Calculate the FFT of the input sequence x consisting of n samples: $X =$`fft`(x)

2. Create a vector h:

$$h(i) = \begin{cases} 1 & \text{for } i = 1, (n/2)+1 \\ 2 & \text{for } i = 2, 3, \ldots, (n/2) \\ 0 & \text{for } i = (n/2)+2, \ldots, n \end{cases} \tag{2.65}$$

3. Calculate the element-wise product of X and h: $Y = X \cdot h$

4. Calculate the inverse FFT of Y.

This algorithm is implemented in MATLAB Signal Processing Toolbox as `hilbert`. The analytic signal can be presented in the form:

$$x_a(t) = A(t)\cos(\phi(t)) \tag{2.66}$$

where $A(t) = |x_a(t)|$ is called the instantaneous amplitude, $\phi(t) = \arg(x_a(t))$ and $f(t) = \frac{1}{2\pi}\frac{d\arg(x_a(t))}{dt}$ is called the instantaneous frequency[1].

The concepts of instantaneous amplitude and frequency are useful in case of simple signals without overlapping frequency components, e.g., in Figure 2.12 a-c; the instantaneous amplitude recovers the modulation of the chirp, and instantaneous frequency recovers its linearly changing frequency. However, the instantaneous measures turn out to be meaningless, if the signal is more complicated, e.g., if there are two structures with different frequencies present at the same moment. Figure 2.12 d-f illustrates such a case. The signal in panel d) is composed of two signals analogous to that in panel a), but shifted in frequency by a constant. Its instantaneous amplitude (panel e) and frequency (panel f) display fast oscillatory behavior.

2.4.2 Analytic tools in the time-frequency domain

2.4.2.1 Time-frequency energy distributions

Signal energy can be computed in time or in frequency domain as:

$$E_x = \int_{-\infty}^{\infty} |x(t)|^2 dt = \int_{-\infty}^{\infty} |X(f)|^2 df \tag{2.67}$$

and the quantities $|x(t)|^2$ or $|X(f)|^2$ are interpreted as energy density. The idea of signal energy density can be extended to the time-frequency space. The time-frequency energy density $\rho_x(t,f)$ should represent the amount of signal energy assigned to a given time-frequency point (t,f), and fulfill the conditions:

1. The total energy is conserved:

$$E_x = \int_{-\infty}^{\infty}\int_{-\infty}^{\infty} \rho_x(t,f)\,dt\,df \tag{2.68}$$

[1]Note, that in case of a stationary signal, e.g., $x = B\cos(2\pi ft)$ these definitions retrieve amplitude $A(t) = B$ and frequency $\phi(t) = f$ of the signal.

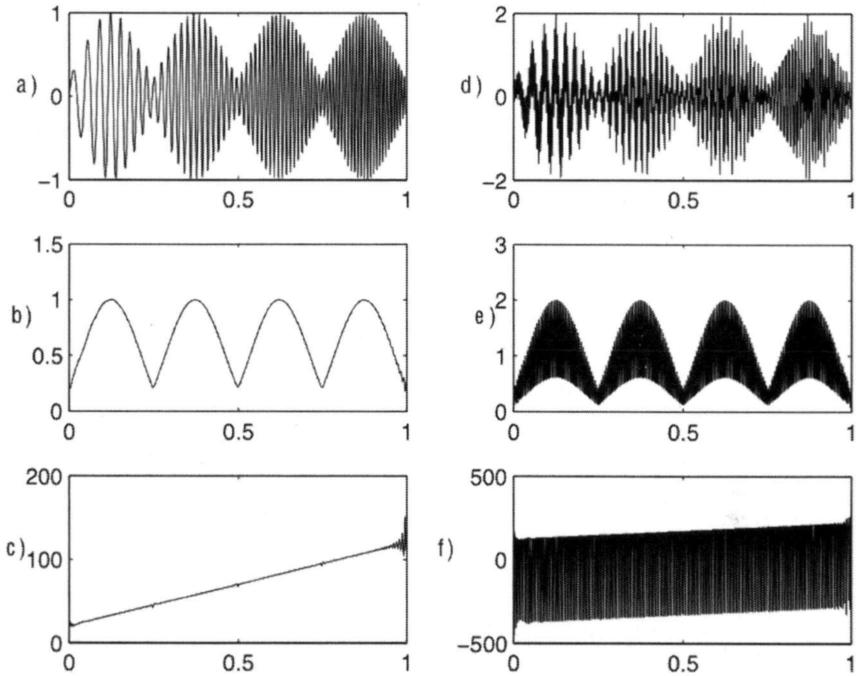

FIGURE 2.12: Instantaneous amplitude and frequency. a) Signal: modulated linear chirp; b) instantaneous amplitude of signal in panel (a); c) instantaneous frequency of signal in panel (a); d) signal composed of two modulated linear chirps with the same slope as in (a) but different intercept frequency; e) instantaneous amplitude of signal in panel (d); f) instantaneous frequency of signal in panel (d).

2. The marginal distributions are conserved:

$$\int_{-\infty}^{\infty} \rho_x(t,f)dt = |X(f)|^2 \tag{2.69}$$

$$\int_{-\infty}^{\infty} \rho_x(t,f)df = |x(t)|^2 \tag{2.70}$$

Convenient MATLAB toolbox for analysis of signals in time-frequency domain is the `tftb` available under the terms of the GNU Public License at `http://tftb.nongnu.org/`

2.4.2.1.1 Wigner-Ville distribution The basic time-frequency energy density distribution is the Wigner-Ville distribution (WVD) defined as (for signals represented in time domain):

$$W_x(t,f) = \int_{-\infty}^{\infty} x(t+\tau/2)x^*(t-\tau/2)e^{-i2\pi f\tau}d\tau \tag{2.71}$$

or equvialently (for signals represented in the frequency domain):

$$W_x(t,f) = \int_{-\infty}^{\infty} X(f+\xi/2)X^*(f-\xi/2)e^{i2\pi\xi t}d\xi \tag{2.72}$$

The $W_x(t,f)$ is implemented as `tfrwv` function in the Matlab toolbox `tftb`. WVD has the following properties:

- Energy conservation equation (2.68)
- Conservation of marginal distributions equations: (2.69) and (2.70)
- Conservation of time and frequency shifts:

$$y(t) = x(t-t_0) \Rightarrow W_y(t,f) = W_x(t-t_0,f) \tag{2.73}$$

$$y(t) = x(t)e^{i2\pi f_0 t} \Rightarrow W_y(t,f) = W_x(t,f-f_0) \tag{2.74}$$

- Conservation of scaling

$$y(t) = \sqrt{k}x(kt) \Rightarrow W_y(t,f) = W_x(kt,f/k) \tag{2.75}$$

WVD is a quadratic representation, so an intrinsic problem occurs when the signal contains more than one time-frequency component. Let's assume that we analyze a signal y composed of two structures x_1 and x_2:

$$y(t) = x_1(t) + x_2(t) \tag{2.76}$$

Then, the WVD can be expressed as:

$$W_y(t,f) = W_{x_1}(t,f) + W_{x_2}(t,f) + 2Re\{W_{x_1,x_2}(t,f)\} \tag{2.77}$$

where $W_{x_1,x_2}(t,f) = \int_{-\infty}^{\infty} x_1(t+\tau/2)x_2^*(t-\tau/2)e^{-i2\pi f\tau}d\tau$ is called a cross-term.

WVD has many desired properties mentioned above and optimal time-frequency resolution, but the presence of cross-terms in real applications can make the interpretation of results difficult. The problem is illustrated in Figure 2.13.

2.4.2.1.2 Cohen class The cross-terms oscillate in the time-frequency space with relatively high frequency, as can be observed in Figure 2.13. The property can be used to suppress the influence of the cross-terms simply by applying a low-pass spatial filter on the WVD. A family of distributions obtained by filtering WVD is called Cohen's class:

$$C_x(t,f,\Pi) = \int_{-\infty}^{\infty}\int_{-\infty}^{\infty} \Pi(s-t,\xi-f)W_x(s,\xi)dsd\xi \tag{2.78}$$

where

$$\Pi(t,f) = \int_{-\infty}^{\infty}\int_{-\infty}^{\infty} f(\xi,\tau)e^{-i2\pi(f\tau+\xi t)}d\tau d\xi \tag{2.79}$$

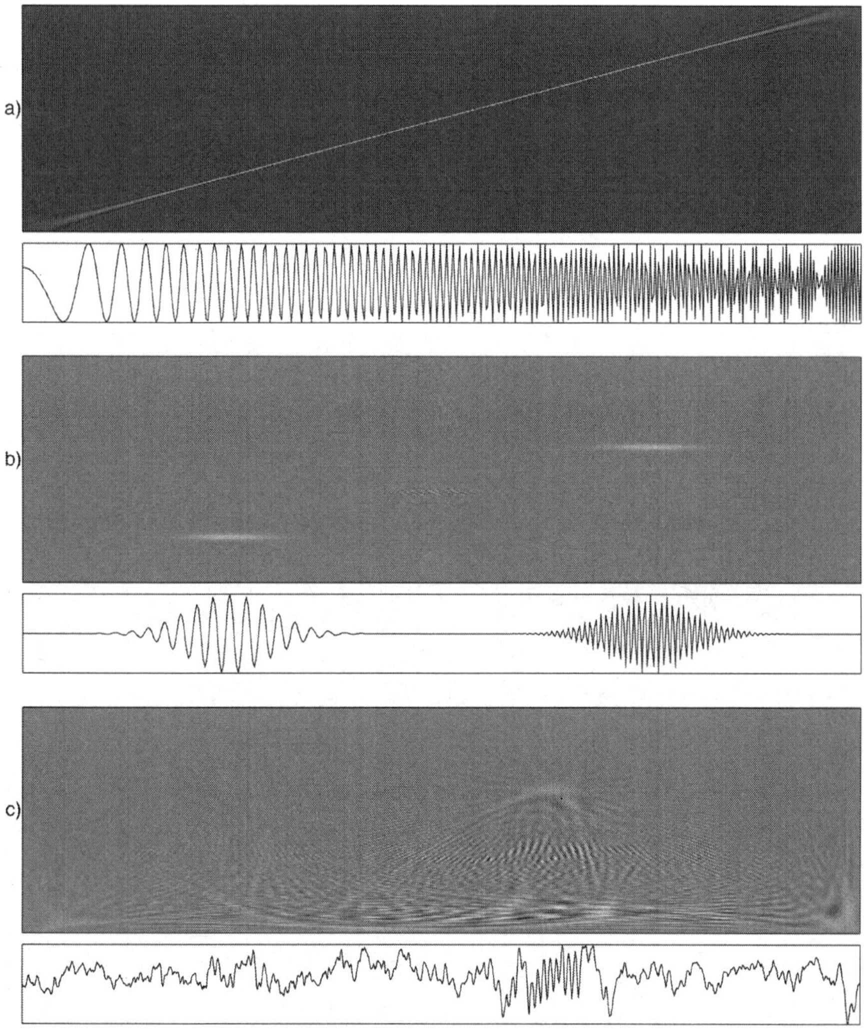

FIGURE 2.13: Illustration of cross-terms in WVD. a) Linear chirp—its time-frequency representation is a straight line, b) two Gabor functions—note the cross-term structure in the time-frequency representation just in the middle of the line connecting the representations of the individual components. c) A real world example: 10 sec. of EEG recorded during sleep stage 2, with a clear sleep spindle visible as a transient oscillatory structure.

and $f(\xi,\tau)$ is the filter kernel. WVD is a member of Cohen's class with $f(\xi,\tau) = 1$. Another very popular member of the class is the Choi-Williams distribution with the

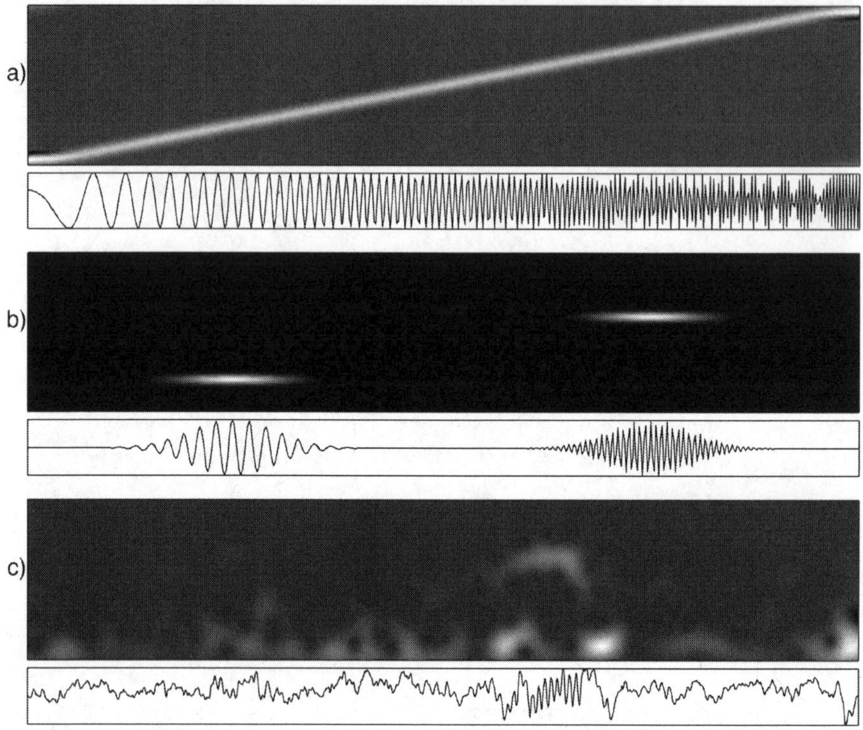

FIGURE 2.14: Ilustrative application of Choi-Williams distribution to the same set of signals as in Figure 2.13. a) Linear chirp—its time-frequency representation is a straight line, b) two Gabor functions—note the cross-term structure in the time-frequency representation is highly reduced. c) A real world example: 10 sec. of EEG recorded during sleep stage 2, with a clear sleep spindle visible as a transient oscillatory structure.

filtering kernel given by two-dimensional Gaussian function:

$$f(\xi,\tau) = \exp\left[-\alpha(\xi\tau)^2\right]. \tag{2.80}$$

A proper selection of the kernel parameter can significantly reduce cross-terms, as illustrated in Figure 2.14.

2.4.2.2 Time-frequency signal decompositions

2.4.2.2.1 Short time Fourier transform and spectrogram Time-frequency representation of the signal can also be obtained in a different manner. In this approach one extracts successive short pieces of signal with a window function and computes its frequency representation with a short time Fourier transform (STFT):

$$F_x(t,f;h) = \int_{-\infty}^{\infty} x(u)h^*(u-t)e^{-i2\pi uf}du \tag{2.81}$$

where h is the window. If the window has a finite energy then the STFT can be reversed:

$$x(t) = \frac{1}{E_h}\int_{-\infty}^{\infty}\int_{-\infty}^{\infty} F_x(u,f;h)h(t-u)e^{i2\pi tf}\,du\,df \tag{2.82}$$

where $E_h = \int_{-\infty}^{\infty}|h(t)|^2\,dt$. Thus we can view the STFT as a tool that decomposes the signal into waveforms, *time-frequency atoms*, of the form:

$$h_{t,f}(u) = h(u-t)e^{i2\pi fu} \tag{2.83}$$

This concept is illustrated in Figure 2.15. Each atom is obtained by translation of a single window h and its modulation with frequency f.

STFT can be used to obtain the distribution of energy in the time-frequency space. If the window function has a unit energy (i.e., $E_h = 1$) then a squared modulus of STFT, a *spectrogram*, is an estimator of energy density (Figure 2.16):

$$S_x(t,f) = \left|\int_{-\infty}^{\infty} x(u)h^*(u-t)e^{-i2\pi fu}du\right|^2 \tag{2.84}$$

Spectrogram conserves the translations in time:

$$y(t) = x(t-t_0) \Rightarrow S_y(t,f;h) = S_x(t-t_0,f;h) \tag{2.85}$$

and in frequency:

$$y(t) = x(t)e^{i2\pi f_0 t} \Rightarrow S_y(t,f;h) = S_x(t,f-f_0;h) \tag{2.86}$$

Spectrogram is a quadratic form; thus in the representation of multi-component signals the cross-terms are also present:

$$y(t) = x_1(t)+x_2(t) \Rightarrow S_y(t,f) = S_{x_1}(t,f)+S_{x_2}(t,f)+2Re\{S_{x_1,x_2}(t,f)\} \tag{2.87}$$

where

$$S_{x_1,x_2}(t,f) = F_{x_1}(t,f)F^*_{x_2}(t,f) \tag{2.88}$$

The formula (2.88) shows that the cross-terms are present only if the signal components are close enough in the time-frequency space, i.e., when their individual STFTs overlap. The time and frequency resolution are determined by the properties of the window h. The time resolution can be observed as the width of the spectrogram of the Dirac's delta:

$$x(t) = \delta(t-t_0) \Rightarrow S_x(t,f;h) = e^{-i2\pi t_0 f}h(t-t_0). \tag{2.89}$$

The frequency resolution can be observed as the width of the spectrogram of the complex sinusoid.

$$x(t) = e^{i2\pi f_0 t} \Rightarrow S_x(t,f;h) = e^{-i2\pi t f_0}H(f-f_0) \tag{2.90}$$

It is clear from (2.89) that the shorter duration of time window h, the better is the time resolution. But due to the uncertainty principle (1.41) the shorter the window in time, the broader is its frequency band; thus from (2.90) the poorer the frequency resolution.

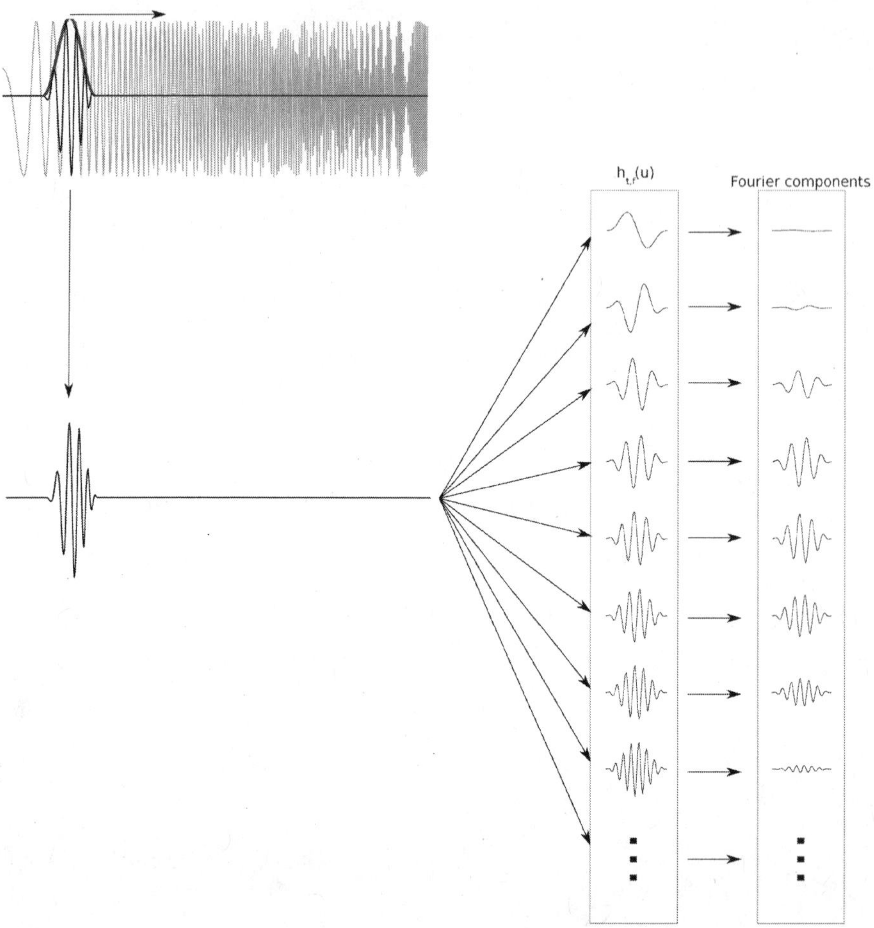

FIGURE 2.15: STFT as atomic decomposition. A sliding window (thick line) extracts a portion of signal roughly localized in time. This windowed signal is "compared" with each of the time frequency atoms of the form $h_{t,f}(u) = h(u-t)e^{i2\pi fu}$, giving the Fourier components ($h_{t,f}(u)$ weighted by intensities).

2.4.2.2.2 Continuous wavelet transform and scalogram The continuous wavelet transform (CWT) is defined as

$$T_x(t,a;\Psi) = \int_{-\infty}^{\infty} x(s)\Psi^*_{t,a}(s)ds \tag{2.91}$$

where $\Psi_{t,a}(s)$ is obtained from the mother wavelet Ψ by time translation t and dilation by scale a:

$$\Psi_{t,a}(s) = \frac{1}{\sqrt{|a|}}\Psi\left(\frac{s-t}{a}\right) \tag{2.92}$$

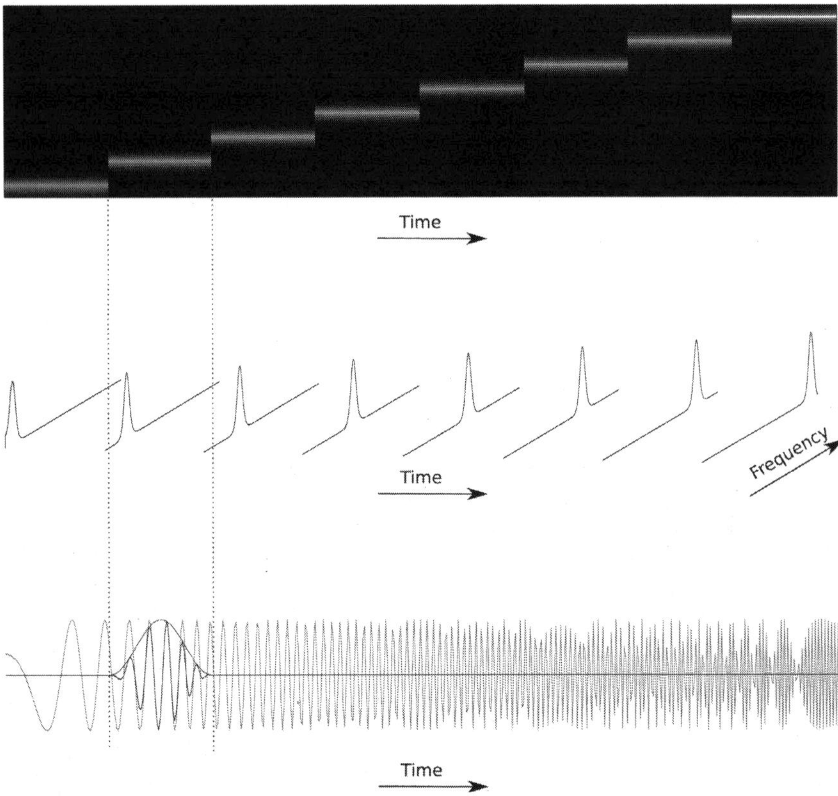

FIGURE 2.16: Illustration of a spectrogram construction. STFT is used to compute the spectrum of a short windowed fragment of the signal. The consecutive spectra are presented as a gray-scale time-frequency map. From the bottom: signal of increasing frequency, spectra in perspective, time-frequency representation.

Please note that $\Psi_{t,a}$ conserves the overall shape of Ψ, in the sense of number of zero-crossings, however it is dilated or contracted. The wavelet Ψ should have the zero mean value. If for scale $a_0 = 1$ the wavelet is concentrated in frequency around f_0, we can bind the scale and frequency by:

$$f_a = \frac{f_0}{a} \tag{2.93}$$

The CWT can be interpreted as decomposition of the signal into the time-frequency atoms, obtained through projection of the signal on the set of waveforms derived from a single wavelet Ψ by time translations and scaling. In this sense the CWT is very similar to STFT (2.81), the main difference being that the window length changes with the scale (or due to (2.93) with the frequency). This property results in different compromise between the time and frequency resolutions in different frequency bands, i.e., the frequency resolution is fine at low frequencies at the expense of poor time resolution, and frequency resolution gradually deteriorates for high frequency bands but the time resolution for these bands improves.

The CWT can be used to form a time-frequency representation of the energy distribution—scalogram, in a way analogous to the way in which a spectrogram is obtained from STFT. The time-scale representation is:

$$S_x(t,a;\Psi) = |T_x(t,a;\Psi)|^2 \tag{2.94}$$

Due to equation (2.93) it can be understood as time-frequency representation:

$$S_x(t,f;\Psi) = |T_x(t,f_0/f;\Psi)|^2 \tag{2.95}$$

Analogously to the spectrogram, the scalogram is disturbed by cross-terms of multi-component signals in the regions of time-frequency space where the CWT of individual components overlap.

2.4.2.2.3 Discrete wavelet transform The CWT in its theoretical form (2.91) operates on continuous time-scale (time-frequency) space. In any practical application the time and scale dimensions have to be sampled. The sampling, in general, can be performed in any way resulting in different approximations of CWT. The common way is to select $t_0 > 0$ and $a_0 > 0$ and then generate a grid of points $\{t = nt_0 {a_0}^m, \quad a = a_0^m\}$ for $m,n \in \mathbb{Z}$

The discrete wavelet transform can be defined then as:

$$T_x(n,m;\Psi) = \int_{-\infty}^{\infty} x(s)\Psi^*_{n,m}(s)ds; \quad m,n \in \mathbb{Z} \tag{2.96}$$

where $\Psi^*_{n,m}$ is a scaled and translated version of Ψ:

$$\Psi_{n,m}(s) = \frac{1}{\sqrt{a_0^m}}\Psi\left(\frac{s - nt_0 a_0^m}{a_0^m}\right) \tag{2.97}$$

2.4.2.2.4 Dyadic wavelet transform—multiresolution signal decomposition However, there is one special manner of sampling which, for some forms of wavelets, creates the orthonormal basis in the time-scale space. This is the *dyadic sampling* obtained for $t_0 = 1$ and $a_0 = 2$. The dyadic sampled wavelets are related to multiresolution analysis and the transform can be efficiently computed using filter banks [Mallat, 1989].

To better understand the multiresolution signal decomposition we should start with considering the so-called *scaling function*. The scaling function is a unique measurable function[2] $\phi(t) \in L^2(\mathbb{R})$ such that, if the dilation of $\phi(t)$ by 2^j is expressed by:

$$\phi_{2^j}(t) = 2^j\phi(2^j t) \tag{2.98}$$

[2] i.e., the integral of squared function is finite.

then $\sqrt{2^{-j}}\phi_{2^j}(t-2^{-j}n)$ for $n, j \in \mathbb{Z}$ is the orthonormal basis in vector space V_{2^j}. The vector space V_{2^j} is the set of all possible approximations of a signal at resolution 2^j. The signal $x(t)$ can be projected on the orthonormal basis:

$$A_{2^j}(t) = 2^{-j} \sum_{n=-\infty}^{\infty} \langle x(u), \phi_{2^j}\left(u - 2^{-j}n\right)\rangle \, \phi_{2^j}\left(t - 2^{-j}n\right) \tag{2.99}$$

The *discrete approximation of signal* $x(t)$ at resolution 2^j is defined by a set of inner products:

$$A^d_{2^j}x(n) = \langle x(u), \phi_{2^j}(u - 2^{-j}n)\rangle \tag{2.100}$$

The above formula is equivalent to the convolution of signal $x(t)$ with the dilated scaling function $\phi_{2^j}(t)$ which is a low-pass filter. The computation of $A^d_{2^j}x$ for consecutive resolutions 2^j can be interpreted as a low-pass filtering of $x(t)$ followed by uniform sampling at rate 2^j. In practice, the computations of $A^d_{2^j}x$ can be realized iteratively by convolving $A^d_{2^{j+1}}x$ with $\hat{h}(n)$, such that $\hat{h}(n) = h(-n)$ is a mirror filter of:

$$h(n) = \langle \phi_{2^{-1}}(u), \phi(u-n)\rangle \tag{2.101}$$

and downsampling the result by 2. This algorithm can be expressed by the formula:

$$\begin{cases} A^d_{2^0} = x(n) \\ A^d_{2^j}x(n) = \mathcal{D}_2\left(\sum_{k=-\infty}^{\infty} \hat{h}(2n-k) A^d_{2^{j+1}}x(n)\right) \end{cases} \tag{2.102}$$

where $\mathcal{D}_2$ is the operator of downsampling by factor 2.

The low-pass filtering used in the algorithm has the effect that after each iteration the consecutive approximations contain less information. To describe the residual information lost in one step and measure the irregularities of signal at resolution 2^j a function Ψ_{2^j} is constructed, such that:

$$\Psi_{2^j}(t) = 2^j \Psi\left(2^j t\right) \tag{2.103}$$

and the functions $\Psi_{2^j}\left(t - 2^{-j}n\right)$ are a basis in a vector space O_{2^j}. The vector space O_{2^j} is orthonormal to V_{2^j} and together they span the vector space $V_{2^{j+1}}$:

$$O_{2^j} \perp V_{2^j} \text{ and } O_{2^j} \oplus V_{2^j} = V_{2^{j+1}} \tag{2.104}$$

The function $\Psi(t)$ is called a wavelet and corresponds to the mother wavelet function in equation 2.92. The projection of a signal $x(t)$ on the space O_{2^j} is called the *discrete detail signal* and is given by the product:

$$D^d_{2^j}x(n) = \langle x(u), \Psi_{2^j}\left(u - 2^j n\right)\rangle \tag{2.105}$$

These inner products can be computed conveniently from higher resolution discrete approximation by the convolution:

$$D^d_{2^j}x(n) = \sum_{k=-\infty}^{\infty} \hat{g}(2n-k) A^d_{2^{j+1}}(k) \tag{2.106}$$

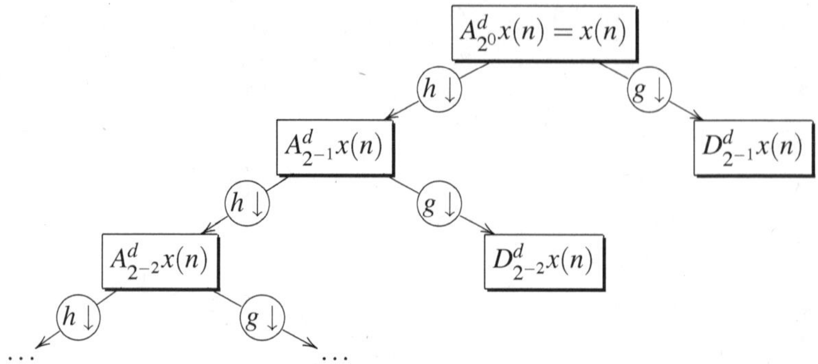

FIGURE 2.17: Cascade of filters producing the discrete wavelet transform coefficients. In this algorithm the signal is passed through the cascade, where at each step the approximation of signal from the previous step is low-pass and high-pass filtered, by convolving with h and g, respectively. Each of the filtered sequences is downsampled by factor 2 (on the scheme it is marked as $\downarrow$). Next the approximation is passed to the following step of the cascade.

with the mirror filter $\hat{g}(n) = g(-n)$ of the filter $g(n)$ defined by:

$$g(n) = \langle \Psi_{2^{-1}}(u), \phi(u-n) \rangle \tag{2.107}$$

The filter $g(n)$ can also be computed from filter $h(n)$ with the formula:

$$g(n) = (-1)^{1-n} h(1-n) \tag{2.108}$$

Filter $g(n)$ is a high-pass filter. The $A^d_{2^j}$ and $D^d_{2^j}$ can be obtained from $A^d_{2^{j+1}}$ by convolving with $h(n)$ and $g(n)$, respectively, and downsampling by factor 2. Computation of discrete approximation by repeating the process for $j < 0$ is called the pyramidal transform. It is illustrated in Figure 2.17.

For a continuous signal the cascade could be infinite. In practical cases of sampled signals, the process of consecutive downsamplings has the natural stop. The coefficients of wavelet transform are the outputs of all the detail branches and the last approximation branch.

For correctly designed filters (quadrature mirror filters) the process of decomposition can be reversed yielding the inverse discrete wavelet transform. A technical discussion of how to design these filters is beyond the scope of this book and is available, e.g., in [Strang and Nguyen, 1996]. If we implement the wavelet transform as an iterated filter bank we do not have to specify the wavelet explicitly.

There is one important consequence of discretization of the wavelet transform. The transform is not shift invariant, which means that the wavelet transform of a signal and of the wavelet transform of the same time-shifted signal are not simply shifted versions of each other. An illustration of the structure of the time-frequency space related to the discrete wavelet representation is shown in Figure 2.18.

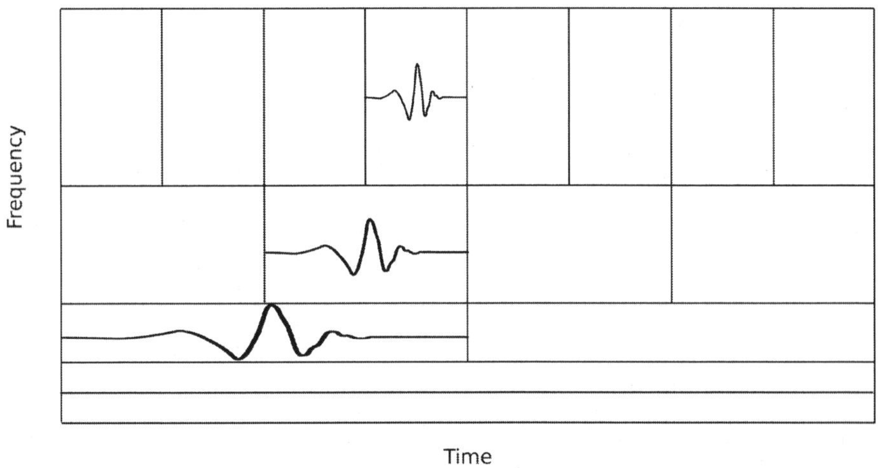

FIGURE 2.18: Structure of the time-frequency plane for discrete wavelets. The wavelet function is dilated and translated.

2.4.2.2.5 Wavelet packets Wavelet packets (WP) are a modification of the discrete wavelet decomposition. The decomposition into WP is achieved in two steps. First is the modification of the filter/downsampling cascade. In the WP scheme at each level of the cascade both branches—the approximation and the detail coefficients—are further filtered and downsampled. As a result, a complete tree is obtained as illustrated in Figure 2.19. The second modification is that from that tree we can select the most suitable decomposition of a given signal, e.g., with respect to an entropy-based criterion [Coifman and Wickerhauser, 1992]. This procedure is called pruning of a decomposition tree.

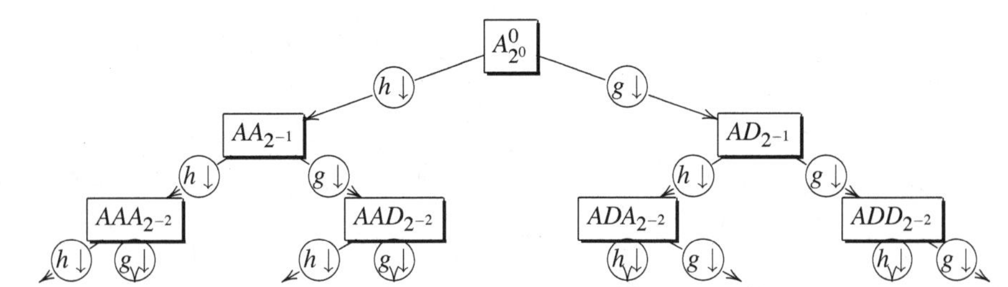

FIGURE 2.19: Three levels of WP decomposition tree.

More detailed information on different aspects of wavelet analysis may be found in [Mallat, 1999].

2.4.2.2.6 Wavelets in MATLAB In MATLAB wavelets analysis can be conveniently performed with the Wavelet Toolbox distributed by MathWorks. The collection of functions delivered by this toolbox can be obtained by command `help wavelet`.

Another MATLAB toolbox for wavelet analysis is WaveLab `http://www-stat.stanford.edu/~wavelab/Wavelab_850/index_wavelab850.html`. WaveLab is a collection of MATLAB functions that implement a variety of algorithms related to wavelet analysis. The techniques made available are: orthogonal and biorthogonal wavelet transforms, translation-invariant wavelets, interpolating wavelet transforms, cosine packets, wavelet packets.

2.4.2.2.7 Matching pursuit—MP The atomic decompositions of multicomponent signals have the desired property of explaining the signal in terms of time-frequency localized structures. The two methods presented above: spectrogram and scalogram are working well but they are restricted by the a priori set trade-off between the time and frequency resolution in different regions of the time-frequency space. This trade-off does not follow the structures of the signal. In fact interpretation of the spectrogram or scalogram requires understanding which aspects of the representation are due to the signal and which are due to the properties of the methods.

A time-frequency signal representation that adjusts to the local signal properties is possible in the framework of matching pursuit (MP) [Mallat and Zhang, 1993].

In its basic form MP is an iterative algorithm that in each step finds an element g_{γ_n} (atom) from a set of functions D (dictionary) that best matches the current residue of the decomposition $R^n x$ of signal x; the null residue being the signal:

$$\begin{cases} R^0 x = x \\ R^n x = \langle R^n x, g_{\gamma_n} \rangle g_{\gamma_n} + R^{n+1} x \\ g_{\gamma_n} = \arg\max_{g_{\gamma_i} \in D} |\langle R^n x, g_{\gamma_i} \rangle| \end{cases} \tag{2.109}$$

where: $\arg\max_{g_{\gamma_i} \in D}$ means the atom g_{γ_i} which gives the highest inner product with the current residue: $R^n x$. Note, that the second equation in (2.109) leads to orthogonality of g_{γ_n} and $R^{n+1}x$, so :

$$||R^n x||^2 = \left|\langle R^n x, g_{\gamma_n} \rangle\right|^2 + \left|\left|R^{n+1} x\right|\right|^2 \tag{2.110}$$

The signal can be expressed as:

$$x = \sum_{n=0}^{k} \left\langle R^n x, g_{\gamma_n} \right\rangle g_{\gamma_n} + R^{k+1} x \tag{2.111}$$

It was proved [Davis, 1994] that the algorithm is convergent, i.e., $\lim_{k\to\infty} \left|\left|R^k s\right|\right|^2 = 0$. Thus in this limit we have:

$$x = \sum_{n=0}^{\infty} \left\langle R^n x, g_{\gamma_n} \right\rangle g_{\gamma_n} \tag{2.112}$$

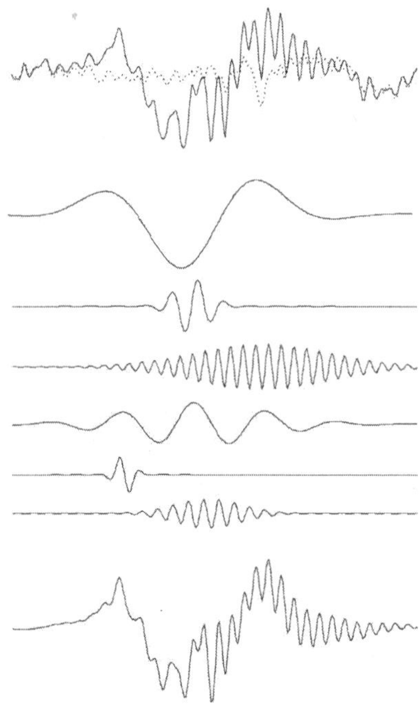

FIGURE 2.20: Illustration of the iterative MP decomposition. a) Signal to be decomposed (black) and the residue after six iterations (dotted line), b) from top to bottom the six consecutive selected atoms, c) the sum of the six atoms.

In this way signal x is represented as a weighted sum of atoms (waveforms) from the dictionary D. The iterative procedure is illustrated in Figure 2.20. Taking into account (2.110) we can see that the representation conserves the energy of the signal:

$$||x||^2 = \sum_{n=0}^{\infty} |\langle R^n x, g_{\gamma_n} \rangle|^2 \tag{2.113}$$

The MP decomposition can be considered as an extension of the atomic decompositions offered by STFT or CWT. The main advantage of the MP paradigm is the relaxation of constraint between the frequency band and frequency resolution. The MP algorithm performs decomposition in an extremely redundant set of functions, which results in a very flexible parametrization of the signal structures.

In principal the dictionary D can be any set of functions. In practical implementations (e.g., `http://eeg.pl/mp`) the dictionary contains a base of Dirac deltas, a base of sinusoids and a set of Gabor functions:

$$g_\gamma(t) = K(\gamma) e^{-\pi\left(\frac{t-u}{\sigma}\right)^2} \sin\left(2\pi f(t-u) + \phi\right) \tag{2.114}$$

with $K(\gamma)$ normalization factor such that $||g_\gamma|| = 1$, and $\gamma = \{u, f, \sigma, \phi\}$ are the parameters of functions in the dictionary (u –time translation, f –frequency, σ – time width, ϕ – phase).

The parameters γ can be sampled in various ways. The original idea of Mallat [Mallat and Zhang, 1993] was to follow a dyadic scheme that mimics the oversampled discrete wavelets. For applications where the parameters are used to form statistics of the atoms the dyadic sampling produced estimators that were biased by the structure of the dictionary. Introduction of stochastic dictionaries relied on randomization of the time-frequency coordinates and time width of atoms [Durka et al., 2001a]. This allowed to obtain a bias free implementation. Further extensions of the MP algorithm allow for analysis of multivariate signals i.e multichannel and multitrial decompositions [Durka et al., 2005b, Sielużycki et al., 2009a] (Sect. 3.6.3).

The atomic decomposition of signal can be used to produce the time-frequency energy distribution. In this approach the best properties of energy distributions (WVD) and atomic decompositions can be joined. The WVD of the whole decomposition is:

$$W_x(t,f) = \sum_{n=0}^{\infty} \left|\left\langle R^n x, g_{\gamma_n}\right\rangle\right|^2 W_{g_{\gamma_n}}(t,f) + \sum_{n=0}^{\infty}\sum_{m=0,m\neq n}^{\infty} \left\langle R^n x, g_{\gamma_n}\right\rangle \left\langle R^m x, g_{\gamma_m}\right\rangle^* W_{g_{\gamma_n} g_{\gamma_m}}(t,f) \tag{2.115}$$

where

$$W_{g_{\gamma_n}}(t,f) = \int g_{\gamma_n}\left(t+\frac{\tau}{2}\right) g^*_{\gamma_n}\left(t-\frac{\tau}{2}\right) e^{-i2\pi f\tau} d\tau \tag{2.116}$$

is WVD of individual atoms. The double sum in equation (2.115) corresponds to the crossterms, but since it is given explicitly, it can be omitted yielding the estimator of the energy density distribution in the form:

$$E_x^{MP}(t,f) = \sum_{n=0}^{M} |\langle R^n x,\ g_{\gamma_n}\rangle|^2\, W_{g_{\gamma_n}}(t,f) \tag{2.117}$$

This interpretation is valid since normalization of atoms (2.114):

$$\int_{-\infty}^{+\infty}\int_{-\infty}^{+\infty} W_g(t,f)\,dt\,df = \|g\|^2 = 1 \tag{2.118}$$

leads to:

$$\int_{-\infty}^{+\infty}\int_{-\infty}^{+\infty} E_x^{MP}(t,f)\,dt\,df = \|x\|^2 \tag{2.119}$$

This representation has implicitly no cross-terms and for Gabor atoms offers the highest time-frequency resolution (Sect. 1.4.7).

MP dictionary can be adjusted to be coherent with particular structures in the signal. The dictionary containing asymmetric functions was designed and proved to be useful in description of components with different time courses of the rising and decaying parts [Jedrzejczak et al., 2009] (Sect. 4.5.2).

2.4.2.2.8 Comparison of time-frequency methods The complex character of biomedical signals and their importance in health research and clinical practice brought a wide variety of signal analysis methods into applications in biomedical research. The most widespread are the spectral methods which make possible the identification of the basic rhythms present in the signal. Conventional methods of the analysis assumed stationarity of the signal, in spite of the fact that interesting processes are often reflected in fast dynamic changes of signal. This implied the application to the analysis of the signals methods operating in time-frequency space.

The available time-frequency methods can be roughly divided into two categories:

- Those that give directly continuous estimators of energy density in the time-frequency space
- Those that decompose the signal into components localized in the time-frequency space, which can be described by sets of parameters, and at the second step the components can be used to create the estimators of time-frequency energy density distribution

An example of time-frequency energy distribution obtained by means of different methods is shown in Figure 2.21.

In the first category—the Cohen's class of time-frequency distributions—one obtains directly the time-frequency estimators of energy density without decomposing the signal into some predefined set of simple elements. This allows for maximal flexibility in expressing the time frequency content of the signal. However, there are two consequences:

- The first consequence is the lack of parametric description of the signals structures
- The second consequence is that, no matter how much the signal structures are separated in time or frequency, they interfere and produce cross-terms.

The problem of compromise between time and frequency resolution manifests when one selects the proper filter kernel to suppress the cross-terms.

In the second category the most natural transition from spectral analysis to the analysis in time-frequency space is the use of short time Fourier transform (STFT) and a representation of energy density derived from it—the spectrogram. The positive properties of this approach are the speed of computations and the time and frequency shift invariance, which makes the interpretation of the resulting time-frequency energy density maps easy to interpret. The main drawbacks are: (1) the a priori fixed compromise between time and frequency resolution in the whole time-frequency space, which results in smearing the time-frequency representation, (2) the presence of cross-terms between the neighboring time-frequency structures.

Another common choice in the second category is the CWT. From the practical point of view the main difference from the STFT relies on another compromise between the time and frequency resolution. In case of CWT, one sacrifices the time resolution for the better frequency resolution of low frequency components and vice

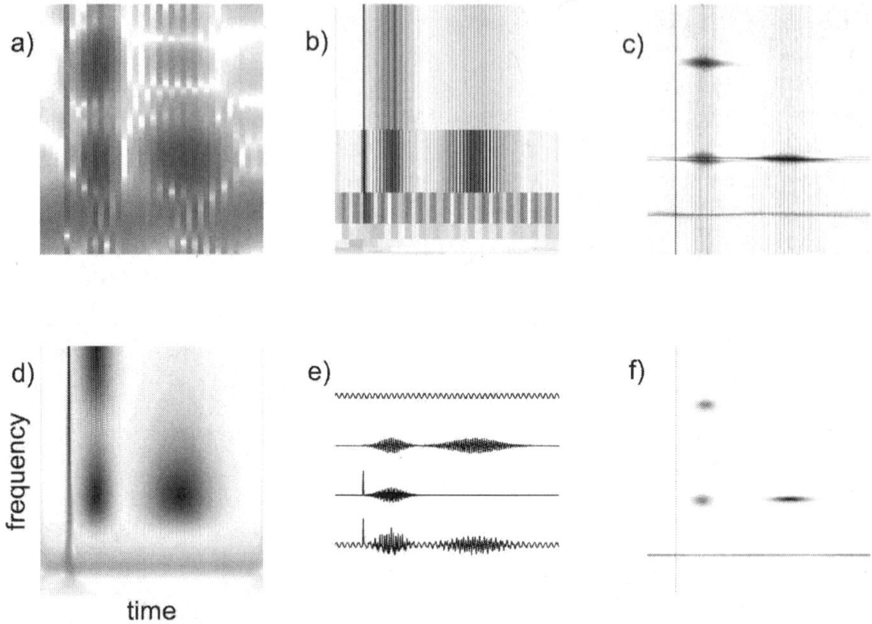

FIGURE 2.21: Comparison of energy density in the time-frequency plane obtained by different estimators for a signal e): a) spectrogram, b) discrete wavelet transform, c) Choi-Williams transform, d) continuous wavelets, f) matching pursuit. Construction of the simulated signal shown in (e), the signal consisting of: a sinusoid, two Gabor functions with the same frequency but different time positions, a Gabor function with frequency higher than the previous pair, an impulse. From [Blinowska et al., 2004b].

versa for higher frequency components; also the change of the frequency of a structure leads to the change of the frequency resolution.

STFT and CWT can be considered as atomic representations of the signal, and as such give a certain parametric description of the signals. However, the representation in not sparse; in other words there are too many parameters; hence they are not very informative.

The sparse representation of the signal is provided by DWT and MP, which leads to efficient parameterization of the time series. The DWT can decompose the signal into a base of functions, that is a set of waveforms that has no redundancy. There are fast algorithms to compute the DWT. Similar to CWT, the DWT has poor time resolution for low frequencies and poor frequency resolution for high frequencies. The DWT is very useful in signal denoising or signal compression applications. The lack of redundancy has a consequence in the loss of time and frequency shift invariance. DWT may be appropriate for time-locked phenomena, but much less for transients appearing in time at random, since parameters describing a given structure depend on its location inside the considered window.

The decomposition based on the matching pursuit algorithm offers the step-wise adaptive compromise between the time and frequency resolution. The resulting decomposition is time and frequency invariant. The time-frequency energy density estimator derived from the MP decomposition has explicitly no cross-term, which leads to clean and easy-to-interpret time-frequency maps of energy density. The price for the excellent properties of the MP decomposition is the higher computational complexity.

The sparsity of the DWT and MP decompositions has a different character which has an effect on their applicability. DWT is especially well suited to describing time locked phenomena since it provides the common bases. MP is especially useful for structures appearing in the time series at random. The sparsity of MP stems from the very redundant set of functions, which allows to represent the signal structures as a limited number of atoms. The MP decomposition gives the parameterization of the signal structures in terms of the amplitude, frequency, time of occurrence, time, and frequency span which are close to the intuition of practitioners.

2.4.2.2.9 Empirical mode decomposition and Hilbert-Huang transform The Hilbert-Huang transform (HHT) was proposed by Huang et al. [Huang et al., 1998]. It consists of two general steps:

- The empirical mode decomposition (EMD) method to decompose a signal into the so-called intrinsic mode function (IMF)
- The Hilbert spectral analysis (HSA) method to obtain instantaneous frequency data

The HHT is a non-parametric method and may be applied for analyzing non-stationary and non-linear time series data.

Empirical mode decomposition (EMD) is a procedure for decomposition of a signal into so called intrinsic mode functions (IMF). An IMF is any function with the same number of extrema and zero crossings, with its envelopes being symmetric with respect to zero. The definition of an IMF guarantees a well-behaved Hilbert transform of the IMF. The procedure of extracting an IMF is called sifting. The sifting process is as follows:

1. Between each successive pair of zero crossings, identify a local extremum in the signal.
2. Connect all the local maxima by a cubic spline line as the upper envelope $E_u(t)$.
3. Repeat the procedure for the local minima to produce the lower envelope $E_l(t)$.
4. Compute the mean of the upper and lower envelope: $m_{11}(t) = \frac{1}{2}(E_u(t)+E_l(t))$.
5. A candidate h_{11} for the first IMF component is obtained as the difference between the signal $x(t)$ and $m_{11}(t)$: $h_{11}(t) = x(t) - m_{11}(t)$.

In a general case the first candidate h_{11} doesn't satisfy the IMF conditions. In such case the sifting is repeated taking h_{11} as the signal. The sifting is repeated iteratively:

$$h_{1k}(t) = h_{1(k-1)}(t) - m_{1k}(t) \tag{2.120}$$

until the assumed threshold for standard deviation SD computed for the two consecutive siftings is achieved. The SD is defined as:

$$SD = \sum_{t=0}^{T} \frac{|h_{1(k-1)}(t) - h_{1k}(t)|^2}{h_{1(k-1)}^2(t)} \tag{2.121}$$

Authors of the method suggest the SD of 0.2–0.3 [Huang et al., 1998]. At the end of the sifting process after k iterations the first IMF is obtained:

$$c_1 = h_{1k} \tag{2.122}$$

The c_1 mode should contain the shortest period component of the signal. Subtracting it from the signal gives the first residue:

$$r_1 = x(t) - c_1 \tag{2.123}$$

The procedure of finding consecutive IMFs can be iteratively continued until the variance of the residue is below a predefined threshold, or the residue becomes a monotonic function—the trend (the next IMF cannot be obtained). The signal can be expressed as a sum of the n-empirical modes and a residue:

$$x(t) = \sum_{i=1}^{n} c_i - r_n \tag{2.124}$$

Each of the components can be expressed by means of a Hilbert transform as a product of instantaneous amplitude $a_j(t)$ and an oscillation with instantaneous frequency $\omega_j(t)$ (Sect. 2.4.1): $c_j = a_j(t)e^{i\int \omega_j(t)dt}$. Substituting this to (2.124) gives representation of the signal in the form:

$$x(t) = \sum_{i=1}^{n} a_j(t)e^{i\int \omega_j(t)dt} \tag{2.125}$$

Equation (2.125) makes possible construction of time-frequency representation—the so-called Hilbert spectrum. The weight assigned to each time-frequency coordinate is the local amplitude.

2.5 Non-linear methods of signal analysis

Non-linear methods of signal analysis were inspired by the theory of non-linear dynamics— indeed the biomedical signal may be generated by the non-linear process. Dynamical systems are usually defined by a set of first-order ordinary differential equations acting on a phase space. The phase space is a finite-dimensional vector space $\mathbb{R}^m$, in which a state $x \in \mathbb{R}^m$ is defined. For the deterministic system we can describe the dynamics by an explicit system of m first-order ordinary differential equations :

$$\frac{dx(t)}{dt} = f(t, x(t)), \qquad x \in \mathbb{R}^m \tag{2.126}$$

If the time is treated as a discrete variable, the representation can take a form of an m-dimensional map:

$$x_{n+1} = F(x_n), \qquad n \in \mathbb{Z} \tag{2.127}$$

A sequence of points x_n or $x(t)$ solving the above equations is called a trajectory of the dynamical system. Typical trajectories can run away to infinity or can be confined to the certain space, depending on F (or f). An attractor is a geometrical object to which a dynamical system evolves after a long enough time. Attractor can be a point, a curve, a manifold or a complicated object called a *strange attractor*. Attractor is considered strange if it has non-integer dimension. Non-linear chaotic systems are described by strange attractors.

Now we have to face the problem that, what we observe is not a phase space object but a time series and, moreover, we don't know the equations describing the process that generates them. The delay embedding theorem of Takens [Takens, 1981] provides the conditions under which a smooth attractor can be reconstructed from a sequence of observations of the state of a dynamical system. The reconstruction preserves the properties of the dynamical system that do not change under smooth coordinate changes. A reconstruction in d dimensions can be obtained by means of the retarded vectors:

$$\xi(t) = x(t_i), x(t_i + \tau), \ldots, x(t_i + (m-1)\tau) \tag{2.128}$$

Number m is called the embedding dimension, τ is called delay time or lag.

According to the above formula almost any scalar observation (e.g., time series), is sufficient to learn about the evolution of a very complex high-dimensional deterministic evolving system. However, we don't know in advance how long the retarded vector must be and we don't know the delay τ.

Choosing a too small value of τ would give a trivial result and too large τ would hamper the information about the original system. Usually the time coordinate of the first minimum of the autocorrelation function is taken as τ. The phase portrait of an ECG obtained by embedding in three-dimensional space is shown in Figure 2.22. Please note (picture b) the distortion of the phase portrait for too large τ. Finding m is even more complex. The method of false neighbors [Kantz and Schreiber, 2000] is difficult to apply and doesn't give unequivocal results. Usually the embedding dimension is found by increasing the dimension step by step.

2.5.1 Lyapunov exponent

Lyapunov exponents describe the rates at which nearby trajectories in phase space converge or diverge; they provide estimates of how long the behavior of a mechanical

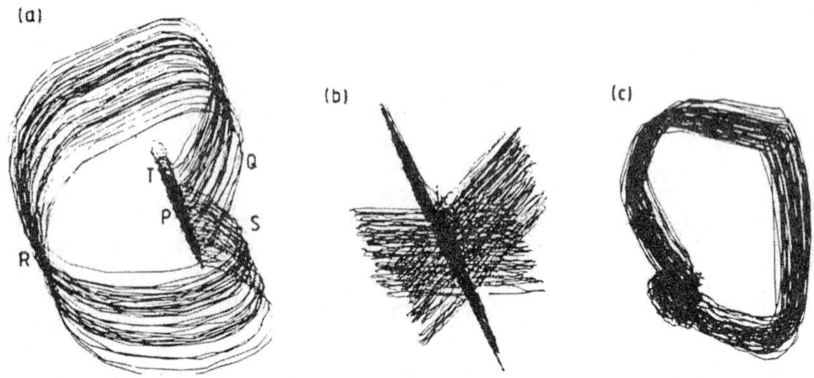

FIGURE 2.22: Phase portraits of human ECG in three-dimensional space. A two-dimensional projection is displayed for two values of the delay τ: (a) 12 ms and (b) 1200 ms. (c) represents the phase portrait constructed from ECG of simultaneously recorded signals from three ECG leads. From [Babloyantz and Destexhe, 1988].

system is predictable before chaotic behavior sets in. The *Lyapunov exponent* or *Lyapunov characteristic exponent* of a dynamical system is a quantity that characterizes the rate of separation of infinitesimally close trajectories. Quantitatively, the separation of two trajectories in phase space with initial distance ΔZ_0 can be characterized by the formula:

$$|\Delta Z(t)| \approx e^{\lambda t}|Z_0| \tag{2.129}$$

where λ is the Lyapunov exponent. Positive Lyapunov exponent means that the trajectories are diverging which is usually taken as an indication that the system is chaotic. The number of Lyapunov exponents is equal to the number of dimensions of the phase space.

2.5.2 Correlation dimension

The concept of generalized dimension (special cases of which are: correlation dimension and Hausdorff dimension) was derived from the notion that geometrical objects have certain dimensions, e.g., a point has a dimension 0, a line—1, a surface—2; in case of chaotic trajectories dimension is not an integer.

The measure called *correlation dimension* was introduced by [Grassberger and Procaccia, 1983]. It involves definition of the correlation sum $C(\varepsilon)$ for a collection of points x_n in some vector space to be the fraction of all possible pairs of points which are closer than a given distance ε in a particular norm. The basic formula for

$C(\varepsilon)$ is:

$$C(\varepsilon) = \frac{2}{N(N-1)} \sum_{i=1}^{N-1} \sum_{j=i+1}^{N} \Theta(\varepsilon - ||x_i - x_j||) \tag{2.130}$$

where Θ is the Heaviside step function,

$$\Theta(x) = \begin{cases} 0 \text{ for } x \leq 0 \\ 1 \text{ for } x > 0 \end{cases} \tag{2.131}$$

The sum counts the pairs (x_i, x_j) whose distance is smaller than ε. In the limit $N \to \infty$ and for small ε, we expect C to scale like a power law $C(\varepsilon) \propto \varepsilon^{D_2}$, so the correlation dimension D_2 is defined by:

$$D_2 = \lim_{\varepsilon \to 0} \lim_{N \to \infty} \frac{\partial \log C(\varepsilon, N)}{\partial \log \varepsilon} \tag{2.132}$$

In practice, from a signal $x(n)$ the embedding vectors are constructed by means of the Takens theorem for a range of m values. Then one determines the correlation sum $C(\varepsilon)$ for the range of ε and for several embedding dimensions. Then $C(m, \varepsilon)$ is inspected for the signatures of self-similarity, which is performed by construction of a double logarithmic plot of $C(\varepsilon)$ versus ε. If the curve does not change its character for successive m we conjecture that the given m is a sufficient embedding dimension and D_2 is found as a slope of a plot of $\log(C(\varepsilon, N))$ versus $\log(\varepsilon)$.

The Haussdorf dimension D_H may be defined in the following way: If for the set of points in M dimensions the minimal number of N spheres of diameter l needed to cover the set increases like:

$$N(l) \propto l^{-D_H} \text{ for } l \to 0, \tag{2.133}$$

D_H is a Hausdorff dimension. $D_H \geq D_2$, in most cases $D_H \approx D_2$.

We have to bear in mind that the definition (2.132) holds in the limit $N \to \infty$, so in practice the number of data points of the signal should be large. It has been pointed out by [Kantz and Schreiber, 2000] that $C(\varepsilon)$ can be calculated automatically, whereas a dimension may be assigned only as the result of a careful interpretation of these curves. The correlation dimension is a tool to quantify self-similarity (fractal—non-linear behavior) when it is known to be present. The correlation dimension can be calculated for any kind of signal, also for a purely stochastic time series or colored noise, which doesn't mean that these series have a non-linear character. The approach which helps in distinguishing non-linear time series from the stochastic or linear ones is the method of surrogate data (Sect. 1.6).

2.5.3 Detrended fluctuation analysis

Detrended fluctuation analysis (DFA) quantifies intrinsic fractal-like correlation properties of dynamic systems. A fundamental feature of a fractal system is scale-invariance or self similarity in different scales. In DFA the variability of the signal is analyzed in respect to local trends in data windows. The method allows to detect long-range correlations embedded in a non-stationary time series. The procedure relies on the conversion of a bounded time series $x_t (t \in \mathbb{N})$ into an unbound process: X_t:

$$X_t = \sum_{i=1}^{t} (x_i - \langle x_i \rangle) \tag{2.134}$$

where X_t is called a cumulative sum and $\langle x_i \rangle$ is the average in the window t. Then the integrated time series is divided into boxes of equal length L, and a local straight line (with slope and intercept parameters a and b) is fitted to the data by the least squares method:

$$E^2 = \sum_{i=1}^{L} (X_i - a_i - b)^2 \tag{2.135}$$

Next, the fluctuation—the root-mean-square deviation from the trend is calculated over every window at every time scale:

$$F(L) = \sqrt{\frac{1}{L} \sum_{i=1}^{L} (X_i - a_i - b)^2} \tag{2.136}$$

This detrending procedure followed by the fluctuation measurement process is repeated over the whole signal over all time scales (different box sizes L). Next, a $\log - \log$ graph of L against $F(L)$ is constructed.

A straight line on this graph indicates statistical self-affinity, expressed as $F(L) \propto L^{\alpha}$. The scaling exponent α is calculated as the slope of a straight line fit to the $\log - \log$ graph of L against $F(L)$.

The fluctuation exponent α has different values depending on the character of the data. For uncorrelated white noise it has a value close to $\frac{1}{2}$, for correlated process $\alpha > \frac{1}{2}$, for a pink noise $\alpha = 1$ (power decaying as $1/f$), $\alpha = \frac{3}{2}$ corresponds to Brownian noise.

In the case of power-law decaying autocorrelations, the correlation function decays with an exponent γ: $R(L) \sim L^{-\gamma}$ and the power spectrum decays as $S(f) \sim f^{-\beta}$. The three exponents are related by relations: $\gamma = 2 - 2\alpha$, $\beta = 2\alpha - 1$, and $\gamma = 1 - \beta$.

As with most methods that depend upon line fitting, it is always possible to find a number α by the DFA method, but this does not necessarily mean that the time series is self-similar. Self-similarity requires that the points on the $\log - \log$ graph are sufficiently collinear across a very wide range of window sizes. Detrended fluctuation is used in HRV time series analysis and its properties will be further discussed in Sect. 4.2.2.4.3.

2.5.4 Recurrence plots

Usually the dimension of a phase space is higher than two or three which makes its visualization difficult. Recurrence plot [Eckmann et al., 1987] enables us to investigate the m-dimensional phase space trajectory through a two-dimensional representation of its recurrences; namely recurrence plot (RP) reveals all the times when the phase space trajectory visits roughly the same area in the phase space. In this way distinct recurrent behavior, e.g., periodicities and also irregular cyclicities can be detected. The recurrence of states, in the meaning that states become arbitrarily close after some time, is a fundamental property of deterministic dynamical systems and is typical for non-linear or chaotic systems.

Recurrence of a state at time i against a different state at time j is marked within a two-dimensional squared matrix with ones and zeros (black and white dots in the plot), where both axes are time axes. More formally RP can be expressed as:

$$R_{i,j} = \Theta(\varepsilon_i - ||x_i - x_j||), \quad x_i \in R^m, \quad i,j = 1\ldots,N \tag{2.137}$$

where N is the number of considered states x_i, ε_i is a threshold distance, $||\cdot||$ a norm, and $\Theta(\cdot)$ the Heaviside function. The visual appearance of an RP (Figure 2.23) gives

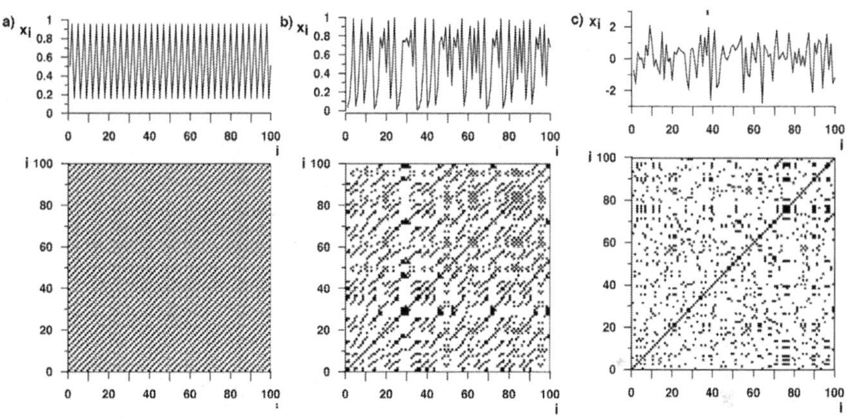

FIGURE 2.23: Examples of the recurrence plots (lower panels) for different signals shown above: a) a periodic signal obtained from the logistic equation for the control parameter r = 3.829, b) a chaotic signal obtained from the logistic equation for r = 3.9999, c) white noise. Note that in the RP for the deterministic chaotic signal short lines parallel to the diagonal are present. Adapted from [Klimaszewska and Zebrowski, 2009].

hints about the dynamics of a system. Uncorrelated white noise results in uniformly distributed dots, periodic patterns are connected with cyclicities in the process—time distance between patterns (e.g., lines) corresponds to period, diagonal lines mean that the evolution of states is similar at different times—the process could be deterministic; if these diagonal lines occur beside single isolated points, the process could be chaotic.

The visual interpretation of RPs requires some experience. Their quantification offers a more objective way of investigating the considered system. With this quantification, the RPs have become a tool for analysis of certain biomedical processes, e.g., cardiac activity. The main advantage of recurrence plots is that they provide information even for short and non-stationary data, where other non-linear methods fail.

2.5.5 Poincaré map

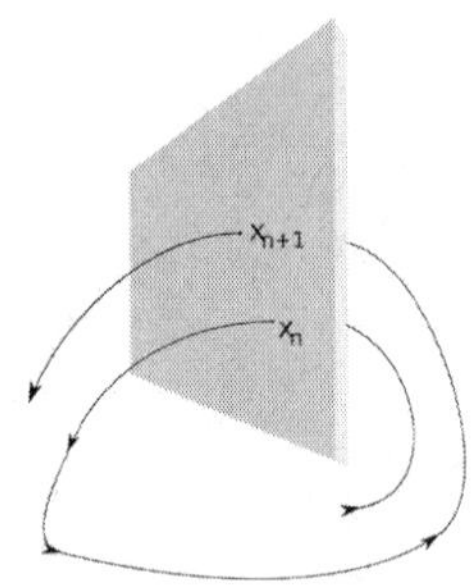

FIGURE 2.24: Construction of the Poincaré map. The Poincaré map relates two consecutive intersection points x_{n+1} and x_n, which come from the same side of the plane.

Another way to visualize the dynamical system evolution in a phase space is a Poincaré map. A Poincaré section is the intersection of a trajectory of a dynamical system in the state space with a certain lower dimensional subspace, transversal to the flow of the system. The Poincaré map relates two consecutive intersection points x_{n+1} and x_n, which come from the same side of the plane (Figure 2.24). A Poincaré map differs from a recurrence plot because it is defined in a phase space, while a recurrence plot is defined in a time space (points on this plot depict pairs of time moments when the system visits roughly the same region of phase space). By means of the Poincaré map it is possible to reduce the phase space dimensionality, at the same time turning the continuous time flow into a discrete time map [Kantz and Schreiber, 2000].

2.5.6 Approximate and sample entropy

Even for low-dimensional chaotic systems, a huge number of points is required to achieve convergence of the algorithms estimating dimension or entropy of the process. In order to overcome this difficulty and provide a measure capable of quantifying the changes of process complexity the modifications of the entropy measure were proposed. Approximate entropy (ApEn) was introduced by [Pincus, 1991]. ApEn measures the (logarithmic) likelihood that trajectories that are close to each other remain close on the next incremental comparison. However, methodological pitfalls in ApEn were pointed out in [Richman and Moorman, 2000] and [Costa et al., 2002] and another measure called sample entropy (SaEn) was introduced in [Richman and Moorman, 2000].

In the SaEn method the vectors (blocks) $\xi(i)$ of differing dimensions k are created by embedding a signal in a way similar to that given in formula (2.128). The reconstructed vectors are the state space (k-dimensional) representation of the dynamics of the system. SaEn is defined as the logarithmic difference between the probability(density) of occurrence of the vector $\xi(i)$ within the chosen distance r in k dimensions and the probability of occurrence of the vector $\xi(i)$ within the same chosen distance in $k+1$ dimension:

$$\mathrm{SaEn} = \log \frac{\rho^k(r)}{\rho^{k+1}(r)}, \tag{2.138}$$

where $\rho^k(r)$, $\rho^{k+1}(r)$—densities of occurrence in k, $k+1$ dimensions, respectively.

(MATLAB code for calculating Sample Entropy is available at: `http://www.physionet.org/physiotools/sampen/matlab/`). SaEn met some criticism, e.g., that it does not characterize the complexity of the signal completely [Govindan et al., 2007] and soon another modifications appeared.

In the approach called multiscale entropy (MSE) [Costa et al., 2002] the procedure involves partitioning the signal into disjoined windows and the data are averaged inside each time window, which in fact is equivalent to downsampling, or low-pass filtering. In the case of a continuous system the same signal sampled at two different sampling rates would show different behavior in MSE analysis, so the method will quantify the system behavior in different frequency ranges.

Yet in another approach SaEn was modified by introducing the time delay δ between the successive components of blocks (or subseries) [Govindan et al., 2007]; δ was chosen as the time point at which autocorrelation function falls below $1/e$. In the density estimation in order not to account for the temporally correlated points, for each center i , the $i+\delta$ surrounding points were discarded.

The above described entropy measures and detrended fluctuation analysis found applications mainly in heart rate variability analysis and their utility will be discussed in the section concerning this signal.

2.5.7 Limitations of non-linear methods

There are several pitfalls in the application of non-linear methods. One of them is the fact that even for infinite-dimensional, stochastic signals low-dimensional estimates may be obtained, which was pointed out by [Theiler et al., 1992]. The temporal coherence of the data may be mistaken for the trace of non-linearity. Therefore before applying non-linear methods one should first check, if indeed the signals have traces of non-linearity, since the non-linear process needs not to produce non-linear time series. The null hypothesis about the stochastic character of the signal may be checked by means of surrogate data technique (Sect. 1.6).

The non-linear methods rely to a large extent on the reconstruction of a phase space. In order to construct the phase space by embedding, rather long stationary data epochs are needed. In case of biomedical signals long stationary data segments are often hard to find. Another problem is connected with the fact, that the non-linear methods are prone to systematic errors connected with the arbitrary choices (e.g., of the bin length in estimating probabilities, or lag of embedding procedure).

However, the biggest problem in application of non-linear methods is the fact that they are very sensitive to noise. As was pointed out by [Kantz and Schreiber, 2000] for a low-dimensional deterministic signal with a noise component of the order 2-3%, it will be difficult to reasonably estimate correlation dimension, since in this case a significant scaling region of $C(\varepsilon)$ could not be found; the reason being that the artifacts of the noise meet the artifacts of the overall shape of the attractor. Since biomedical signals contain a significant noise component one should approach non-linear methods with caution and apply them only in cases where the non-linearity of a signal is well established and linear methods seem not to work properly.

3

Multiple channels (multivariate) signals

The technological progress in the biomedical field has led to the construction of recording equipment which allows recording of activity from multiple sites. Today a typical dataset contains not only two or four but dozens or hundreds of channels. This is especially the case for EEG, MEG, and sometimes also for ECG and EMG signals. Analysis of multichannel data can give a better insight into the relations between the investigated sites, but it is a challenging task. Besides many experimental and computational difficulties, the problem quite often lies in the proper application of existing mathematical tools. The techniques capitalizing on the covariance structure of the multichannel (multivariate) data are especially useful in this respect. In this chapter an introduction to basic aspects of multichannel data processing will be presented.

3.1 Cross-estimators: cross-correlation, cross-spectra, coherence (ordinary, partial, multiple)

Joint moments of the order two: cross-correlation R_{xy} and cross-covariance C_{xy}, were defined in Sect. 1.1 (equations 1.11 and 1.12) by means of ensemble averaging formalism. Under the assumption of ergodicity they can be expressed by the formulas:

$$R_{xy}(\tau) = \int_{-\infty}^{\infty} x(t)y(t+\tau)dt \tag{3.1}$$

$$C_{xy}(\tau) = \int_{-\infty}^{\infty} (x(t) - \mu_x)(y(t+\tau) - \mu_y)\,dt \tag{3.2}$$

Cross-correlation and cross-spectrum are bound by means of Fourier transform and inverse Fourier transform (cf. Sect. 2.3.2.1.3):

$$S_{xy}(f) = \int_{-\infty}^{\infty} R_{xy}(\tau)e^{-i2\pi f\tau}d\tau \tag{3.3}$$

$$R_{xy}(\tau) = \int_{-\infty}^{\infty} S_{xy}(f)e^{i2\pi f\tau}df \tag{3.4}$$

Similarly to the power spectrum, cross-spectrum is usually computed by means of the Fourier transforms X and Y of the signals x and y :

$$S_{xy}(f) = \lim_{T\to\infty} \frac{1}{T} X(f,T)Y^*(f,T) \tag{3.5}$$

The fact that we have only a limited data window of length T has the same consequences as in cases of power spectra estimation (see Sect. 2.3.2.1). Usually non-rectangular window functions are used and smoothing is applied. Computation of cross-spectrum is implemeted in MATLAB Signal Processing Toolbox as `cpsd` function. It estimates the cross power spectral density of the discrete-time signals using Welch's averaged, modified periodogram method of spectral estimation.

$S_{xy}(f)$ is a complex value consisting of real and imaginary parts:

$$S_{xy}(f) = Re(S_{xy})(f) + iIm(S_{xy})(f) \tag{3.6}$$

In polar coordinates it can be expressed by the formula:

$$S_{xy}(f) = |S_{xy}(f)|e^{i\Phi xy(f)}, \tag{3.7}$$

where $|S_{xy}(f)| = \sqrt{Re(S_{xy})^2(f) + Im(S_{xy})^2(f)}$ is a modulus and $\Phi_{xy}(f) = \tan^{-1}\frac{Im(S_{xy})}{Re(S_{xy})}$ is a phase.

Coherence is a measure which is often used in biomedical application. It is expressed by the formula:

$$\gamma_{xy}(f) = \frac{S_{xy}(f)}{\sqrt{S_x(f)S_y(f)}} \tag{3.8}$$

where S_x and S_y are spectra of signals x and y. Since $S_{xy}(f) \leq \sqrt{S_x(f)S_y(f)}$, function $\gamma_{xy}(f) \leq 1$. Coherence shows the relation between two signals in frequency and phase. The square of the coherence measures the spectral power in a given frequency common to both signals.

The above formula defines ordinary (bivariate) coherence. If a data set contains more than two channels, the signals can be related with each other in different ways. Namely, two (or more) signals may simultaneously have a common driving input from the third channel. Depending on the character of relations between channels, some of them may be connected directly with each other and some connections can be indirect (through other channels). To distinguish between these situations partial and multiple coherences were introduced.

The construction of partial coherence relies on subtracting influences from all other processes under consideration. For three channels partial coherence is defined as a normalized partial cross spectrum:

$$\kappa_{xy|z}(f) = \frac{|S_{xy|z}(f)|}{\sqrt{(S_{xx|z}(f)S_{yy|z}(f)}} \tag{3.9}$$

where partial cross-spectrum is defined as:

$$S_{xy|z}(f) = S_{xy}(f) - S_{xz}(f)S_{zz}^{-1}(f)S_{zy}(f) \tag{3.10}$$

For an arbitrary number of channels partial coherence may be defined in terms of the minors of spectral matrix $\mathbf{S}(f)$, which on the diagonal contains spectra and off-diagonal cross-spectra:

$$\kappa_{ij}(f) = \frac{\mathbf{M}_{ij}(f)}{\sqrt{\mathbf{M}_{ii}(f)\mathbf{M}_{jj}(f)}} \tag{3.11}$$

where $\mathbf{M}_{ij}$ is a minor of $\mathbf{S}$ with the ith row and j^{th} column removed. Its properties are similar to ordinary coherence, but it is nonzero only for direct relations between channels. If a signal in a given channel can be explained by a linear combination of some other signals in the set, the partial coherence between them will be low.

Multiple coherence is defined by:

$$G_i(f) = \sqrt{1 - \frac{\det \mathbf{S}(f)}{S_{ii}(f)\mathbf{M}_{ii}(f)}} \tag{3.12}$$

Its value describes the amount of common components in the given channel and the other channels in the set. If the value of multiple coherence is close to zero then the channel has no common components with any other channel of the set. The high value of multiple coherence for a given channel means that a large part of the variance of that channel is common to all other signals; it points to the strong relation between the signals. Partial and multiple coherences can be conveniently found by means of the autoregressive parametric model MVAR.

3.2 Multivariate autoregressive model (MVAR)

3.2.1 Formulation of MVAR model

MVAR model is an extension of the one channel AR model for an arbitrary number of channels. In the MVAR formalism, sample $x_{i,t}$ in channel i at time t is expressed not only as a linear combination of p previous values of this signal, but also by means of p previous samples of all the other channels of the process. For k channels model of order p may be expressed as:

$$\vec{x}_t = A_1\vec{x}_{t-1} + A_2\vec{x}_{t-2} + \ldots A_p\vec{x}_{t-p} + \vec{E}_t \tag{3.13}$$

where : $\vec{x}_t = [x_{1,t}, x_{2,t}, \ldots x_{k,t}]$ is a vector of signal samples at times t in channels $\{1,2,\ldots k\}$. $\vec{E}_t = [E_{1,t}, E_{2,t}, \ldots E_{k,t}]$ is a vector of noise process samples at time t. The covariance matrix $\mathbf{V}$ of a noise process is expressed as:

$$\vec{E}_t\vec{E}_t^{\mathsf{T}} = \mathbf{V} = \begin{pmatrix} \sigma_1^2 & 0 & \ldots & 0 \\ 0 & \sigma_2^2 & \ldots & 0 \\ \vdots & & & \vdots \\ 0 & \ldots & 0 & \sigma_k^2 \end{pmatrix} \tag{3.14}$$

where T denotes transposition.

$\mathbf{V}$ is a matrix representing residual variances of noises assumed in the model. In the process of model estimation we obtain an estimate of the variance matrix as $\hat{\mathbf{V}}$, which accounts for the variance not explained by the model coefficients. Matrix $\hat{\mathbf{V}}$ usually contains small non-diagonal elements; their value informs us how well the model fits the data. The MVAR model coefficients for each time lag l are $k \times k$-sized matrices:

$$\mathbf{A}_t = \begin{pmatrix} A_{11}(l) & A_{12}(l) & \dots & A_{1k}(l) \\ A_{21}(l) & A_{22}(l) & \dots & A_{2k}(l) \\ \vdots & & & \vdots \\ A_{k1}(l) & \dots & A_{kk-1}(l) & A_{kk}(l) \end{pmatrix} \tag{3.15}$$

Before starting a fitting procedure, certain preprocessing steps are needed. First, the temporal mean should be subtracted for every channel. Additionally, in most cases normalization of the data by dividing each channel by its temporal variance is recommended. This is especially useful when data channels have different amplification ratios.

The estimation of model parameters in the case of a multivariate model is similar to the one channel model. The classical technique of AR model parameters estimation is the Yule-Walker algorithm. It requires calculating the correlation matrix $\mathbf{R}$ of the system up to lag p. The model equation (3.13) is multiplied by $\vec{x}_{t+s}^{\mathrm{T}}$, for $s = 0, \dots, p$ and expectations of both sides of each equation are taken:

$$R_{ij}(s) = \frac{1}{N_s} \sum_{t=1}^{N_s} x_{i,t} x_{j,t+s} \tag{3.16}$$

Assuming that the noise component is not correlated with the signals, we get a set of linear equations to solve (the Yule-Walker equations):

$$\begin{pmatrix} \mathbf{R}(0) & \mathbf{R}(-1) & \dots & \mathbf{R}(p-1) \\ \mathbf{R}(1) & \mathbf{R}(0) & \dots & \mathbf{R}(p-2) \\ \vdots & \vdots & & \vdots \\ \mathbf{R}(1-p) & \mathbf{R}(2-p) & \dots & \mathbf{R}(0) \end{pmatrix} \begin{pmatrix} \mathbf{A}(1) \\ \mathbf{A}(2) \\ \vdots \\ \mathbf{A}(p) \end{pmatrix} = \begin{pmatrix} \mathbf{R}(-1) \\ \mathbf{R}(-2) \\ \vdots \\ \mathbf{R}(-p) \end{pmatrix} \tag{3.17}$$

and

$$\hat{\mathbf{V}} = \sum_{j=0}^{p} \mathbf{A}(j)\mathbf{R}(j) \tag{3.18}$$

This set of equations is similar to the formula (2.50) for one channel model, however the elements $\mathbf{R}(i)$ and $\mathbf{A}(i)$ are $k \times k$ matrices. Other methods of finding MVAR coefficients are the Burg (LWR) recursive algorithm and covariance algorithm. The Burg algorithm produces high resolution spectra and is preferred when closely spaced spectral components are to be distinguished. Sinusoidal components in a spectrum are better described by the covariance algorithm or its modification. Recently, a Bayesian approach has been proposed for estimating the optimal model order and model parameters. In most cases, however, the spectra produced by different algorithms are very similar to each other. MVAR model and methods of determination of its coefficients are described in: [Priestley, 1981, Kay, 1988, Marple, 1987, Lutkepohl, 1993].

Several criteria of the determination of the MVAR model order were proposed in [Lutkepohl, 1993]. Similarly to the one channel case we seek the minimum of the function consisting of two terms: the first one is a reward for minimizing the residual variance, the second one is a punishment for a too high model order. The first term depends on the estimated residual variance $\hat{\mathbf{V}}(p)$ for a given p, the second one is a function of model order, number of channels k, and number of data points N. The criteria presented below differ in respect to the second term:

AIC criterion:

$$AIC(p) = \log[\det(\hat{\mathbf{V}})] + 2\frac{pk^2}{N} \tag{3.19}$$

Hannan-Quin criterion:

$$HQ(p) = \log[\det(\hat{\mathbf{V}})] + 2\log(\log(N))\frac{pk^2}{N} \tag{3.20}$$

Schwartz criterion:

$$SC(p) = \log[\det(\hat{\mathbf{V}})] + \log(N)\frac{pk^2}{N} \tag{3.21}$$

AIC criterion is the one which is mostly used, but some authors [Kay, 1988, Marple, 1987] claim that it sometimes gives a too high model order.

3.2.2 MVAR in the frequency domain

In analogy to the procedure described in Sect. 2.3.2.2.3 equation (3.13) can be easily transformed to describe relations in the frequency domain:

$$\mathbf{E}(f) = \mathbf{A}(f)\mathbf{X}(f) \tag{3.22}$$

$$\mathbf{X}(f) = \mathbf{A}^{-1}(f)\mathbf{E}(f) = \mathbf{H}(f)\mathbf{E}(f) \tag{3.23}$$

where

$$\mathbf{H}(f) = \left(\sum_{m=0}^{p} \mathbf{A}(m)e^{-2\pi imf\Delta t}\right)^{-1} \tag{3.24}$$

and $\Delta t = \frac{1}{F_s}$; F_s is the sampling frequency

From the form of that equation we see that the model can be considered as a linear filter with white noises $\mathbf{E}(f)$ on its input (flat dependence on frequency) and the signals $\mathbf{X}(f)$ on its output. The transfer matrix $\mathbf{H}(f)$ contains information about all relations between channels of a process. From the transfer matrix, spectra and cross-spectra may be calculated:

$$\mathbf{S}(f) = \mathbf{X}(f)\mathbf{X}^*(f) = \mathbf{H}(f)\mathbf{E}(f)\mathbf{E}^*(f)\mathbf{H}^*(f) = \mathbf{H}(f)\mathbf{V}\mathbf{H}^*(f) \tag{3.25}$$

The matrix $\mathbf{S}(f)$ contains auto-spectra of each channel on the diagonal and cross - spectra off the diagonal. The simple rule connecting the model order with the number of spectral peaks does not hold in the case of MVAR. Usually the model order is lower, since there are more coefficients to describe spectral properties; their number is equal k^2p.

3.3 Measures of directedness

3.3.1 Estimators based on the phase difference

In the study of biomedical signals a problem of particular interest is finding the direction of the influence exerted by one channel on the others, or in other words finding the directedness. The most straightforward way to determine the directedness seems the application of cross measures: cross-correlation or coherence. Cross-correlation and coherence are statistically equivalent, since they are bound by the Fourier transform, but cross-correlation operates in time domain and coherence in the frequency domain. Hence these measures emphasize different features of signals.

The amplitude of the cross-correlation describes the similarity of the signals and, if for certain delay τ, function $R_{xy}(\tau)$ has maximum, we can assume that signal y is delayed by τ in respect to signal x, so the direction may be inferred from the delay between signals found for the maximum of the function. It gives information concerning the most pronounced frequency component. When there are different frequency components the signal has to be filtered first (by means of procedure not disturbing phases (in MATLAB `filtfilt`).

For signals containing different rhythmical components coherence is usually a method of choice. Coherence contains phase information (Sect. 3.1) which can be translated into a delay between signals. The delay in samples Δx can be computed from the formula: $\Delta x = \frac{\phi}{\pi}\frac{F_s}{f}$, where f is the frequency for which the phase ϕ is calculated. Since phase is dependent on frequency, the delays may be found for particular frequencies. However, we have to bear in mind that phase is determined modulo 2π; therefore the result may be ambiguous.

In the case of measures of directedness the problems connected with driving the two channels by a third one are critical and can lead to false patterns of propagation. Therefore bivariate correlation and bivariate coherence are hardly applicable for multichannel systems with the number of channels higher than two. This problem may be alleviated by application of partial correlations; however the estimation of directedness by means of phases of partial correlations is prone to high statistical errors. The asymptotic variance for the phase of partial coherence in the case of three channels is:

$$\mathrm{var}\left(\phi_{xy|z}(f)\right) = \frac{1}{n}\left(\frac{1}{\kappa^2_{xy|z}(f)} - 1\right) \tag{3.26}$$

where n is the number of degrees of freedom dependent on the method of estimation of cross-spectra.

Therefore, for low values of coherence the errors will be large. Also in the case of estimation of coherence by means of Fourier transform, even with smoothing, the number of degrees of freedom is low. Partial correlations found by means of MVAR are computed under the assumption of minimum phase and they are not suitable for determination of directedness. By means of cross-correlation or coherence it is not possible to detect reciprocal connections. This type of connection may be found by means of measures based on the causality principle.

3.3.2 Causality measures

3.3.2.1 Granger causality

In biomedical studies finding causality between channels of multivariate processes is of particular interest. The testable definition of causality was introduced by Granger[1] (1969) in the field of economics [Granger, 1969]. Granger defined causality in terms of predictability of time series; namely if some series $y(t)$ contains information in past terms that helps in the prediction of series $x(t)$, then $y(t)$ is said to cause $x(t)$. More specifically, if we try to predict a value of $x(t)$ using p previous values of the series x only, we get a prediction error ε:

$$x(t) = \sum_{j=1}^{p} A'_{11}(j)x(t-j) + \varepsilon(t) \tag{3.27}$$

If we try to predict a value of $x(t)$ using p previous values of the series x and p previous values of y we obtain another prediction error ε_1:

$$x(t) = \sum_{j=1}^{p} A_{11}(j)x(t-j) + \sum_{j=1}^{p} A_{21}(j)y(t-j) + \varepsilon_1(t) \tag{3.28}$$

If the variance ε_1 (after including series y to the prediction) is lower than the variance ε we say that y causes x in the sense of Granger causality. Similarly we can say that x causes y in the sense of Granger causality when the variance ε_2 is reduced after including series x in the prediction of series y:

$$y(t) = \sum_{j=1}^{p} A_{22}(j)y(t-j) + \sum_{j=1}^{p} A_{21}(j)x(t-j) + \varepsilon_2 \tag{3.29}$$

Three different measures of causality or directionality based on the Granger causality concept were introduced: Granger causality index (GCI) [Geweke, 1982], directed transfer function (DTF) [Kaminski and Blinowska, 1991] and partial directed coherence (PDC) [Baccalá and Sameshima, 2001]. GCI operates in the time domain, DTF and PDC in the frequency domain. Granger causality was originally defined for a two channel system. The above mentioned estimators were extended to multichannel systems.

[1] Nobel Laureate in economics.

3.3.2.2 Granger causality index

For a two channels system GCI is based directly on the principle formulated by Granger. Namely we check, if the information contributed by the second signal improves the prediction of the first signal. In order to find out, we have to compare the variance of univariate AR model (equation 3.27) with a variance of the model accounting for the second variable (equation 3.28).

Granger causality index showing the driving of channel x by channel y is defined as the logarithm of the ratio of residual variance for a two channel model to the residual variance of the one channel model:

$$\mathrm{GCI}_{y \to x} = \log \frac{\varepsilon_1}{\varepsilon} \tag{3.30}$$

This definition can be extended to the multichannel system by considering how the inclusion of the given channel changes the residual variance ratios. To quantify directed influence from a channel x_j to x_i for n-channel autoregressive process in time domain we consider n- and $n-1$ dimensional MVAR models. First, the model is fitted to a whole n-channel system, leading to the residual variance $\hat{V}_{i,n}(t) = \mathrm{var}(E_{i,n}(t))$ for signal x_i. Next, a $n-1$ dimensional MVAR model is fitted for $n-1$ channels, excluding channel j, which leads to the residual variance $\hat{V}_{i,n-1}(t) = \mathrm{var}(E_{i,n-1}(t))$. Then Granger causality is defined as:

$$\mathrm{GCI}_{j \to i}(t) = \log \frac{\hat{V}_{i,n}(t)}{\hat{V}_{i,n-1}(t)} \tag{3.31}$$

GCI is smaller or equal to 1, since the variance of the n-dimensional system is lower than the residual variance of a smaller $n-1$ dimensional system. GCI(t) is used to estimate causality relations in time domain. When the spectral content of the signals is of interest, which is frequent for biomedical signals, the estimators defined in the frequency domain: DTF or PDC are appropriate choices.

3.3.2.3 Directed transfer function

Directed transfer function (DTF) introduced by Kaminski and Blinowska, [Kaminski and Blinowska, 1991] is based on the properties of the transfer matrix $\mathbf{H}(f)$ of MVAR which is asymmetric and contains spectral and phase information concerning relations between channels. DTF describes the causal influence of channel j on channel i at frequency f:

$$\mathrm{DTF}_{ij}^2(f) = \frac{|H_{ij}(f)|^2}{\sum_{m=1}^{k} |H_{im}(f)|^2} \tag{3.32}$$

The above equation defines a normalized version of DTF, which takes values from 0 to 1 giving a ratio between the inflow from channel j to channel i to all the inflows to channel i. Sometimes one can abandon the normalization property and use values of elements of transfer matrix which are related to causal connection strength. The non-normalized DTF can be defined as:

$$\theta_{ij}^2(f) = |H_{ij}(f)|^2 \tag{3.33}$$

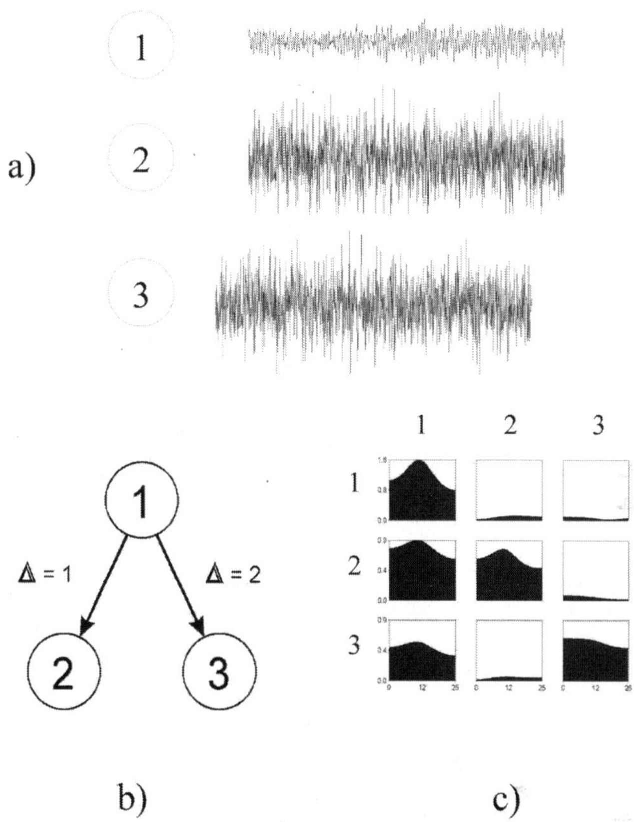

FIGURE 3.1: Performance of DTF for noisy signals. a) 1—stochastic signal, 2 and 3— signals delayed by, respectively, one and two samples with white noise of variance 9 times greater than the signal variance added. b) Simulation scheme, c) results of DTF; in each box DTF as a function of frequency, on the diagonal power spectra; propagation from channels labeled above to the channels labeled at the left of the picture.

It was shown by means of modeling that non-normalized DTF is related to the causal connection strength between channels [Kaminski et al., 2001]. DTF is an estimator which provides spectral information on the transmitted activity and is very robust in respect to noise. In Figure 3.1 an example is given where the direction of flow is determined correctly in case of signals where variance of noise was 9 times as high as that of the signal. However, DTF shows direct as well as indirect transmissions; that is in case of cascade flow $a \rightarrow b \rightarrow c$, it shows also flow: $a \rightarrow c$.

3.3.2.3.1 dDTF In order to distinguish direct from indirect flows direct directed transfer function (dDTF) was introduced [Korzeniewska et al., 2003]. The dDTF is defined as a multiplication of a modified DTF ($F_{ij}^2(f)$) by partial coherence ($C_{ij}^2(f)$). The modification of DTF concerned normalization of the function in such a way as to make the denominator independent of frequency. The dDTF ($\chi_{ij}(f)$) showing direct propagation from channel j to i is defined as:

$$\chi_{ij}^2(f) = F_{ij}^2(f)\kappa_{ij}^2(f) \tag{3.34}$$

$$F_{ij}^2(f) = \frac{|H_{ij}(f)|^2}{\sum_f \sum_{m=1}^{k} |H_{im}(f)|^2} \tag{3.35}$$

$\chi_{ij}(f)$ has a nonzero value when both functions $F_{ij}^2(f)$ and $\kappa_{ij}^2(f)$ are nonzero; in that case there exists a causal relation between channels $j \rightarrow i$ and that relation is direct. The DTF and dDTF for the same simulation scheme are presented in Figure 3.2. One can see that dDTF shows only the direct flows, whereas in the case of DTF

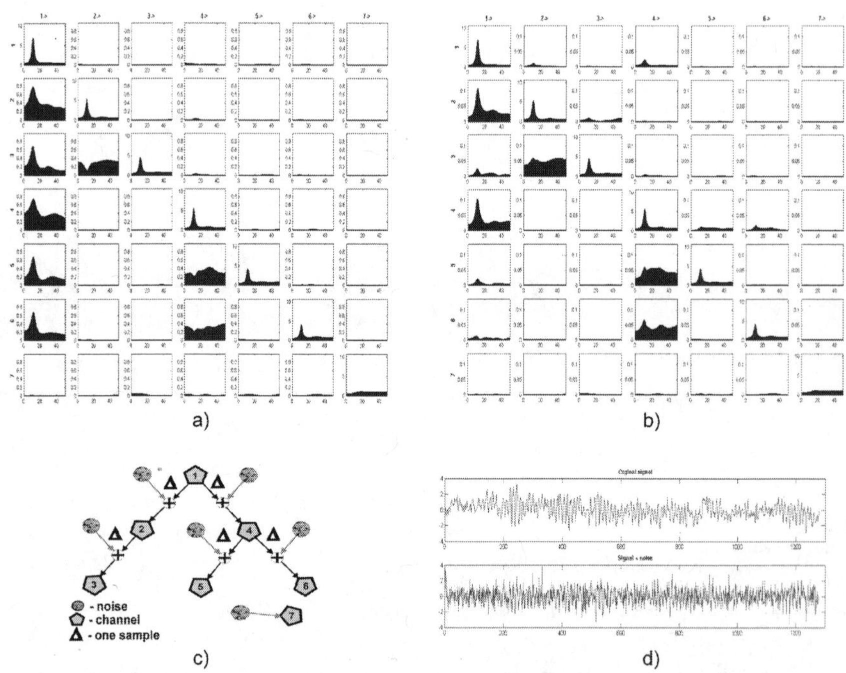

FIGURE 3.2: a) In each box DTF as a function of frequency, on the diagonal power spectra; propagation from channels labeled above to the channels labeled at the left of the picture; b) dDTF is shown using the same convention as in a); c) simulation scheme; d) signals used in the simulation.

cascade flows are also present, e.g., from channel 1 to 3 and 5.

3.3.2.3.2 SDTF In biomedical applications it is often of interest to grasp the dynamical changes of signal propagation, which means that we have to apply a very short measurement window. However, it deteriorates strongly the statistical properties of the estimate. In order to fit the MVAR model properly the number of data points in the window should be bigger than the number of model parameters. For MVAR, number of parameters is pk^2 (where k number of channels, p model order), number of data points is kN, so the condition for proper model fitting is that $\frac{kp}{N}$ should be small, preferably smaller than 0.1. The same rule holds for all estimators based on MVAR. In case of time varying processes, when we would like to follow the dynamics of the transmission the data window has to be short.

To increase the number of the data points we may use multiple repetitions of the experiment. We may treat the data from each repetition as a realization of the same stochastic process. Then the number of data points is kN_SN_T (where N_T is number of realizations, N_S is number of data points in the window), and their ratio to the number of parameters (pk^2) effectively increases. Based on this observation, we can divide a non-stationary recording into shorter time windows, short enough to treat the data within a window as quasi-stationary. We calculate the correlation matrix between channels for each trial separately. The resulting model coefficients are based on the correlation matrix averaged over trials. The correlation matrix has a form:

$$\hat{R}_{ij}(s) = \frac{1}{N_T}\sum_{r=1}^{N_T} R_{ij}^{(r)}(s) = \frac{1}{N_T}\sum_{r=1}^{N_T}\frac{1}{N_S}\sum_{t=1}^{N_S} X_{i,t}^{(r)} X_{j,t+s}^{(r)} \tag{3.36}$$

$R_{ij}^{(r)}$ is the correlation matrix for short windows of N_S points, and r is the index of the repetition.

The averaging concerns correlation matrices for short data windows—data are not averaged in the process. The choice of the window size is always a compromise between quality of the fit (depending on the ratio between number of data points and number of model parameters) and time resolution. By means of the above described procedure for each short data window the MVAR coefficients are obtained and then the estimators characterizing the signals: power spectra, coherences, DTFs are found. By application of the sliding window, multivariate estimators may be expressed as functions of time and frequency and their evolution in time may be followed.

The tutorial and software for calculation of DTF can be found at http://eeg.pl.

3.3.2.4 Partial directed coherence

The partial directed coherence (PDC) was introduced by Baccalá and Sameshima [Baccalá and Sameshima, 2001] in the following form:

$$P_{ij}(f) = \frac{A_{ij}(f)}{\sqrt{a_j^H(f)a_j(f)}} \tag{3.37}$$

In the above equation $A_{ij}(f)$ is an element of $A(f)$—a Fourier transform of MVAR model coefficients $a_j(f)$ is j-th column of $A(f)$, and the $(.)^H$ denotes the Hermitian transpose. The PDC from j to i represents the relative coupling strength of the interaction of a given source (signal j), with regard to some signal i as compared to all of j's connections to other signals. From the normalization condition it follows that PDC takes values from the interval $P_{ij} \in [0,1]$. PDC shows only direct flows between channels. Although it is a function operating in the frequency domain, the dependence of $A(f)$ on the frequency has not a direct correspondence to the power spectrum. Figure 3.3 shows PDC for the simulation scheme presented in Figure 3.2 c).

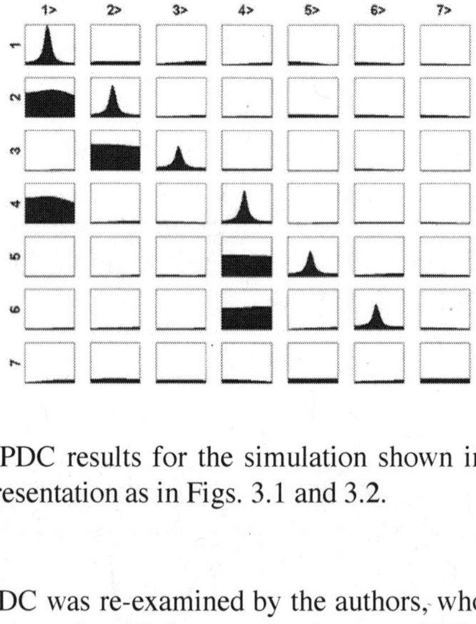

FIGURE 3.3: The PDC results for the simulation shown in Figure 3.2 c). The same convention of presentation as in Figs. 3.1 and 3.2.

The definition of PDC was re-examined by the authors, who considered the fact that PDC dependence on a signal's dynamic ranges, as modified by gains, obscures the PDC ability to correctly pinpoint the direction of information flow [Baccalá et al., 2006]. They proposed the so-called generalized partial directed coherence (GPDC) defined by the formula:

$$\mathrm{GPDC}_{ij}(f) = \frac{\frac{1}{\sigma_i} A_{ij}(f)}{\sqrt{\sum_{k=1}^{N} \frac{1}{\sigma_k^2} A_{ki}(f) A_{ki}^*(f)}} \tag{3.38}$$

Other problems inherent to PDC were pointed out by [Schelter et al., 2009]. The most important one was that PDC measures the strength of influence relative to a given signal source. It means that, in particular, adding further variables that are influenced by the source variable decreases the PDC, although the relationship between source and target processes remains unchanged. To counteract this unfortunate property in [Schelter et al., 2009], another kind of normalization of PDC was proposed. This problem will be considered further in Sect. 3.5.

3.4 Non-linear estimators of dependencies between signals

3.4.1 Non-linear correlation

The general idea of the non-linear correlation $h^2_{y|x}$ [Lopes da Silva et al., 1989] is that, if the value of y is considered to be a function of x, the value of y can be predicted according to a non-linear regression curve. The formula describing the non-linear correlation coefficient is given by the equation:

$$h^2_{y|x} = \frac{\sum_{k=1}^{N} y(k)^2 - \sum_{k=1}^{N} (y(k) - f(x_i))^2}{\sum_{k=1}^{N} y(k)^2} \tag{3.39}$$

where $y(k)$ are samples of the signal y of N points, $f(x_i)$ is the linear piecewise approximation of the non-linear regression curve. In practice to determine $f(x_i)$ a scatter plot of y versus x is studied. Namely the values of signal x are divided into bins; for each bin, the value x of the midpoint (r_i) and the average value of y (q_i) are calculated. The regression curve is approximated by connecting the resulting points (r_i, q_i) by straight lines. By means of $h^2_{y|x}$ estimator the directedness from x to y can be determined. Defined in an analogous way, estimator $h^2_{x|y}$ shows the influence of y on x.

3.4.2 Kullback-Leibler entropy, mutual information and transfer entropy

Kullback-Leibler (KL) entropy introduced in [Kullback and Leibler, 1951] is a non-symmetric measure of the difference between two probability distributions P and Q. KL measures the expected number of extra bits required to code samples from P when using a code based on Q, rather than using a code based on P. For probability distributions $P(i)$ and $Q(i)$ of a discrete random variable i their KL divergence is defined as the average of the logarithmic difference between the probability distributions $P(i)$ and $Q(i)$, where the average is taken using the probabilities $P(i)$:

$$D_{\mathrm{KL}}(P||Q)) = \sum_i P(i) \log \frac{P(i)}{Q(i)} \tag{3.40}$$

If the quantity $0 log 0$ appears in the formula, it is interpreted as zero. For continuous random variable x, KL-divergence is defined by the integral:

$$D_{\mathrm{KL}}(P||Q)) = \int_{-\infty}^{\infty} p(x) \log \frac{p(x)}{q(x)} dx \tag{3.41}$$

where p and q denote the densities of P and Q. Typically P represents the distribution of data, observations, or a precise calculated theoretical distribution. The measure Q typically represents a theory, model, description, or approximation of P. The KL from P to Q is not necessarily the same as the KL from Q to P. KL entropy is also called information divergence, information gain, or Kullback-Leibler divergence. In MATLAB KL divergence between distributions can be computed using function `kldiv`. This function can be downloaded from the MATLAB Central file exchange `http://www.mathworks.com/matlabcentral/fileexchange/`.

Mutual information is based on the concept of the entropy of information (see Sect. 2.3.1.2). For a pair of random variables x and y the mutual information (MI) between them is defined as:

$$\mathrm{MI}_{xy} = \sum p_{ij} \log \frac{p_{ij}}{p_i p_j} \tag{3.42}$$

where p_{ij} is the joint probability that $x = x_i$ and $y = y_j$. This measure essentially tells how much extra information one gets on one signal by knowing the outcomes of the other one. If there is no relation between both signals MI_{xy} is equal to 0. Otherwise, MI_{xy} will be positive attaining a maximal value of I_x (self information of channel x) for identical signals (Sect. 2.3.1.2).

MI is a symmetric measure and it doesn't give any information about directionality. It is possible to get a notion of direction by introducing a time lag in the definition of MI_{xy}. However, it is transfer entropy which is frequently applied as a non-linear directionality measure.

The transfer entropy was introduced by [Schreiber, 2000], who used the idea of finite-order Markov processes to quantify causal information transfer between systems I and J evolving in time. Assuming that the system under study can be approximated by a stationary Markov process of order k, the transition probabilities describing the evolution of the system are: $p(i_{n+1}|i_n,\ldots,i_{n-k+1})$. If two processes I and J are independent, then the generalized Markov property $p(i_{n+1}|i_n^{(k)}) = p(i_{n+1},|i_n^{(k)}, j_n^{(l)})$ holds, where $i_n^{(k)} = i_n,\ldots,i_{n-k+1}$ and $j_n^{(l)} = (j_n,\ldots,j_{n-l+1})$ is the number of conditioning states from process I and J, respectively. Schreiber proposed to use the Kullback- Leibler entropy (equation 3.40) to quantify the deviation of the transition probabilities from the generalized Markov property. This resulted in the definition of transfer entropy (TE):

$$\mathrm{TE}_{J\to I} = \sum (p(i_{n+1}, i_n^{(k)}, j_n^{(l)}) \log \frac{p(i_{n+1}|i_n^{(k)}, j_n^{(l)})}{p(i_{n+1}|i_n^{(k)})}) \tag{3.43}$$

The TE can be understood as the excess amount of bits that must be used to encode the information of the state of the process by erroneously assuming that the actual transition probability distribution function is $p(i_{n+1}|i_n^{(k)})$ instead of $p(i_{n+1}|j_n^{(l)})$.

MI and TE quantify the statistical dependence between two time series, with no assumption about their generation processes, which can be linear or non-linear. Despite the apparent simplicity of the estimators, the practical calculation of MI or TE from experimental signals is not an easy task. In practice estimation of the probabilities involves obtaining the histograms of the series of outcomes and finding, say, p_i as the ratio between the number of samples in the i-th bin of the histogram and the total number of samples. In order to get an accurate estimate of this measure by using histogram-derived probabilities, one needs to have a large number of samples and small bins [Quian Quiroga et al., 2002]. When taking small bins of the same size for each individual variable, it may happen that the values $p_{ij} = 0$ occur. These difficulties and other problems, including high probability of systematic errors in calculation of MI were discussed in [Pereda et al., 2005], where the method of the estimation of MI and its applications are also described.

3.4.3 Generalized synchronization

Generalized synchronization (GS) is a concept rather than a measure, since following the original idea of GS introduced by [Rulkov et al., 1995], different formulas for calculation of GS have been introduced [David et al., 2004, Pereda et al., 2005]. GS measure is based on the theory of chaotic systems and makes use of the embedding theorem (Sect. 2.5) which is applied to the signals x and y. GS quantifies how well one can predict the trajectory in phase space of one of the systems knowing the trajectory of the other; or alternatively, it quantifies how neighborhoods (i.e., recurrences) in one attractor maps into the other [Quian Quiroga et al., 2000]. In brief, one calculates for each sample x_n the squared mean Euclidean distance to its k neighbors $R_n^k(x)$. The y conditioned squared mean Euclidean distance $R_n^k(x|y)$ is defined by replacing the nearest neighbors of x_n by the equal number of time samples of the closest neighborhood of y_n. Then the $\mathrm{GS}^k(x|y)$ interdependence measure may be defined as [Arnhold et al., 1999, Quian Quiroga et al., 2002]:

$$\mathrm{GS}^k(x|y) = \frac{1}{N}\sum_{n=1}^{N}\frac{R_n^k(x)}{R_n^k(x|y)}. \tag{3.44}$$

The GS measures based on the normalized difference of $R_n^k(x|y)$ and $R_n^k(y|x)$ are asymmetric and hence may give the information on the directionality of the interaction. However, one should be careful in drawing the conclusions on the driver and driven system [Pereda et al., 2001].

3.4.4 Phase synchronization

Two non-linear coupled oscillators may synchronize their phases, even if their amplitudes remain uncorrelated. This phenomenon, called phase synchronization, may be detected by analyzing the phases of the time series. Namely the phases may be found from analytical signals obtained by application of the Hilbert transform to the experimental signals (Sect. 2.4.1). Synchronization means that the phase locking condition applies for time t:

$$\phi_{n,m}(t) = |n\phi_x(t) - m\phi_y(t)| \leq constant, \tag{3.45}$$

where $\phi_x(t)$, $\phi_y(t)$ are the phases of signals x and y (equation 2.66, Sect. 2.4.1) and n, m are small natural numbers, most often $n = m = 1$.

Several indexes of phase synchronization were introduced. Mean phase coherence, also called phase locking value (PLV) or synchrony factor [Quian Quiroga et al.,

2002], is defined by the formula:

$$\mathrm{PLV}_{n,m} = \sqrt{\langle \cos \phi_{n,m}(t) \rangle^2 + \langle \sin \phi_{n,m}(t) \rangle^2}, \tag{3.46}$$

where $\langle \cdot \rangle$ denotes averaging over time. From the above formula one can find how the relative phase is distributed over the unit circle. If the two signals are phase synchronized, the relative phase will occupy a small portion of the circle, which means high phase coherence.

Other measures of phase synchronization (PS) are based on the calculation of the distribution of phases [Pereda et al., 2005]. In the so-called stroboscopic approach the phase of one of the oscillators is observed at those instants where that of the other one attains a certain value.

The above methods require bandpass filtering of the signals. When the signals are characterized by modulated natural frequency the relative phase distributions are broad, which makes estimation of PS difficult. The problems connected with the construction of histograms, mentioned in the context of finding generalized synchronization, are also present in case of estimation of phase synchronization based on the consideration of phase distributions.

PS does not allow determination of the direction of interaction between the channels of a process. Some attempts were made to judge the direction from the dynamical model of the process under the assumption of weak dependence on the amplitude.

3.4.5 Testing the reliability of the estimators of directedness

Testing the null hypothesis about the lack or presence of causal relations between time series is not straightforward, since analytical determination of the statistical distributions of the estimators of directedness is difficult. The asymptotic distribution of DTF under the null hypothesis of no information flow was derived by [Eichler, 2006]. He proposed a frequency dependent point-wise significance level that forms an upper bound for the true unknown critical value of the asymptotic distribution of DTF.

However in the absence of a general analytical approach the significance of the directedness estimators is usually tested by means of surrogate data. Namely the obtained results are compared with those computed for the time series with no dependencies between channels.

The simplest approach is to randomly shuffle the samples of the data series, but such procedure does not preserve the autocorrelation structures of the signals. The use of such a white-noise version of data as a control condition turns the data into a very unlikely realization of the physiological process. Therefore this procedure is not recommended.

A better approach is to use the surrogates which preserve the spectral properties of the time series and randomize only the phases of each signal. The procedure of the generation of this type of data was described in (Sect. 1.6). The outcome of this kind of test shows whether there is directedness among the set of signals. A similar approach is used to test for non-linearity of the time series. The problem of construction of bivariate surrogate data was considered in [Andrzejak et al., 2003a]. In case of multichannel data the test indicates presence or lack of dependencies, but it doesn't indicate if the character of dependence is linear or non-linear. The generation of surrogate data for multichannel series destroying only the non-linear part of the interdependence is a problem which is not solved at the moment in a satisfactory way.

In case of estimators operating in time-frequency space such as SDTF, the problem of reliability testing is more difficult, since all realizations of the process are used to compute the estimator. In this case the bootstrap method [Zoubir and Boashash, 1988] is applied. It relies on construction of the new artificial time series consisting of the same number of realizations as the original data. The realizations are created by drawing randomly (with repetitions) trials from the original set. Let us assume that the original ensemble of trials consists of M realizations:

$$[r] = [r^1, r^2, r^3, r^4, r^5, \ldots, r^k, r^{k+1}, \ldots, r^M] \tag{3.47}$$

New ensemble of trials may be:

$$[r_{boot}] = [r^3, r^1, \ldots, r^3, r^7, \ldots, r^5, \ldots, r^5, r^M] \tag{3.48}$$

Repeating such procedure for all k channels L times (in practice 100 or more times) we obtain L ensembles of realizations and we can calculate the average $\overline{SDTF}$ and its variance as:

$$\text{var}(SDTF) = \sum_{i=1}^{L} \frac{(\overline{SDTF} - SDTF)^2}{(L-1)} \tag{3.49}$$

The method may be used for evaluation of the errors of any estimator based on ensemble averaging.

3.5 Comparison of the multichannel estimators of coupling between time series

Among approaches to quantify the coupling between multivariate time series, we can distinguish between linear and non-linear methods as well as approaches based on bivariate and multivariate estimators. For a long time the standard methods of establishing relations between signals have been cross-correlation and coherence. However, in case of these bivariate measures we don't know if the two channels are really coupled or if there is another channel that drives them.

This point may be elucidated by simulation illustrating the situation where the activity is measured in different distances from the source (Figure 3.4). The signals in channels 2-5 were constructed by delaying the signal from channel one (the source) by $1, 2, 3, 4$ samples, respectively, and adding the noise in each step. The coherences

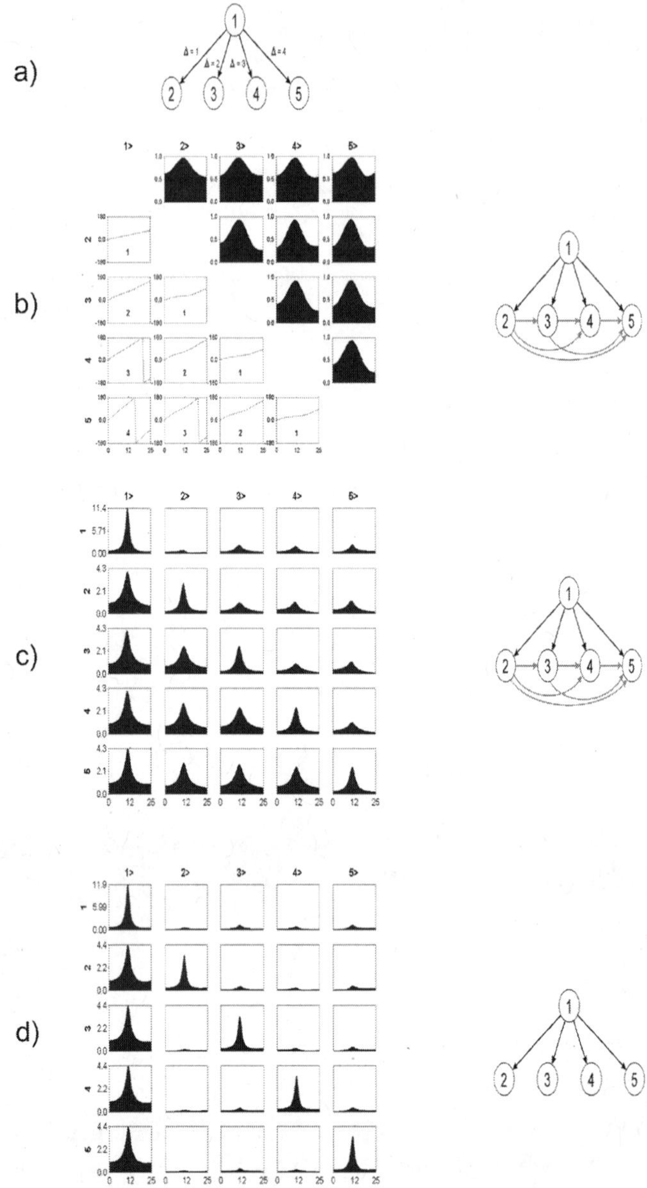

FIGURE 3.4: Propagations estimated for simulation scheme shown in a) by means of: b) bivariate coherences, c) bivariate DTF model, d) multivariate DTF. In panel b) above the diagonal, the magnitudes of coherences, below diagonal, phases of the coherences are shown. From phases, the delays (values printed in each box) were calculated. For c) and d) in each box DTF as a function of frequency, on the diagonal power spectra; propagation from channels labeled above to the channels labeled at the left of the picture. Resulting propagation schemes shown in the right part of each panel.

were calculated between all signals pairwise. For the peak frequency the phases were found and assuming the linear phase relations, the delays were calculated. The resulting scheme of propagation presented in Figure 3.4 b) shows many false flows. Similar results were obtained for bivariate AR model (Figure 3.4 c). This situation is common to all bivariate methods; namely the propagation is found in each case where the phase difference is present [Blinowska et al., 2004a]. Sometimes, for bivariate measures even the reverse propagation may be found [Kus et al., 2004]. The results obtained by means of the multivariate method (DTF) show the correct scheme of propagation (Figure 3.4 d).

Another problem is that, from correlation or coherence (also partial coherence), it is not possible to detect reciprocal flows (interaction in both directions). Such patterns of propagation may be found by means of estimators based on the causality principle.

All non-linear methods described above are bivariate, so they suffer from the disadvantages pointed out above. There were some attempts to apply non-linear measures for number of channels bigger than two. Chen et al. [Chen et al., 2004] proposed conditional extended Granger causality and applied it to 3 channel simulated non-linear time series, but the advantages over the linear approach were not clearly demonstrated and the influence of noise was not studied. The method requires determination of embedding dimension and neighborhood size, which are difficult to find in an objective and optimal way. The same objections concern another methods aiming to extend non-linear estimators for a number of channels higher than two. The problems concerning state space reconstruction and noise sensitivity become even more serious for a higher number of channels.

In the study devoted to comparison of linear and non-linear methods of coupling [Netoff et al., 2006], non-linear estimators: mutual information, phase correlation, continuity measure[2] were compared with correlation in case of non-linear signals in the presence of noise. The authors found that any method that relies on an accurate state space reconstruction will be inherently at a disadvantage over measures which do not rely on such assumptions. Another finding was the high sensitivity to noise of non-linear estimators. The authors conclude: *"We have been as guilty as any of our colleagues in being fascinated by the theory and methods of nonlinear dynamics. Hence we have continually been surprised by robust capabilities of linear CC (correlation) to detect weak coupling in nonlinear systems, especially in the presence of noise."*

We can conclude that non-linear methods are not recommended in most cases. They might be used only when there is clear evidence that there is a good reason to think that there is non-linear structure either in data themselves or in the interdependence between them [Pereda et al., 2005]. We have to bear in mind that many systems composed of highly non-linear components exhibit an overall linear type of behavior; the examples may be some electronic devices, as well as brain signals

[2]Measure similar to GS testing for continuity of mapping between neighboring points in one data set to their corresponding points in the other data set.

such as electroencephalograms or local field potentials (LFP). This problem will be further discussed in Sect. 4.1. In this context it is worth mentioning that some linear methods, e.g., DTF, work quite well also for non-linear time series [Winterhalder et al., 2005].

In the above quoted reference comparison of linear techniques inferring directed interactions in multivariate systems was carried out. Partial phase spectrum (PC) and three methods based on Granger causality: GCI, DTF and PDC, were tested in respect to specificity in absence of influences, correct estimation of direction, direct versus indirect transmissions, non-linearity of data, and influences varying in time. All methods performed mostly well with some exceptions. PC failed in the most important aspect—estimation of direction. It was due to the large errors in phase estimation for particular frequencies and the inability to detect reciprocal interactions. GCI failed in the presence of non-linearities. DTF did not distinguish direct from cascade flows, but when such distinction is important dDTF may be used. PDC for signals with a non-linear component gave correct results only for very high model orders, which makes the method hardly applicable for strongly non-linear signals because of limitations connected with the number of parameters of MVAR mentioned in Sect. 3.3.2.3.2.

More important drawbacks of PDC were pointed out by [Schelter et al., 2009], namely:

i) PDC is decreased when multiple signals are emitted from a given source,

ii) PDC is not scale-invariant, since it depends on the units of measurement of the source and target processes,

iii) PDC does not allow conclusions on the absolute strength of the coupling.

The authors proposed the renormalization of PDC similar to the one used in the definition of DTF, which helped in alleviation of the above problems.

Point iii) is illustrated in Figure 3.5. In a situation when activity is emitted in several directions PDC shows weaker flows than in the situation when the same activity is emitted in one direction only. Another feature of PDC is a weak dependence on frequency. The PDC spectrum is practically flat, whereas DTF spectrum (especially for non-normalized DTF) reflects the spectral characteristics of the signal. An example of application of DTF and PDC to experimental data (EEG) will be given in Sect. 4.1.6.2.2.

We can conclude that the methods estimating direction which perform best are the linear multivariate methods based on the Granger causality concept, and in the frequency domain the methods or choice are DTF (or dDTF) and renormalized PDC.

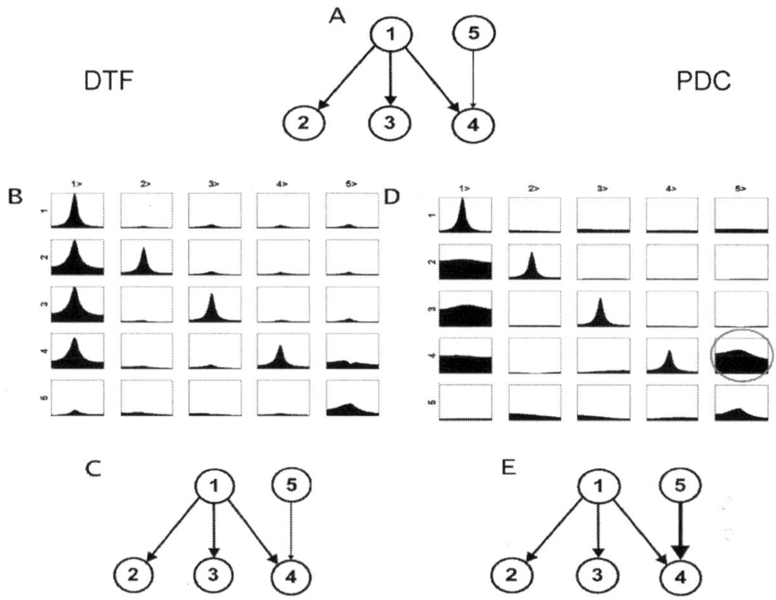

FIGURE 3.5: Comparison of DTF (panel B) and PDC (panel D) for simulation scheme A. Resulting schemes of propagation for DTF and PDC below corresponding panels. Thickness of arrows proportional to the flow intensities. The convention of presenting DTF/PDC is as in Figure 3.4. Note that the weak flow from channel 5 is enhanced by PDC.

3.6 Multivariate signal decompositions

3.6.1 Principal component analysis (PCA)

3.6.1.1 Definition

Principal component analysis (PCA) is a method of decomposition of multichannel epoched data into components that are linearly independent; that is, spatially and temporally uncorrelated. Depending on the field of application, it is also named the discrete Karhunen-Loéve transform (KLT), the Hotelling transform, or proper orthogonal decomposition (POD). The geometrical intuition behind this method is the following. We treat samples from all channels at a given time moment as a point in the space of dimension equal to the number of channels. One epoch of multichannel data forms a cloud of points in that space. This cloud can be spread more in some directions than in the others. The measure of the spread is the variance of the points' locations in a given direction. The PCA finds a set of orthogonal axes, a base, such that each consecutive axis spans the directions with consecutively decreasing variance. The projections of the points onto these axes constitute the components. Each component can be visualized as a time series. On the other hand the original time series can be recovered as a linear combination of these components. Components corresponding to the smallest variance can be neglected and in this way a reduction of data dimensionality can be achieved.

3.6.1.2 Computation

PCA can be performed by means of the singular value decomposition (SVD) algorithm. An epoch of k channels data of length m can be represented as a $m \times k$ matrix $\mathbf{x}$. It is always possible to find three matrices **P**, **A**, and **M** such that:

$$\mathbf{x} = \mathbf{PAM}^{\mathsf{T}} \tag{3.50}$$

where **P** is the $m \times k$ matrix containing k normalized principal component waveforms ($\mathbf{P}^{\mathsf{T}}\mathbf{P} = 1$), **A** is $k \times k$, diagonal matrix of components amplitudes, **M** is a $k \times k$ matrix mapping components to original data such that M_{ij} is the contribution of jth component to i^{th} channel; $\mathbf{M}^{\mathsf{T}}\mathbf{M} = 1$. The SVD algorithm is implemented in MATLAB as a function `svd`; to find the decomposition of $\mathbf{x}$ into **P**, **A**, and **M** execute:

```
[P,A,M] = svd(x);
```

When data concerning multiple conditions or subjects are to be analyzed by PCA the matrix $\mathbf{x}$ is composed by concatenating the matrixes of individual conditions or subjects along the time dimension. Thus for example for N_c conditions we have to decompose matrix $\mathbf{x}$ which has dimension $m \cdot N_c \times k$. The nth row of resulting matrix **P** represents N_c concatenated time courses obtained for each condition for the nth component. The corresponding topographical map is contained in the nth column.

3.6.1.3 Possible applications

PCA can be used as a tool for decomposition of multichannel data into a linear combination of components characterized by different spatial, temporal, and amplitude distribution. These components can be used as a preprocessing step in source localization techniques. However, there is no direct correspondence between the principal components and individual sources. A single source may produce a signal that decomposes into a number of components, while on the other hand multiple sources can contribute to a single component. In general, the individual components should not be ascribed directly to any physiologically meaningful phenomenon [Lamothe and Stroink, 1991].

A similar method, which leads to the reduction of dimensionality of the data is factor analysis (FA). FA is related to PCA but not identical. PCA performs a variance-maximizing rotation (varimax) of the variable space, thus it takes into account all variability in the data. In contrast, factor analysis estimates how much of the variability is due to common factors. The number of common factors is usually less than the dimension of the original variable space [Bryant and Yarnold, 1994]. FA can describe a large set of recorded wave shapes from multiple sensors in terms of a small number of quantitative descriptors. These descriptors, or factors can be conceptualized as basic waveforms produced by hypothetical signal generators which, mixed in correct proportions, would reproduce the original waveforms in the set. FA was used to describe the evoked potentials [John and Thatcher, 1977]. FA can be performed in MATLAB using function `factoran` from the Statistical Toolbox.

3.6.2 Independent components analysis (ICA)

3.6.2.1 Definition

Independent component analysis (ICA) is a statistical signal processing method that decomposes a multichannel signal into components that are statistically independent. In simple words two components s_1 and s_2 are independent when information of the value of s_1 does not give any information about the value of s_2. The ICA can be represented by a simple generative model:

$$\mathbf{x} = \mathbf{D}\mathbf{s} \tag{3.51}$$

where the $\mathbf{x} = \{x^1, x^2, \ldots, x^n\}$ is the measured n channel signal, $\mathbf{D}$ is the mixing matrix, and $\mathbf{s} = \{s^1, s^2, \ldots, s^n\}$ is the activity of n sources. The main assumption about $\mathbf{s}$ is that the s^i are statistically independent. To be able to estimate the model we must also assume that the independent components have *nongaussian* distribution [Hyvärinen and Oja, 2000].

The model implies the following:

- The signal is a linear mixture of the activities of the sources
- The signals due to each source are independent
- The process of mixing sources and the sources themselves are stationary
- The energies (variances) of the independent components cannot be determined unequivocally.[3] The natural choice to solve this ambiguity is to fix the magnitude of the independent components so that they have unit variance: $E[s^i] = 1$.
- The order of the components is arbitrary. If we reorder both in the same way: the components in $\mathbf{s}$, and the columns in $\mathbf{D}$, we obtain exactly the same signal $\mathbf{x}$.

The main computational issue in ICA is the estimation of the mixing matrix $\mathbf{D}$. Once it is known, the independent components can be obtained by:

$$\mathbf{s} = \mathbf{D}^{-1}\mathbf{x} \tag{3.52}$$

[3]This is because the multiplication of the amplitude of the ith source can be obtained either by multiplication of s^i or by multiplication of the ith column of matrix $\mathbf{D}$.

3.6.2.2 Estimation

Finding the independent components can be considered in the light of the central limit theorem as finding components of least gaussian distributions. To comprehend this approach let us follow the heuristic given by [Hyvärinen and Oja, 2000]. For simplicity, let us assume that the sought independent components have identical distributions. Let us define $y = \mathbf{w}^{\mathsf{T}}\mathbf{x}$. Please note, that if $\mathbf{w}^{\mathsf{T}}$ is one of the columns of the matrix $\mathbf{D}^{-1}$, then y is one of the sought components. With the change of variables $\mathbf{z} = \mathbf{D}^{\mathsf{T}}\mathbf{w}$ we can write $y = \mathbf{w}^{\mathsf{T}}\mathbf{x} = \mathbf{w}^{\mathsf{T}}\mathbf{D}\mathbf{s} = \mathbf{z}^{\mathsf{T}}\mathbf{s}$. This exposes the fact that y is a linear combination of the components s^i with the weights given by z_i. It stems from central limit theorem that the sum of independent random variables has more gaussian character than each of the variables alone. The linear combination becomes least gaussian when $\mathbf{z}$ has only one nonzero element. In that case y is proportional to s^i. Therefore the problem of estimation of the ICA model can be formulated as a problem of finding $\mathbf{w}$ which maximizes the nongaussianity of $y = \mathbf{w}^{\mathsf{T}}\mathbf{x}$. Maximizing nongaussianity of y gives one independent component, corresponding to one of the $2n$ maxima[4] in the optimization landscape. In order to find all independent components we need to find all the maxima. Since the components are uncorrelated the search for consecutive maxima can be constrained to the subspace orthogonal to the one already analyzed.

3.6.2.3 Computation

The intuitive heuristics of maximum nongaussianity can be used to derive different functions whose optimization enables the estimation of the ICA model; such a function may be, e.g., kurtosis. Alternatively, one may use more classical notions like maximum likelihood estimation or minimization of mutual information to estimate ICA. All these approaches are approximatively equivalent [Hyvärinen and Oja, 2000].

There are several tools for performing ICA in MATLAB. The EEGLAB toolbox [Makeig, 2000] offers a convenient way to try different algorithms: *runica* [Makeig et al., 1996], *jader* [Cardoso, 1999] and *fastica* [Hyvärinen, 1999]. The *runica* and *jader* algorithms are a part of the default EEGLAB distribution. To use the *fastica* algorithm, one must install the *fastica* toolbox (`http://www.cis.hut.fi/projects/ica/fastica/`) and include it in the MATLAB path. In general, the physiological significance of any differences in the results of different algorithms (or of different parameter choices in the various algorithms) have not been thoroughly tested for physiological signals. Applied to simulated, relatively low dimensional data sets for which all the assumptions of ICA are exactly fulfilled, all three algorithms return near-equivalent components.

Very important note: As a general rule, finding N stable components (from N-channel data) typically requires more than kN^2 data sample points (at each channel), where N^2 is the number of elements in the unmixing matrix that ICA is trying to

[4] Corresponding to s^i and $-s^i$.

estimate and k is a multiplier. As reported in the EEGLAB tutorial[5], the value of k increases as the number of channels increases. In general, it is important to give ICA as much data as possible for successful estimation. ICA works best when a large amount of basically similar and mostly clean data are available.

3.6.2.4 Possible applications

The most often reported applications of ICA in the field of biomedical signal analysis are related to EEG and MEG artifact reduction (Sect. 4.1.5) and feature extraction in studies of event-related brain dynamics [Onton and Makeig, 2006] and as a preprocessing step in the search of sources of EEG and MEG activity localization [Grau et al., 2007]. ICA is applied as well in analysis of heart (Sect. 4.2) and muscle signals (Sect. 4.3).

3.6.3 Multivariate matching pursuit (MMP)

In Sect. 2.4.2.2.7 a single channel version of matching pursuit algorithm was described. The MP algorithm decomposes the signal into waveforms g_γ (time-frequency atoms). Each atom is characterized by a set of parameters γ. In the case of Gabor functions (equation 2.114) these are $\gamma = \{u, f, \sigma, \phi\}$ where u –time translation, f –frequency, σ – time width, ϕ – phase. The MP algorithm can be extended into multivariate cases by introducing multichannel time-frequency atoms. A multichannel time-frequency atom is a set of functions $\mathbf{g}_\gamma = \{g_\gamma^1, g_\gamma^2, \ldots, g_\gamma^n\}$. Each of the g_γ^i functions is a univariate time-frequency atom. Let $\mathbf{x} = \{x^1, x^2, \ldots, x^n\}$ be the n channel signal. The multivariate matching pursuit acts in the following iterative way:

1. The null residue is the signal $\mathbf{R^0x} = \{R^0x^1, R^0x^2, \ldots, R^0x^n\} = \mathbf{x}$.
2. In the k-th iteration of the algorithm a multichannel atom $\mathbf{g}_{\gamma_k}$ is selected that satisfies the optimality criterion.
3. The next residuum is computed as the difference of the current residuum and the projection of the selected atom on each channel. For channel i it can be expressed as: $R^{k+1}x^i = R^kx^i - \langle R^kx^i, g_{\gamma_k}^i \rangle g_{\gamma_k}^i$.

An interesting possibility in the MMP framework is that the optimality criterion can be model driven, i.e., it can correspond to the assumed model of signal generation. As an example we can presume that each given component of the measured multichannel signal results from the activity of an individual source, and that the propagation of signal (e.g., electric or magnetic field) from the source to the sensors is instantaneous. In such a case, the most straightforward optimality condition is that in a given iteration k the atom g_{γ_k} is selected which explains the biggest amount of total energy (summed over the channels) with the constraint that all the univariate atoms in individual channels have all the parameters identical, excluding the amplitude which

[5] `http://sccn.ucsd.edu/wiki/EEGLAB`.

varies topographicaly. Hence atom g_{γ_k} can be expressed as:

$$g_{\gamma_k} = \arg\max_{g_\gamma \in D} \sum_{i=1}^{n} |\langle R^n \mathbf{x}^i, g_\gamma^i \rangle|^2 \quad and \quad \forall_{i,j}\; g_\gamma^i = g_\gamma^j \tag{3.53}$$

Another example of optimality criterion can be formulated when the multivariate signal is obtained as an ensemble of realization of a process, e.g., multiple epochs containing the realizations of an evoked potential. In such a case the criterion could be analogous to the one given by (equation 3.53), but the constraint of the same phase for all univariate atoms can be relaxed.

4

Application to biomedical signals

4.1 Brain signals: local field potentials (LFP), electrocorticogram (ECoG), electroencephalogram (EEG), and magnetoencephalogram (MEG), event-related responses (ERP), and evoked fields (EF)

Electroencephalogram (EEG) is a record of the electric signal generated by the cooperative action of brain cells, or more precisely, the time course of neuronal extracellular field potentials generated by their synchronous action. EEG recorded in the absence of stimuli is called spontaneous EEG; a brain electric field generated as a response to external or internal stimulus is called an event-related potential (ERP). EEG can be measured by means of electrodes placed on the scalp or directly on the cortex. In the latter case it is called an electrocorticogram (ECoG); lately also iEEG (intracranial EEG) abbreviation is used. Electric fields measured intracortically with electrodes implanted in the brain structures were named local fields potentials (LFP). The same electrodes (if small enough) can also record action potentials (spikes). The amplitude of EEG of a normal subject in the awake state, recorded with the scalp electrodes, is 10-100 μV. In case of epilepsy the EEG amplitudes may increase by almost an order of magnitude. Scalp potentials are largely independent of electrode size which is due to severe space averaging by volume conduction between brain and scalp. Intracranial potentials amplitude depend on electrode size (smaller electrode—higher potential) and may vary in size over four orders of magnitude.

Variable electric field generates a magnetic field which follows from the Maxwell equations. The recording of the magnetic field of the brain is called a magnetoencephalogram (MEG). The amplitude of the MEG is less than 0.5 picotesla (pT) and its frequency range is similar to that of the EEG. Since electric and magnetic fields are orthogonal, radial sources (dipoles oriented perpendicular to the skull) are better visible in EEG and tangential sources in MEG. Radial sources, due to geometry, hardly contribute to MEG, but the advantage of this technique is that the magnetic field is weakly influenced by the structures of the head. Figure 4.1 shows schematic representations of the electric and magnetic fields generated by a current dipole.

Applications of EEG. EEG found application in the investigation of the information processing in the brain and in medical diagnosis. In particular it is helpful in

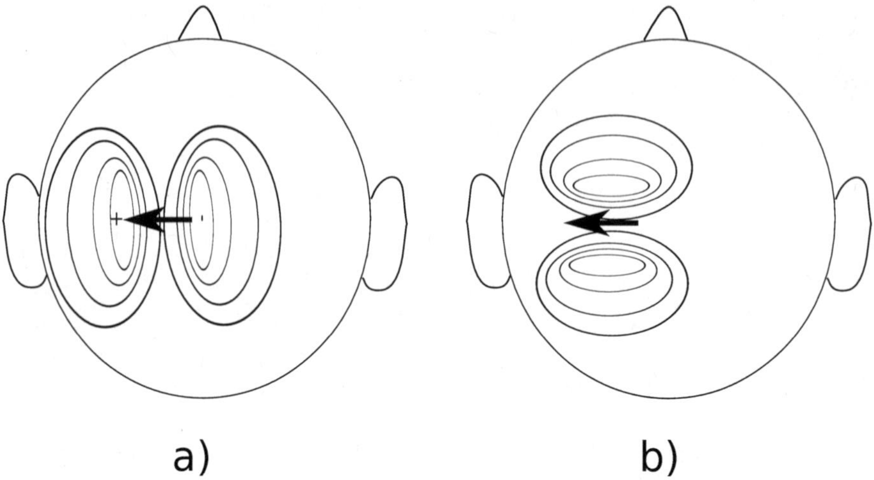

FIGURE 4.1: Isolines of the electric (a) and magnetic (b) fields generated by a current dipole.

solving the following medical problems and diseases:

- Epilepsy
- Brain tumors, head injury, stroke
- Psychiatric diseases
- Sleep disorders
- CNS disorders, e.g., cerebral anoxia, cerebral inflammatory processes, cerebral palsy
- Creutzfeld-Jacob disease, metabolic nervous system disorders
- Developmental disorders
- Testing of psychotropic and anticonvulsant drugs
- Monitoring alertness, e.g., controlling anaesthesia depth
- Testing sensory pathways

Another application of EEG concerns design of brain machine interfaces for direct communication between brain and computer and then possibly with specific devices. In the investigation of brain processes, especially perception and cognitive functions usually ERPs are analyzed, since they supply information on the reaction of the brain to specific external or internal stimuli. ERPs are also used for testing the sensory pathways, in psychiatric diseases, and developmental disorders, e.g., in dyslexia diagnosis.

Applications of MEG. Magnetoencephalography (MEG), in which magnetic fields generated by brain activity are recorded outside of the head, is now in routine clinical practice throughout the world. MEG has become a recognized and vital part of the presurgical evaluation of patients with epilepsy and patients with brain tumors. The big advantage of the MEG technique is that the magnetic field generated inside the brain is affected to a much lesser extent by the conductivities of the skull and scalp than the electric field generated by the same source. In the limiting case of spherical head model, concentric inhomogeneities do not affect the magnetic field at all, whereas they have to be taken into account in the analysis of EEG data [Hämäläinen et al., 1993]. This property of MEG is very advantageous when localizing sources of activity. A current review showing an improvement in the post-surgical outcomes of patients with epilepsy by localizing epileptic discharges by means of MEG can be found in [Stufflebeam et al., 2009]. Magnetic evoked fields (EF) are a counterpart of ERP; they are an effect of event-related brain activity which can be elicited by visual, auditory, sensory, or internal stimuli. Their analysis usually concerns identification of the brain region responsible for their generation.

The MEG sensors are not directly attached to the subject's head. This implies that during the MEG measurement the subject's head should not move in respect to the sensors. This requirement limits the possibility of long-session measurements, or long-term monitoring.

4.1.1 Generation of brain signals

In the brain there are 10^{11} nerve cells. Each of them is synaptically connected with up to 10^4 other neurons. Brain cells can also communicate by means of electrical synapses (gap junctions) transmitting current directly. Electric activity of neurons is manifested by generation of action potentials and post-synaptic potentials (PSP). Action potentials occur when the electrical excitation of the membrane exceeds a threshold. The generation of action potentials is connected with rapid inflow of Na^+ ions to the cell, which changes the polarization of the inside of the neuron from about -80 mV to about +40 mV; the repolarization of the cell is connected with the outflow of K^+ ions to the extracellular space. In this way an action potential of a characteristic spike-like shape (duration about 1 ms) is created. The exact time course of the action potential is shaped by the interplay of many ionic currents, specific for the given neuron type.

Postsynaptic potentials are sub-threshold phenomena connected with the processes occurring on the postsynaptic membrane. When the action potential arrives at the synapse, it secretes a chemical substance called a mediator or transmitter, which causes a change in the permeability of the postsynaptic membrane of the target neuron. As a result, ions traverse the membrane and a difference in potentials across the membrane is created. When the negativity inside the neuron is decreased, an excitatory postsynaptic potential (EPSP) is generated. An inhibitory postsynaptic potential (IPSP) is created when the negativity inside the neuron is increased, and the neuron becomes hyperpolarized. Unlike the action potential, the PSPs are graded potentials; their amplitudes are proportional to the amount of secreted mediator. Postsynaptic potentials typically have amplitudes of 5-10 mV and a time span of 10-50 msec. In order to obtain supra-threshold excitation, the amplitudes of many postsynaptic potentials have to be superimposed.

The electrical activity of neurons, producing electric and magnetic fields conforming approximately to that of a dipole, generates currents along the cell membrane in the intra- and extracellular spaces. Macroscopic observation of these fields requires the synchronization of electrical activity of a large number of dipoles oriented in parallel [Nunez, 1981]. Indeed, pyramidal cells of the cortex are to a large degree parallel and moreover, they are synchronized by virtue of common feeding by thalamocortical connections [Lopes da Silva, 1996]. The condition of synchrony is fulfilled by the PSPs, which are relatively long in duration. The contribution from action potentials to the electric field measured extracranially is negligible. EEG comes from the summation of synchronously generated postsynaptic potentials. The contribution to the electric field of neurons acting synchronously is approximately proportional to their number, and for these firing non-synchronously, as the square root of their number. For example: if an electrode records action of 10^8 neurons (which is typical for a scalp electrode) and 1% of them are acting in synchrony, their contribution will be 100 times bigger than the contribution of neurons acting asynchronously, since $\frac{10^6}{\sqrt{10^8}} = 100$. Scalp potentials are mostly due to sources coherent at the scale of at least several centimeters (roughly 10^8 neurons) with geometries encouraging the superposition of potentials generated by many local sources. Nearly all EEGs are believed to be generated by cortical sources [Nunez, 1981] by virtue of the following reasons: cortical proximity to scalp, relatively large sink-source separation in pyramidal cells constituting cortex dipoles, property of the cortex to produce large cortical layers, high synchrony of pyramidal cells fed by the common thalamic input. MEG is closely related to EEG. The source of the MEG signal is the same electrical activity of the synchronized neural populations as is the case for EEG.

The processes at the cellular level (action potential generation) are non-linear. Non-linear properties can also be observed in the dynamics of well-defined neural populations. However, at the EEG level, non-linearities are the exception rather than the rule. The lack of the traces of non-linear dynamics in EEG was demonstrated by means of surrogate data [Achermann et al., 1994, Stam et al., 1999] and by means of comparison of linear and non-linear forecasting [Blinowska and Malinowski, 1991]. In the latter work it was demonstrated that also in the case of LFP, recorded from implanted electrodes, the linear forecasting gave the same or better results than non-linear. During epileptic seizures some EEG epochs of non-linear character were found [Pijn et al., 1997]. Indeed during seizure large parts of the brain became synchronized.

Since non-linear methods are very sensitive to noise and prone to systematic errors, as was mentioned in Sect. 2.5.7, one should have good reason to apply them to the EEG analysis. It is recommended to check first if the signal has non-linear character and if the linear methods are not sufficient. Most functions can be locally linearized and thus linear methods may also work quite well for non-linear signals.

4.1.2 EEG/MEG rhythms

The problem of the origins of EEG rhythmical activity has been approached by electrophysiological studies on brain nerve cells and by the modeling of electrical activity of the neural populations [Freeman, 1975, Lopes da Silva, 1996, Wright et al., 2000, David et al., 2004]. It is generally accepted that cooperative properties of networks consisting of excitatory and inhibitory neurons connected by feedback loops play a crucial role in establishing EEG rhythms, but there is some evidence that the intrinsic oscillatory properties of some neurons may contribute to the shaping of the rhythmic behavior of networks to which they belong. The frequency of oscillation depends on the intrinsic membrane properties, on the membrane potential of the individual neurons, and on the strength of the synaptic interactions. The following rhythms have been distinguished in EEG (the same rhythms can be observed in MEG) [Niedermayer and Lopes da Silva, 2004]: delta (0.1–4 Hz), theta (4–8 Hz), alpha (8–13 Hz), beta (13–30 Hz), and gamma (above 30 Hz) (Figure 4.2). Frequencies of gamma rhythms measured by scalp electrodes usually did not exceed 40–60 Hz because of strong damping of high frequencies by head tissues. High frequency gamma (60–250) Hz were first observed in ECoG. They may be recorded as well by means of MEG and also in EEG by application of sophisticated modern recording techniques.

The contribution of different rhythms to the EEG/MEG depends on the age and behavioral state of the subject, mainly the level of alertness. There are also considerable inter-subject differences in EEG characteristics. EEG pattern is influenced by neuropathological conditions, metabolic disorders, and drug action [Niedermayer and Lopes da Silva, 2004].

- Delta rhythm is a predominant feature in EEGs recorded during deep sleep. In this stage delta waves usually have large amplitudes (75-200 μV) and show strong coherence all over the scalp.

- Theta rhythms occur in the state of drowsiness and in some emotional states, but they are also involved in cognitive and working memory processes. In the latter case theta waves are associated with gamma activity. Occurrence of theta rhythm can also be connected with the slowing of alpha rhythms due to pathology. Theta waves are predominant in rodents; in this case the frequency range is broader (4-12 Hz) and waves have a high amplitude and characteristic sawtooth shape.

- Alpha rhythms are predominant during wakefulness and are most pronounced in the posterior regions of the head. They are best observed when the eyes are closed and the subject is in a relaxed state. They are blocked or attenuated by attention (especially visual) and by mental effort. Mu waves have a frequency band similar to alpha, but their shape resembles the Greek letter μ. They are prevalent in the central part of the head and are related to the function of motor cortex, namely they are blocked by motor actions.

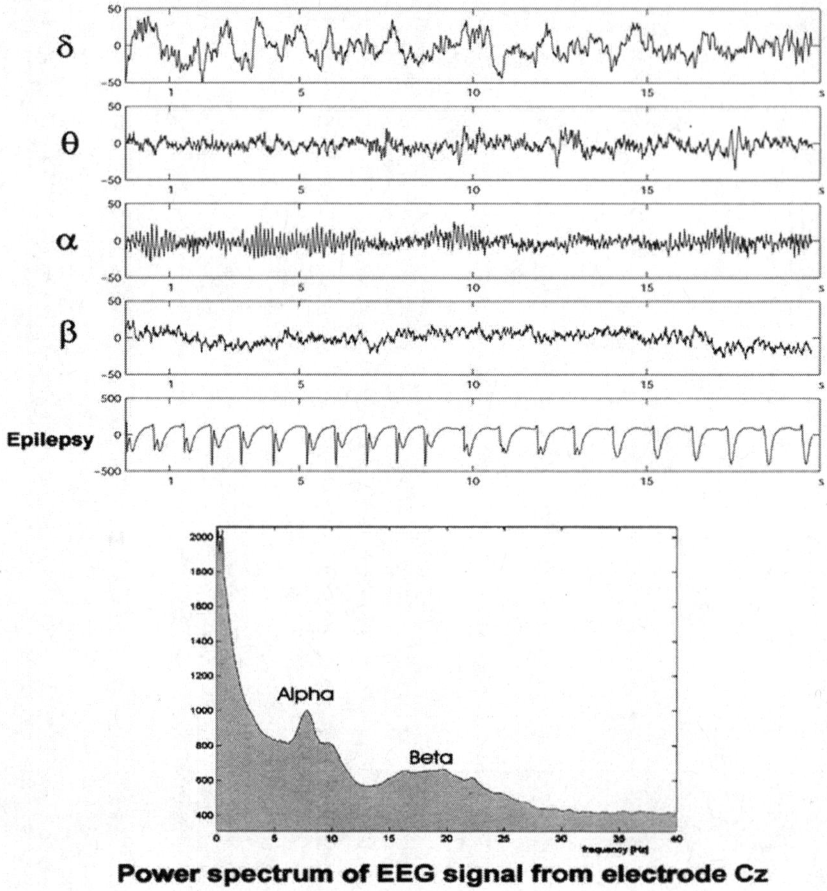

FIGURE 4.2: Example of EEG rhythms and EEG power spectrum.

- Beta activity is characteristic for the states of increased alertness and focused attention, as was shown in several animal and human studies.
- Gamma activity is connected with information processing, e.g., recognition of sensory stimuli and the onset of voluntary movements. High frequency gamma rhythms are linked to neuronal spikes. They are correlated with the degree of functional activation of the brain and are relevant to cortical computations.

The EEG is observed in all mammals, the characteristics of primate EEG being closest to the human. Cat, dog, and rodent EEG also resemble human EEG, but have different spectral content. In lower vertebrates electric brain activity is also observed, but it lacks the rhythmical behavior found in higher vertebrates.

4.1.3 EEG measurement, electrode systems

EEG is usually registered by means of electrodes placed on the scalp. They can be secured by an adhesive or embedded in a special snug cap. A block diagram of the recording setup is shown in Figure 4.3. The first elements of the system are the electrodes and the differential amplifier. The quality of the measurement depends to a large extent on the ratio between the impedance of the electrodes and an input impedance of the amplifier. This requirement is met by modern amplifiers which have an input impedance up to 10^{12} Ω. In such a case the standard requirement that the electrode resistance be less than 5 kΩ may be slightly relaxed.

Prior to sampling, low pass anti-aliasing filters (Sect. 1.2.1.1) are used; high pass filters are applied in order to eliminate artifacts of lowest frequencies. EEG is usually digitized by an analog-to-digital converter (ADC). The digitization of the signal introduces the quantization error (Sect.1.2.2), which depends on the covered voltage range and the number of the ADC's bits (for modern apparatus up to 24 bits). The sampling frequency ranges from 100 Hz for spontaneous EEG and several hundred Hz for ERP to several kHz for recording short latency ERP and for intracranial activity. Knowledge of the exact positions of electrodes is very important for both

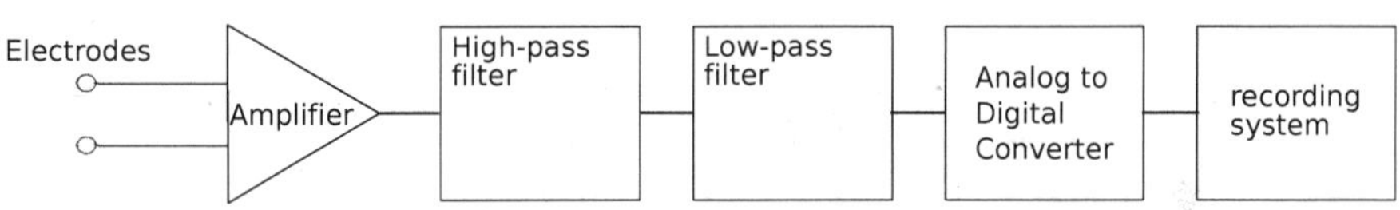

FIGURE 4.3: Block diagram of recording setup.

interpretation of a single recording as well as for comparison of results obtained for a group of subjects; hence the need for standardization. The traditional 10-20 electrode system gives positions of 19 EEG electrodes (and two electrodes placed on earlobes A1/A2) related to specific anatomic landmarks, such that 10-20% of the distance between them is used as the electrode spatial interval. The first part of the derivation's name indexes the array's row—from the front of head: Fp, F, C, P, and O (Figure 4.4). In the extended 10-20 system, electrode sites halfway between those defined by the standard 10-20 system were introduced. Nowadays 100 or even 200 electrode systems are used. The number of electrodes providing adequate space sampling is $\geq$128, since inter-electrode distance should be around 20 mm to prevent spatial aliasing [Nunez and Srinivasan, 2006].

EEG is a measure of potential difference; in the referential (or unipolar) setup it is measured relative to the same electrode for all derivations. There is no universal consent regarding the best position of the reference electrode. Since currents com-

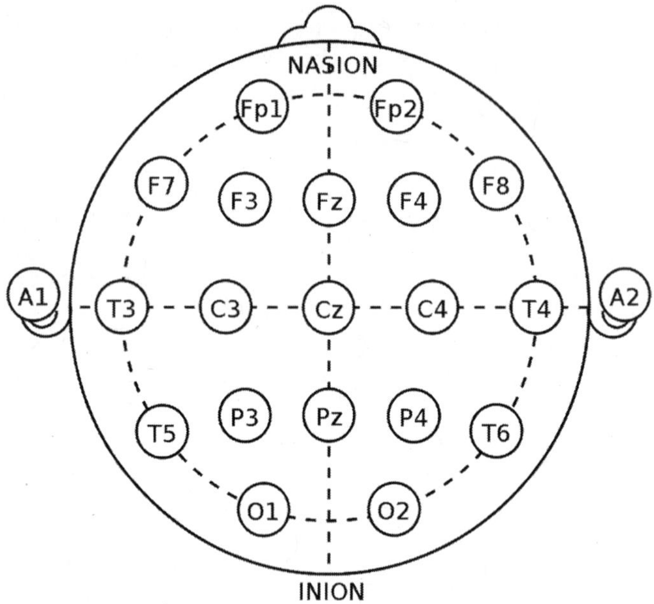

FIGURE 4.4: Electrode locations of the International 10-20 system for EEG recording.

ing from bioelectric activity of muscles, heart, or brain propagate all over the human body the reference electrode has to be placed in proximity of the brain: on the earlobe, nose, mastoid, chin, neck, or scalp center. In the bipolar setup (montage) each channel registers the potential difference between two particular scalp electrodes. Data recorded in a referential setup can be transformed into any bipolar montage, for the sake of display or further processing. The "common average reference" montage is obtained by subtracting from each channel the average activity from all the remaining derivations.

The Laplacian operator method is based on the fact that the Laplacian, calculated as a second spatial derivative of a signal, offers information about vertical current density and cancels common component due to volume conduction [Nunez, 1981, Nunez and Pilgreen, 1991]. Hjorth transform [Hjorth, 1975] is an approximation of the Laplacian operator method. The Hjorth transform references an electrode E_0 to its four closest neighbors:

$$J \sim 4E_0 - E_1 - E_2 - E_3 - E_4 \tag{4.1}$$

where J vertical current density at electrode 0 , E_i—electric potential at electrode i.

There is no optimal reference recommended for all studies. For many applications the Laplace operator method may be advised, but it requires a dense grid of electrodes. For the methods where the correlations between channels are important, as it is for MVAR, the reference electrode should be "neutral," since the calculations present in the common average approach or Hjorth method introduce additional correlations between signals, which disturbs the original correlation structure.

4.1.4 MEG measurement, sensor systems

The MEG measurement is technically much more demanding than the EEG measurement. This stems mainly from the fact that the magnetic fields generated by the brain are on the order of tenths to hundreds of femtotesla, which is orders of magnitude weaker than the Earth's magnetic field and fields generated by, e.g., moving ferromagnetic objects in the environment. Partially the environmental artifacts are removed by special spatial setup of measurement coils and partially by special magnetic shielded chambers (walls made of a few layers of μ-metal with high magnetic permeability separated by pure highly conducting aluminum layers allowing for generation of Eddy currents). The practical measurement of the extremely weak fields of MEG is possible thanks to the superconducting quantum interference device (SQUID) which uses quantum effects in superconducting electrical circuits. The SQUIDs are magnetically coupled to the sensor coils, which in principle can be of three types: magnetometer, axial gradiometer, or planar gradiometer (Figure 4.5). In a gradiometer inverse winding of the sensor coil eliminates the external magnetic fields which are slowly changing in space.

The sensor configuration of modern whole-head MEG systems consists either of magnetometers, or axial or planar gradiometers, or even a combination of them, covering the whole head in a helmet-like fashion.

Magnetometers are superior to gradiometers with regard to the possibility of recording signals from deeper sources. Apart from their different sensitivity to noise, these sensor configurations differ in the location of the signal's amplitude extremum with respect to its source. Unlike planar gradiometers for which the maximum signal is right above the source, magnetometers (as well as axial gradiometers) show extrema on both sides of the underlying source [Hämäläinen et al., 1993]. In the case of signal analysis on the level of sensors, planar gradiometer representation is easier in interpretation than that of axial gradiometers or magnetometers, since the maximum of the averaged signal directly indicates the location of the activated region.

4.1.5 Elimination of artifacts

An artifact in EEG can be defined as any potential difference due to an extracerebral source. The careful identification and elimination of artifacts is of utmost importance for EEG analysis. We can distinguish the artifacts of technical and biological origin. The first ones can be connected with: power supply, spurious electrical noise from engines, elevators, tramway traction, etc., bad electrode contact, or its detachment. Another kind of artifact has its origin within the subject's body. The most common of them are connected with eye blinks, eye movements, muscle activity, electrocardiogram (ECG). Eye movement generates an electric signal called an electrooculogram (EOG) because of potential difference between cornea and the back of the eye. Body and head movements may induce not only muscle electrical activ-

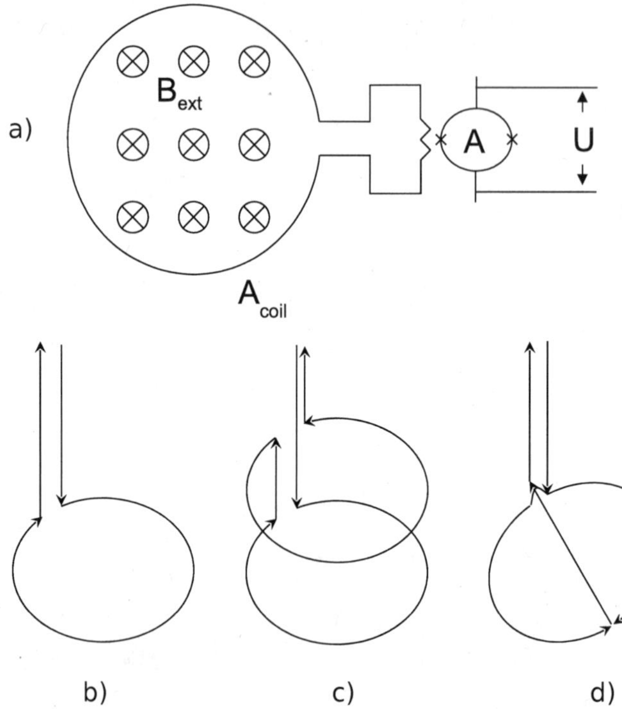

FIGURE 4.5: The schema of the MEG measurement system and types of coils. a) SQUID coupled to the sensor coil: B_{ext} external magnetic field, A_{coil} area of the sensor coil, A area of the SQUID coil , b) magnetometer coil, c) axial gradiometer coil, d) planar gradiometer coil.

ity, but also slow potential shifts, which can disturb low frequency EEG rhythms. Changes in scalp potentials may also be caused by respiration which causes changes in scalp impedance and by the electrodermographic activity (changes in skin potential connected with electric activity of sweat glands). Additionally the secretion of sweat can affect the resistance of electrode contacts. Figure 4.6 shows examples of different kinds of artifacts. Artifact elimination, when performed on the segmented data, usually involves rejection of the whole segment (epoch). If continuous data are considered, the criteria concerning the segment length to be rejected should be set in a way to add ample time before and after the artifact cut-off in order to avoid the sub-threshold but significant contamination.

Several techniques have been developed for artifact elimination. Some of them concern particular kinds of artifacts, e.g., muscular [Gevins and Morgan, 1986, van de Velde et al., 1998, Moretti et al., 2003, Mammone and Morabito, 2008], heart [Jiang et al., 2007], or ocular [Gevins and Morgan, 1986, Moretti et al., 2003, LeVan et al., 2006].

These systems have been evaluated on a limited number of epochs from one

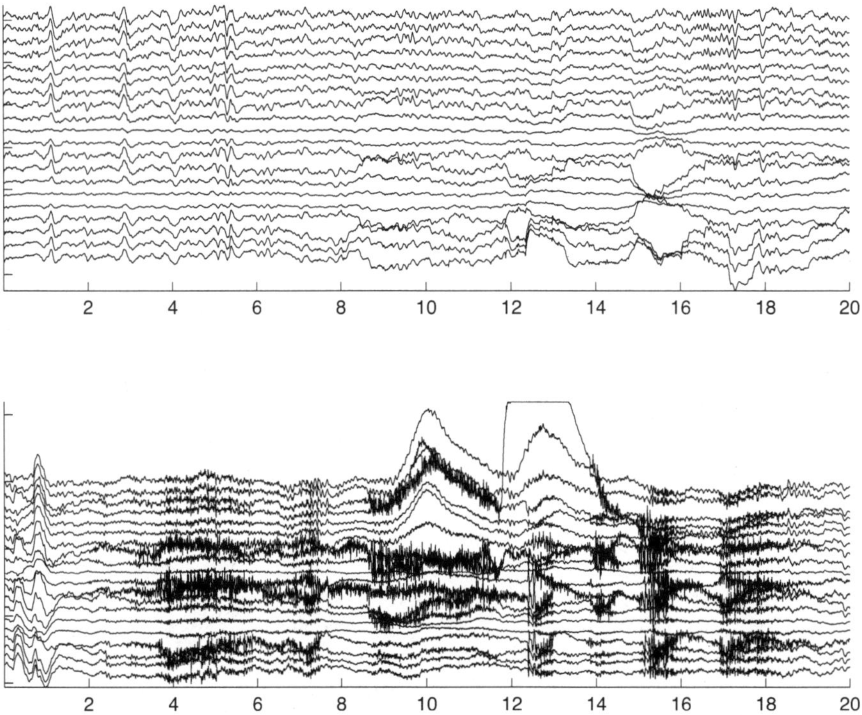

FIGURE 4.6: Examples of different kinds of artifacts. Upper panel, 20 s epoch of EEG contaminated with ECG artifact around seconds 1 and 3, and the eye movement artifact around second 16. Lower panel, muscle and movement artifact causing overflow around second 13.

[Gevins and Morgan, 1986, Mammone and Morabito, 2008] or more subjects [van de Velde et al., 1998, Moretti et al., 2003, LeVan et al., 2006, Jiang et al., 2007]. The methods aiming at elimination of different kinds of artifacts in the framework of the same system are relatively few. Among them we may distinguish semi-automatic [Nakamura et al., 1996, Schlögl et al., 1999, Delorme and Makeig, 2004] and automatic methods [Durka et al., 2003, Klekowicz et al., 2009].

The different approaches to identification of artifacts are described in the review by [Anderer et al., 1999]. The simplest methods were based on amplitude thresholds (overflow check), with different scenarios for setting the thresholds. Further works used time-varying autoregressive modeling and slope detection [van de Velde et al., 1999], wavelets [Jiang et al., 2007], and also artificial neural networks [Schaltenbrand et al., 1993]. There were also attempts to separate the artifacts based on the modeling of the sources of extra-cerebral activity. These methods usually modeled EOG generators as dipoles and estimated the influence of their activity on EEG.

However the general nonuniqueness of source localization solutions and the poor spatial resolution afforded by EEG data limit applicability of this approach [Achim et al., 1991].

Among the methods, which do not rely on specific models is a class of blind source separation algorithms (BSS), which includes PCA and ICA approaches (Sect. 3.6.1 and 3.6.2) They are based on the assumption that by projecting the data onto orthogonal (PCA) or statistically independent components (ICA) the components corresponding to artifacts can be isolated.

The rejection of all artifact contaminated epochs may result in severe loss of data. Therefore methods for correcting the artifacts rather than only detecting the artifact contaminated epochs were devised. They were based on filtering, regression analysis, PCA, and more recently on ICA [Jung et al., 2000].

Simple filtering of signals to eliminate, e.g., muscle or ECG artifacts relies on the assumption that the interesting part of EEG signal is expressed in a different frequency band (usually lower) than the artifacts. However, the low-pass filtering leads to excessive smoothing of data and usually disturbs the higher frequency EEG rhythms.

The regression method is based on the assumptions that: 1) artifact activity may be measured concurrently with EEG, 2) there is a linear relationship between EEG and activity connected with the artifact source, 3) there is no temporal delay between both signals. These assumptions are to a large degree valid for eye movements. The algorithms for EOG artifact corrections were developed by [Gratton et al., 1983, Semlitsch et al., 1986]. They are based on the mechanism of finding blink-affected data epochs in the EOG channels. Then the regression of the eye channel(s) with each individual EEG channel is performed. The corrected EEG_{cor} series are given by the formula:

$$EEG_{cor} = EEG - w \cdot EOG. \tag{4.2}$$

where w is the regression weight for the given channel. There are several problems connected with this approach discussed in [Croft et al., 2005]; the most important one is connected with the fact that not only does EOG influence EEG, but also, vice versa, EEG contributes to EOG signal. In the process of correction some EEG activity is also removed. For this reason nowadays the methods based on ICA are favored [Makeig et al., 1996, Delorme et al., 2007].

The procedure of finding the independent components was described in Sect. 3.6.2. For multichannel EEG or ERP the input matrix $\mathbf{x}$ (of k rows corresponding to the sensors) is used to train the ICA algorithm, which estimates an unmixing matrix $\mathbf{D}^{-1}$ that minimizes statistical dependence of the outputs $\mathbf{s}$. We can expect that the activities generated by different sources, e.g., eye or muscle artifacts, will be isolated as separate ICA components. Brain activities of interest can be obtained by projecting selected ICA components back on the scalp. In the process of back-projection the components corresponding to artifacts can be eliminated by setting the corresponding rows of the source matrix $\mathbf{s}$ to zero. In this way the corrected signals may be obtained. However, one should keep in mind that the "unmixing" is usually not perfect, and there may be some crosstalk between the components; thus in removing the artifact component, one may also remove some part of EEG signal.

James and Gibson [James and Gibson, 2003] proposed temporally constrained ICA which extracts signals that are statistically independent, yet which are constrained to be similar to some reference signal which can incorporate a priori information related to a particular kind of artifact. The authors argued that the relative morphology of the reference is relatively unimportant as long as the temporal features of interest are captured. However, the phase of the reference must be closely matched to that of the desired underlying component. The method was applied to remove artifacts from EEG and MEG signals. An example of removal of different kinds of artifacts form MEG signals by means of the above described method is shown in Figure 4.7.

The BSS approach for artifact rejection, called second order blind inference (SOBI), was proposed by [Joyce et al., 2004]. ICA algorithm assumes that the components are statistically independent at each time point, which is not a well-founded assumption. SOBI considers the relationship between component values at different time lags and insists that these values be decorrelated as much as possible. The remaining correlated components can isolate highly temporally correlated sources, which according to the authors, is a crucial feature for ocular artifact detection. Another advantage of SOBI put forth in [Joyce et al., 2004] is that it uses only second order statistics that can be more reliably estimated than higher order statistics used in the ICA approach. It seems that different methods devised in the framework of the BSS approach require further tests.

In spite of the large number of works devoted to the methodology of artifact removal very few of them were tested on sufficient experimental material and were developed into fully working systems. The semi-automatic system for rejecting artifacts is implemented in EEGLAB `http://sccn.ucsd.edu/eeglab/` [Delorme and Makeig, 2004]. It includes several routines to identify artifacts based on detection of: extreme values, abnormal trends, improbable data, abnormally distributed data, abnormal spectra. It also contains routines for finding independent components.

Please note that if the aim of the further analysis is finding the causal relations (Sect. 3.3.2) between the signals from a multichannel set, the subtraction of the artifacts cannot be used as a preprocessing step, since it disturbs the phase relations between the channels of the process.

The elimination of artifacts in the overnight polysomnographic sleep recordings is of especial importance because of the very large volume of data to be inspected—whole night sleep recording as printed on standard EEG paper would be over half a kilometer long. There was an attempt to design the system of artifact elimination in sleep EEG in the framework of the EC SIESTA project [Schlögl et al., 1999, Anderer et al., 1999], but this system was far from being robust and automatic and it was tested only on selected 90 min. epochs taken from 15 recordings.

A fully automatic system of artifact detection in polysomnographic recordings devised by [Klekowicz et al., 2009] was tested on over 100 polysomnographic recordings of overnight sleep. The parameters reflecting explicitly the likelihood that a given epoch is contaminated by a certain type of artifact were calculated for the following types of artifacts: eye movements, eye blinks, muscle artifacts, ECG, low

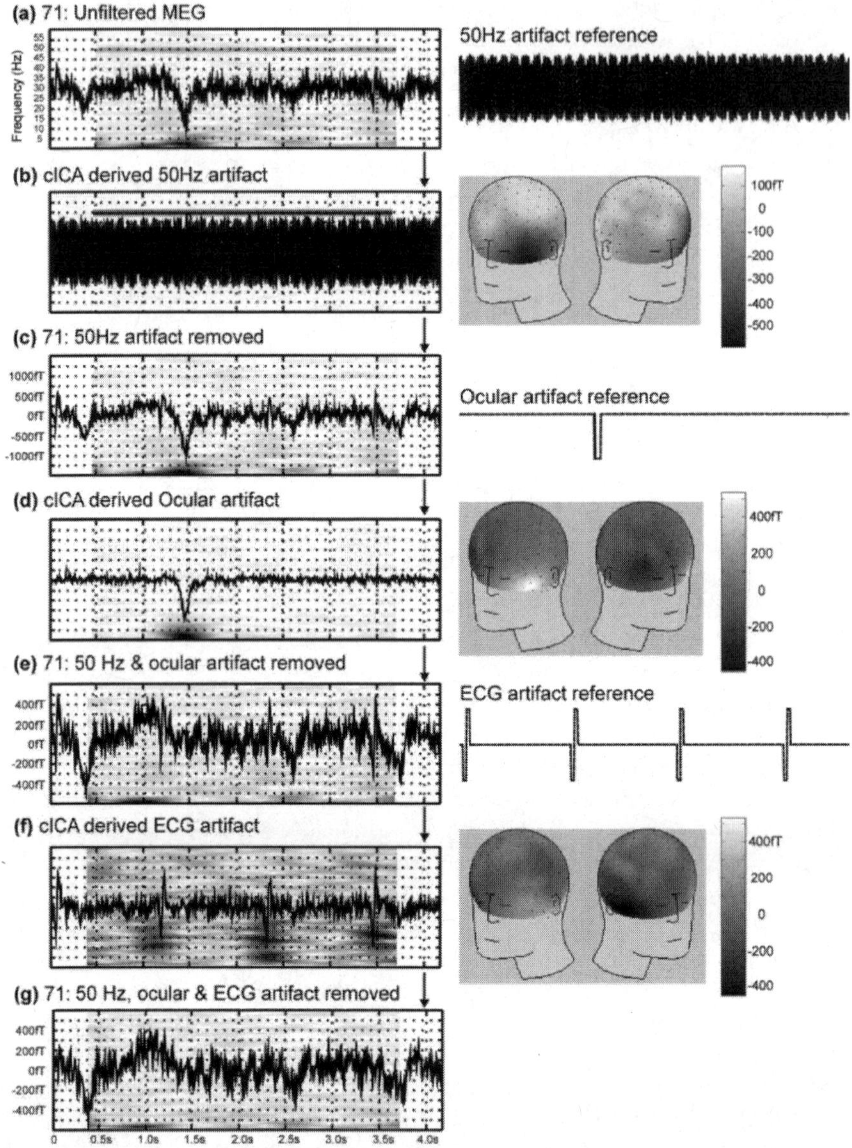

FIGURE 4.7: (a) 4 s epoch of MEG contaminated with 50 Hz, ocular and ECG artifact. (b) ICA output after using 50-Hz reference depicted. (c) The MEG signal with 50 Hz artifact subtracted. (d) ICA output after using ocular artifact reference depicted. (e) The signal with 50 Hz and ocular artifact subtracted. (f) ICA output after using ECG reference depicted. (g) The signal with 50 Hz, ocular and ECG artifact subtracted. The topographies for each ICA derived component are shown on the right. From [James and Gibson, 2003].

frequency disturbances, power supply interference, abrupt slopes, electrode pop artifacts. The eye movements and eye blink artifacts identification was based on the correlation between EOG electrodes and/or certain EEG derivations. For muscle and low frequency artifacts criteria based on spectral content of the signals were used. The thresholds for each kind of artifact were set separately. Sensitivity of the system to each type of artifact can be a priori adjusted or can be set by means of an automatic procedure for thresholds optimization based on minimalization of the cost function. The agreement with the visual scoring was on the level of inter-expert agreement. The MATLAB version of the above described software is available from `http://eeg.pl/Members/hubert/artefakt`. It includes a complete source code for all the computations, the user interface as well as input/otput routines for some formats.

4.1.6 Analysis of continuous EEG signals

Since the first recording of EEG from the human scalp by Hans Berger in 1929, clinicians and scientists have investigated EEG patterns by visual inspection of signals recorded on paper charts and, up to the present time, visual analysis is still in use, but the signals are displayed on a computer screen. The complicated character of EEG consisting of rhythmical components and transient structures has promoted early attempts for an automatic EEG analysis. Berger assisted by [Dietch, 1932] applied Fourier analysis to EEG series. The first digital analysis of EEG was performed in the time domain [Brazier and Casby, 1956]. The development of theoretical background of spectral analysis by [Blackman and Tukey, 1958] promoted the development of signal analysis in frequency domain and in 1963 the first paper on digital spectral analysis of EEG appeared [Walter and Adey, 1963]. With the development of computer techniques and implementation of fast Fourier transform, spectral analysis became a basic digital tool for EEG analysis, since the contribution of characteristic rhythms in brain activity has an important impact on clinical diagnosis.

Due to its complexity, the EEG time series can be treated as a realization of a stochastic process, and its statistical properties can be evaluated by typical methods based on the theory of stochastic signals. These methods include: probability distributions and their moments (means, variances, higher order moments), correlation functions, and spectra. Estimation of these observables is usually based on the assumption of stationarity. While the EEG signals are ever changing, they can be subdivided into quasi-stationary epochs when recorded under constant behavioral conditions. On the basis of empirical observations and statistical analysis performed by several authors, quasi-stationarity can be assumed for EEG epochs of 10 s length approximately, measured under constant behavioral conditions [Niedermayer and Lopes da Silva, 2004].

EEG signal can be analyzed in the time or frequency domain, and one or several channels can be analyzed at a time. The applied methods involve spectral analysis by Fourier transform (FT), autoregressive (AR) or MVAR parametric models, time-frequency and time-scale methods (Wigner distributions, wavelets, matching pursuit). The most common methods used for post-processing include: cluster analysis, discriminant analysis, or artificial neural networks (ANN).

4.1.6.1 Single channel analysis

Estimation of power spectra is one of the most frequently used methods of EEG analysis. It provides information about the basic rhythms present in the signal and can be easily and rapidly calculated by means of the fast Fourier transform (FFT). Maximum entropy power spectrum may be obtained by means of the autoregressive model, which can be recommended for the EEG analysis. Notwithstanding the advantages of AR spectral estimate mentioned already (Sect. 2.3.2.2.3) the transfer function of the AR model represented as a system of parallel filters is compatible with the presence of EEG rhythms.

The impulse response function of the AR model is a sum of damped sinusoids (Sect. 2.3.2.2.4) as was demonstrated by [Franaszczuk and Blinowska, 1985], who introduced the method of parametrization of EEG rhythms called FAD (frequency, amplitude, damping). These parameters determined from the AR model give directly frequency and amplitude of the rhythms without the need of plotting the spectra. The parameter concerning the damping of rhythms is useful in modeling of EEG time series [Wright, 1989]. FAD method was used, e.g., for a study of stability of electrocortical rhythm generators [Mitraszewski et al., 1987]. It was demonstrated that the transfer function of the AR model is compatible with the transfer function of the physiological model of rhythms generation. The FAD parameters were connected with the parameters describing action of neural populations connected in a feedback loop [Blinowska and Franaszczuk, 1989].

The spectral content of EEG changes during childhood and adolescence, attains a stable character about 21 years of age, and then may be subject to changes in old age. In the immature brain there is a preponderance of low frequency rhythms and progressively the role of higher rhythms increases. The enhancement of the low frequency activity in the adult EEG may be a symptom of neuropathology and also learning and performance difficulties. This observation lead [Matousek and Petersen, 1973] to introduce the so-called age quotient based on the contribution to the EEG, rhythms appropriate for each age (determined by regression analysis on a large pool of subjects). The discrepancy above 0.8 in age quotient (a ratio between the age estimated from EEG and the actual one) was considered as a sign of possible pathology. In the computer assisted diagnostic system for neurological and psychiatric diseases—Neurometric [John and Thatcher, 1977] the ratio of the slow rhythms delta and theta to the alpha rhythm computed for each derivation by means of spectral analysis was one of the markers of possible disturbances. Another marker in this system was the degree of asymmetry of spectral power between hemispheres.

The application of the psychoactive or anti-epileptic drugs causes pronounced changes in EEG spectral characteristics. The influence of drugs and the effects of medication are evaluated by means of calculation of spectral power in the characteristic frequency bands and their ratios. In psychopharmacology the topographical distribution of spectral power is also considered. Usually the statistically significant differences between spectral power in frequency bands before and after medication are mapped. The average power values in each frequency band and their ratios are also useful markers in depression [Cook et al., 2009].

In some applications not only rhythmic activity, but also transient structures of EEG are of importance. In this case the methods of non-stationary signal analysis have to be used. The examples of such transients are sleep spindles or epileptic spikes, which are important, e.g., for assessment of the influence of sleep-inducing or anti-epileptic drugs. The optimal method for detection of transients occurring randomly in signals is matching pursuit. The advantages of the method in estimation of influence of sleep-inducing drugs was demonstrated in [Durka et al., 2002], for identification of spindles in [Zygierewicz et al., 1999], and for detection of epileptic spikes in [Durka, 2004].

4.1.6.2 Multiple channel analysis

4.1.6.2.1 Mapping One of the commonly used methods of presenting multichannel brain activity is the topographical display of signals features called mapping. A map may help to make a direct comparison between the topographic distribution of EEG features and an anatomic image given, e.g., by the tomographic brain scan. Three types of features are most commonly mapped for clinical applications: 1) direct variable such as amplitude, 2) transformed variable such as total spectral power or relative spectral power in frequency band, 3) the result of statistical test applied to a given EEG feature. The amplitude values of the signals or the spectral power in selected bands are frequently displayed. To obtain a map composed of hundreds of pixels spatial interpolation between the discrete electrode positions is necessary.

Different interpolation algorithms are possible, ranging from the simple N-nearest neighbors electrode algorithm to the more complex spline approximation. The advantage of spline interpolation methods is the easy computation of the second spatial derivative and thus application of the Laplacian operator. In MATLAB the interpolation of maps can be done, e.g., with the `interp2` function, which supports nearest neighbor, linear, and cubic spline interpolation. For more sophisticated spline interpolations the Spline Toolbox may be useful.

One should not assume that maximum of signal power visible on the map corresponds to the underlying source, since it can come from superposition of activity of coherent sources. The character of the map is strongly dependent on the reference. As was already mentioned, there is no optimal reference system. In case of mapping one has to be aware of this limitation. Common average reference system is liable to serious ambiguities in cases when some of the electrodes pick up similar signals, e.g., occipital alpha rhythm may appear in frontal derivations and eye movement artifacts may affect posterior electrodes [Pfurtscheller, 1991].

The recommended representation involves surface Laplacians (Sect. 4.1.3). However, a reliable computation of surface Laplacian requires at least 64 electrodes and adequate spatial sampling is obtained for more than 128 electrodes. Therefore, quite frequently an approximation of the Laplacian operator by Hjorth transform is applied. Results obtained by application of the Laplacian operator may be further ameliorated by deblurring; that is, using a mathematical model of volume conduction through the skull and scalp to downwardly project scalp-recorded potentials, which provides a computational estimate of the electrical potentials, that would be recorded near the superficial cortical surface [Lee and Gevins, 1993].

Mapping is often performed for the sake of comparison, e.g., to detect changes connected with medication, or to find out possible difference between groups. For the comparison of one map with a group of maps the z statistics can be used. The transformation: $Z = \frac{X-\bar{X}}{\sigma_X}$ of a map derived from an individual subject (X) is calculated for each pixel of a map of an individual subject in comparison to the mean ($\bar{X}$) and standard deviation σ_X for a group of maps. The t-statistics can be used to differentiate between maps, but to use this statistics, the normal distribution of the group data has to be assumed. In many cases the distribution can be transformed to an approximately normal one by Box-Cox transformation [Box and Cox, 1964]. A special case of this transformation is subjecting the data to logarithmic transform of the type: $y = \log(x)$ which is often used to normalize the values of absolute power spectra. The second assumption of t-test is the homoscedasticity (the assumption that the variances in both compared groups are equal). In cases where the variances can be different, one should use the Welch's t test [Welch, 1947]. For comparison of values that cannot be transformed to normal distribution, non-parametric tests should be used. The problem of choice of the appropriate test is discussed in section 1.5.2. To make statistical inference on the maps pixel by pixel one has to take into account the multiple comparison problem. This problem and its possible solutions are discussed in Sect. 1.5.3.

4.1.6.2.2 Measuring of dependence between EEG signals Interdependence between two EEG signals can be found by a cross-correlation function or its analogue in the frequency domain—coherence. Cross-correlation can be used for comparison of EEGs from homologous derivations on the scalp. A certain degree of difference between these EEGs may be connected with functional differences between brain hemispheres, but a low value of cross-correlation may also indicate a pathology.

In EEG studies, not correlations but rather coherences are usually estimated, since they provide the information about the rhythm synchronization between channels. Conventionally, ordinary coherences calculated pair-wise between two signals have been used for EEG as well as for ERP studies. However, for the ensemble of channels taken from different derivations the relationship between two signals may result from the common driving from another site. This is often a case for EEG, the signals recorded from the scalp are strongly interdependent. Therefore the patterns of bivariate coherences are usually complicated and not consistent. To obtain a complete pattern of coherence structure of multichannel EEG the estimation not only of ordinary, but also of partial and multiple coherences is recommended [Franaszczuk et al., 1985].

In Figure 4.8 ordinary (bivariate), partial, and multiple coherences are shown for sleep EEG stage 2 [Kaminski et al., 1997]. Multiple coherences have high amplitudes for all derivations and whole frequency range which means that the system is strongly interconnected. Ordinary coherences decrease monotonically with a distance, which is a rather trivial observation. Partial coherences show only direct connections between channels and the information provided by them is much more selective. The computation of ordinary, partial, and multiple coherences can be easily achieved in the framework of MVAR.

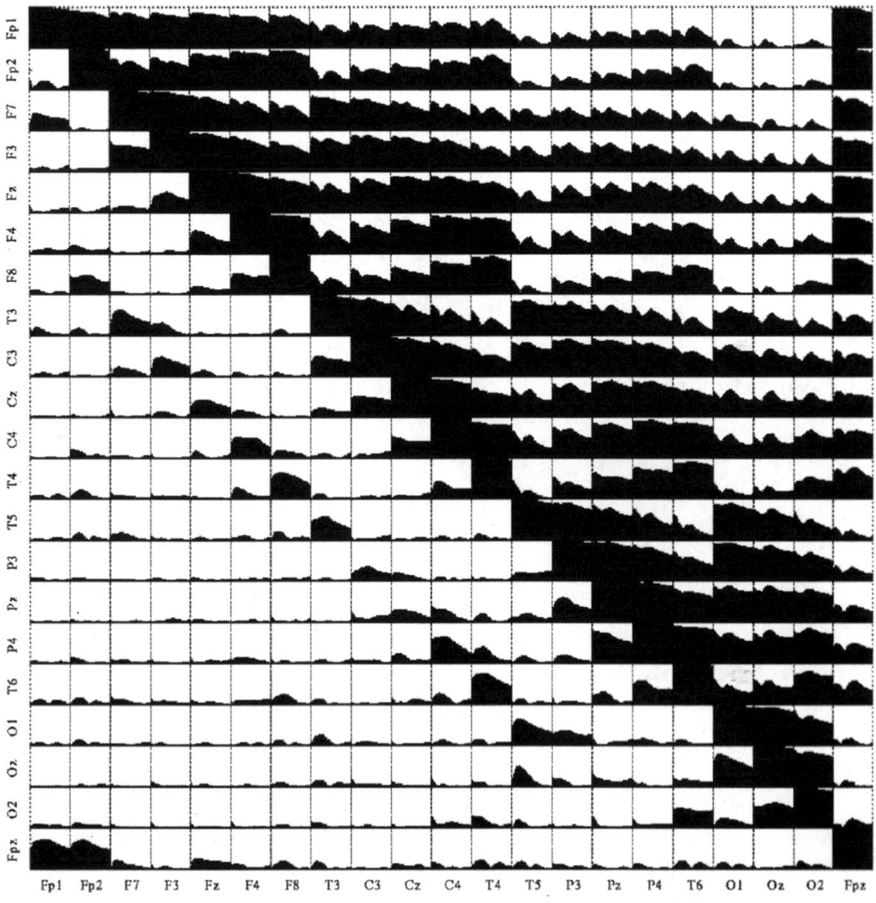

FIGURE 4.8: Coherences: ordinary—above diagonal, partial—below diagonal and multiple—on the diagonal for stage 2 of Sleep EEG. From [Kaminski et al., 1997].

The important problem in brain research is the estimation of functional connectivity and the determination of the direction of the propagation of EEG signals. It was approached by the estimation of phase difference between channels, but as was explained in Sect.3.5, the method gives ambiguous and sometimes misleading results.

The progress in estimation of directionality was connected with introduction of the measures based on the Granger causality principle: DTF and PDC.

DTF was successfully used, e.g., for estimation of EEG transmission during overnight sleep [Kaminski et al., 1997], for finding patterns of LFP propagation during locomotion in behaving animals [Korzeniewska et al., 1997], in epiloptegenesis [Medvedev and Willoughby, 1999], for finding interhemispheric dependencies in presleep/wake states [Bertini et al., 2007], for determination of directionality of EEG synchronization in Alzheimer disease [Babiloni et al., 2009], assessing the functional connectivity [Astolfi et al., 2005].

Comparison of the different methods of directionality estimation for the same experimental data, namely EEG recorded in the awake state eyes closed, is illustrated in Figure 4.9. It is known that the main sources of activity in such a case are located in the posterior parts of cortex and some weaker sources occur more frontally.

One can see that the clearest picture indicating the localizations of the sources of activity (back of the head and some frontal localizations) was revealed by DTF. dDTF showed transmissions determined by the anatomical tracts. In the bivariate case the picture was quite confusing; namely one can even observe the reversal of propagation, namely from C3 to P3 instead of P3 to C3. The propagation structure obtained by means of PDC shows rather the sinks not the sources, so the results are more difficult to interpret, which is connected with the normalization of the measure (Sect. 3.3.2.4).

Inspecting Figure 4.9 it is easy to see that in the case of scalp EEG it is not important to use an estimator showing only direct (not cascade) flows. It is more important which parts of the cortex are communicating than what is the exact "wiring" scheme. Therefore for estimation of transmissions of EEG registered by scalp electrodes DTF is recommended. dDTF should be used for the intracranial electrodes where the direct connections between brain structures are important [Korzeniewska et al., 2003].

In Figure 4.10 the results of the study of activity propagation during overnight sleep are shown. The obtained patterns of EEG transmissions averaged over nine subjects show a high degree of consistency. The most complicated pattern is visible for wakefulness since subjects differ in exact locations of sources. This pattern becomes more organized during sleep. The results have a plausible physiological explanation [Kaminski et al., 1997].

DTF is based on the phase differences between channels. It has non-zero value only when there is a phase difference between signals from different derivations. This fact was confirmed in the simulation studies and by the observation that even strong artifacts, but with no phase difference between channels, did not influence results at all. Volume conduction is a zero phase propagation; therefore no phase difference between channels is generated. Thus in theory volume conduction shouldn't have any influence on DTF results. In practice, it may have some influence, e.g., increasing the noise level. However, this influence is not critical; it is much less important than in case of other methods.

In some publications before the application of DTF the transformation to the source space was performed, e.g., [Astolfi et al., 2005]. We find this procedure unnecessary, since the influence of volume conduction is not significant in the case

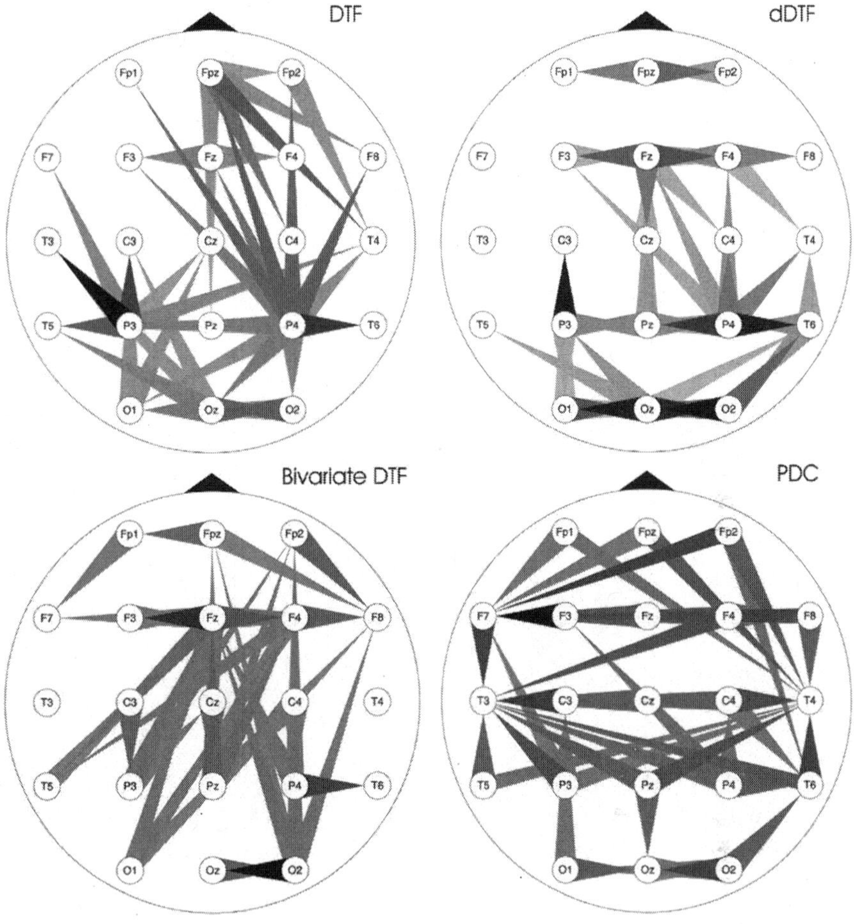

FIGURE 4.9: EEG transmission in awake state eyes closed obtained by means of different methods.

of DTF. On the other hand making the transformation to the source space involves calculations, which can change the delicate balance in phase dependencies between channels. Determination of MVAR model coefficients (and hence DTF) is based on the correlation matrix between signals. The transformation changes this correlation matrix in such a way that the correlations may be disturbed. It is recommended to use as input to DTF the signals referenced to "neutral" electrode, e.g., "linked ears," "nose," etc. Before fitting the MVAR model mean value of the signal should be subtracted, and the data should be normalized dividing by the variance. Preprocessing influencing the phase relation between channels (e.g., artifact subtraction) should be avoided. The PDC method similar to DTF is robust in respect to noise and insen-

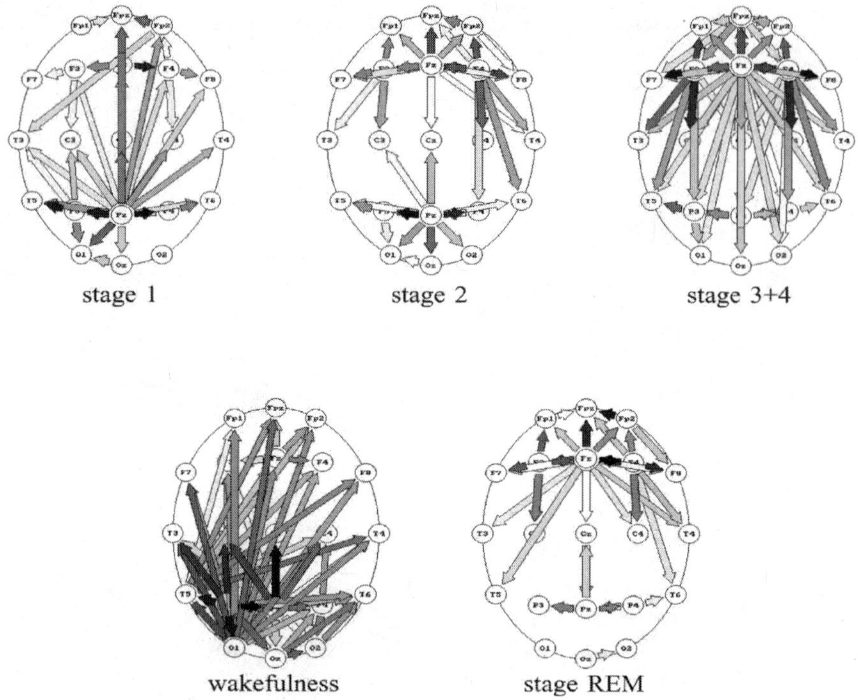

FIGURE 4.10: The propagation of EEG activity during overnight sleep averaged over nine subjects, estimated by means of DTF integrated in the 0–30 Hz band. Arrows are drawn for DTFs greater than 0.06. Their intensities are reflected by the shade of arrows. From [Kaminski et al., 1997].

sitive to volume conduction. In case of PDC the same provisions concerning signal preprocessing have to be made.

PDC have been used, e.g., by [Fanselow et al., 2001] for the study of thalamic bursting in rats during different awake behavioral states, by [Astolfi et al., 2006] for assessing functional connectivity, and for investigation of epileptic activity by [Baccalá et al., 2004, Takahashi et al., 2007]. In the last paper the statistical properties of PDC were analyzed. The properties of PDC were considered also in [Schelter et al., 2009], where the problems connected with interpretation of PDC were raised and a new normalization similar to the one applied in DTF was proposed.

4.1.6.3 Sleep EEG analysis

Sleep EEG reveals a characteristic pattern called a sleep architecture, which is classically described by division into stages originally defined by [Rechtschaffen and Kales, 1968] (R&K): stage 1 (drowsiness), stage 2 (light sleep), stage 3 (deep sleep), stage 4 (very deep sleep), and REM (dreaming period accompanied by rapid eye movements). Recently new criteria for sleep scoring were proposed by the American Academy of Sleep Medicine [Ibert et al., 2007], however R&K criteria are still the golden rule for sleep staging. The differentiation of the sleep stages involves measurement of several signals: not only EEG, but also electrooculogram (EOG), muscular activity—electromyogram (EMG), and respiration, sometimes also measurement of blood flow, electrocardiogram (ECG), and oxygen level in blood. This kind of recording is called a polysomnogram. EOG is measured by means of electrodes placed at the canthi of the eyes. The EOG and EMG help to distinguish Rapid Eye Movement (REM) state, which is connected with dreaming. During the night there are usually four or five cycles of sleep consisting of a period of NREM (non-REM) sleep followed by REM. The sequence of sleep stages is usually illustrated in the form of the hypnogram (hypnogram is shown in Figure 4.13). The recognition of stages is based on the contribution of the different rhythms and the occurrence of characteristic signal transients absent in wake EEG, namely: sleep spindles, vertex waves, and K-complexes.

Sleep spindles are rhythmic waves of frequency 11-15 Hz and duration 0.5-2 s; characteristic increase and then gradual decrease of amplitude is not always observed. Vertex wave is a compound potential: a small spike discharge of positive polarity preceding a large spike and followed by a negative wave of latency around 100 ms and often another small positive spike. The K-complex consists of an initial sharp component, followed by a slow component that fuses with a superimposed fast component. The sharp component may be biphasic or multiphasic. There is hardly universal consent among encephalographers concerning definition of K-complexes, which makes their automatic detection difficult. The most precise definition of K-complexes may be found in [Paiva and Rosa, 1991]. K-complexes and vertex waves are presumably a kind of evoked responses. Sleep stages may be briefly characterized as follows:

Stage 1 (drowsiness) is associated with decrease of alpha rhythm, rhythms in 2-7 Hz frequency band, and low amplitude rhythms of 15-25 Hz band. Sometimes occurrence of vertex waves.

Stage 2 is characterized by appearance of K-complexes and sleep spindles; the slow wave activity (SWA) should be less than 20% of the epoch.

Stage 3 is scored when 20–50% of the epoch contains delta waves (SWA) of peak-to-peak amplitude greater than 75 μV.

Stage 4 is dominated by slow wave activity of high amplitude. Stage 4 is scored when more than 50% of the epoch contains delta activity.

REM is characterized by decrease of EEG amplitude, occurrence of faster rhythms, rapid eye movements and a loss of muscular activity. Spectral characteristics in REM is polyrhythmic and on the basis of EEG only, it is difficult to distinguish REM from stage1.

There is evidence that when sleep becomes deeper the sources that drive EEG activity move from the posterior regions of the head (prevalent during awake state with eyes closed) to the centro-frontal regions [Kaminski et al., 1997].

The sleep pattern changes very much during childhood and adolescence. For newborn babies REM takes most of the sleep time, and in young children only REM and non-REM stages can be distinguished. Diminution of deep, slow wave sleep and increase in wakefulness continues through the entire life span after 1 year of age. In old age the contribution of stages 3 and 4 decreases markedly and first REM stage appears later in the night.

A hypnogram describes sleep macrostructure and relies on division of the time axis into fixed time epochs (20 or 30 s) —this naturally implies some limitations. In the alternative approaches treating sleep EEG as a continuous process, its microstructure is described in terms of: evolution of spindles and SWA activity or occurrence of transient arousals. They are defined as an abrupt shift in EEG frequency, which may include theta, alpha, and/or frequencies greater than 16 Hz, but not spindles. A certain number of spontaneous arousals seems to be an intrinsic component of physiological sleep, but their frequent occurrence may be connected with respiratory sleep disorders, nocturnal myoclonus, and other clinical conditions. The method providing the continuous description of sleep, called the cyclic alternating pattern (CAP), was proposed by [Rosa et al., 1999].

Sleep EEG analysis finds application in diagnosis of several disorders: insomnias, somnambulism, narcolepsy, epilepsy (some epileptic conditions appear mainly during sleep), depression, dementia, drug withdrawal. In sleep scoring, usually central EEG derivations C3 or C4 (referenced to electrodes A1, A2 placed at mastoids), and additionally two EOG channels and an EMG channel are considered. Sometimes occipital electrodes are also taken into account.

The evaluation of sleep pattern connected with construction of hypnogram, or finding arousals, is a very tedious and time consuming job involving analysis of a large volume of polysomnographic recordings. Therefore as early as the 1970s attempts were made to design a system for automatic sleep scoring. The early works dealing with this problem included hybrid systems [Smith and Karacan, 1971] and pattern recognition techniques [Martin et al., 1972]. Later, expert systems [Ray et al., 1986, Kubat et al., 1994] and artificial neural networks (ANN) found applications in sleep analysis [Roberts and Tarassenko, 1992]. ANNs with input coming from an AR model were proposed for sleep staging by [Pardey et al., 1996]. The problem was also attacked by model based approaches [Kemp et al., 1985] and independent component analysis [McKeown et al., 1998]. Wavelet analysis was used for the detection of transient structures in EEG [Jobert, 1992] and for the identification of microarousals [Carli et al., 1999]. The review of the studies concerning digital sleep analysis may be found in [Penzel et al., 2007].

There are several problems in automatic sleep analysis; one of them is poor consent between electroencephalographers concerning the classification of stages [Kubicki et al., 1989, Noman et al., 2000]. Another obstacle is the difficulty in the identification of the transient structures such as sleep spindles, K-complexes, and vertex waves. The detection of these waveforms is best performed by the methods working in the time-frequency domain. The method providing parametrization in the time-frequency space is wavelet analysis, which was applied, e.g., by [Jobert, 1992]. However the parametrization of signal structures in wavelet analysis is bound by the predetermined framework (usually dyadic) in the time-frequency space. Therefore, a given structure may be described by wavelet coefficients belonging to several scale levels in time and frequency (Figure 4.11). The description is not sparse and is not related to the physiological characteristics of the transient. The method, which

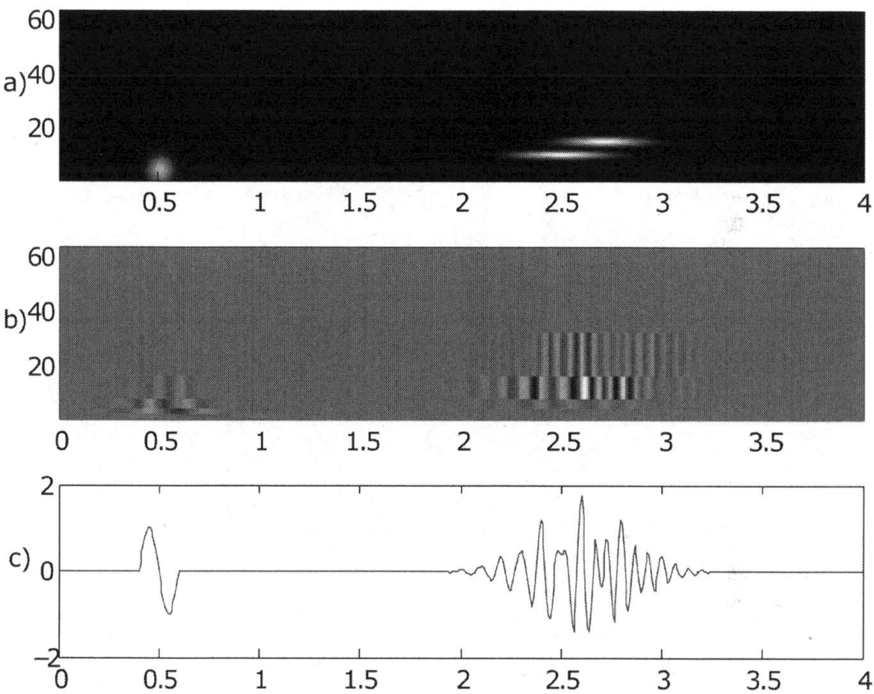

FIGURE 4.11: Comparison of time-frequency representation of sleep structures of a signal (c) consisting of idealized K-complex and two spindles. a) and b) representations obtained by means of MP and WT (Coiflet3), respectively. Note that MP separates the two spindles overlapping in time and close in frequency.

is maximally adaptive, supplies the parametrization of the rhythmical components and transients in the same framework and characterizes the structures of the signal by means of parameters compatible with the clinical definitions is matching pursuit. Therefore, we shall discuss this method in a larger extent. The t-f distribution of EEG from different sleep stages obtained by means of MP are shown in Figure 4.12; data structures corresponding to rhythmic and transient components can be easily distinguished. In MP the signal structures are characterized by the frequency, time span,

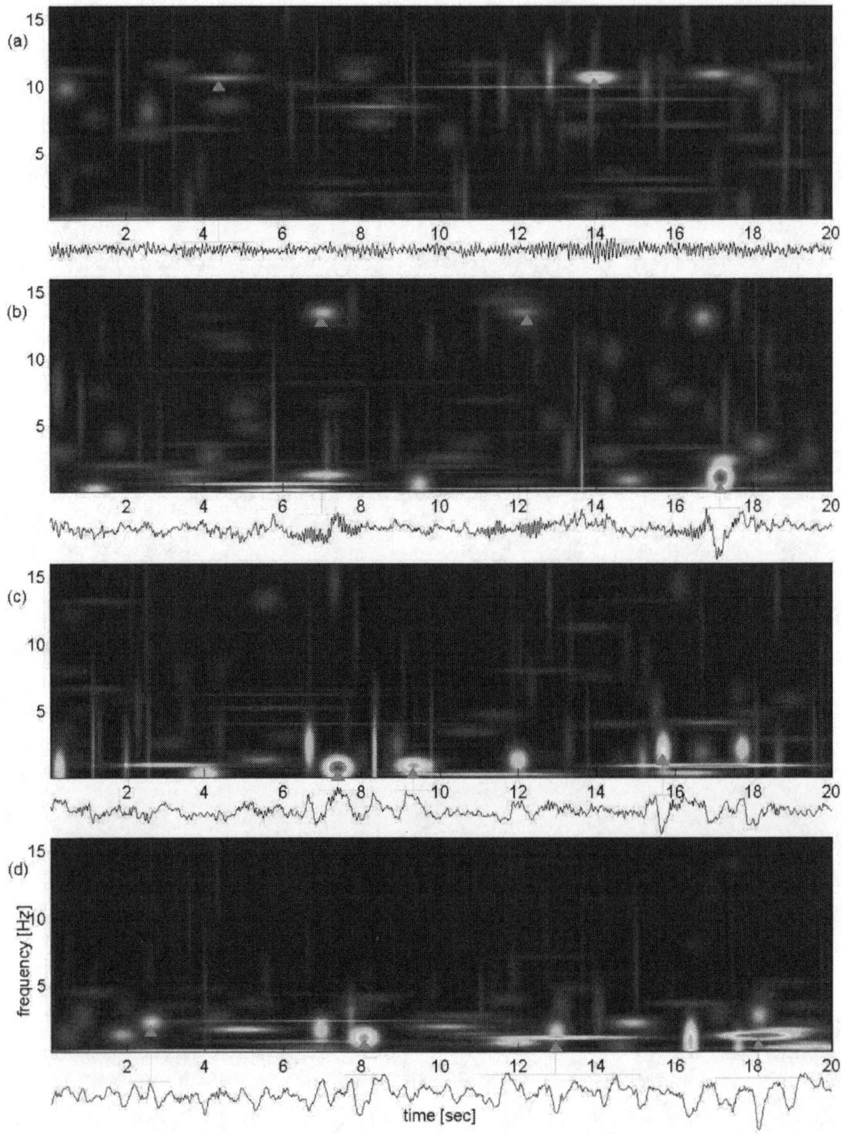

FIGURE 4.12: (SEE COLOR INSERT) Time frequency maps of EEG signal energy for 20-second epochs from different sleep stages shown above the corresponding signals. a) Wakefulness, alpha waves marked by arrows; b) stage 2, two spindles and K-complex marked; c) stage 3, two slow waves and a K-complex marked; d) stage 4, slow waves marked. From [Malinowska et al., 2006].

time occurrence amplitude, and phase. This allows to construct explicit filters, based on clinical definition, for finding in EEG specific structures of both transient and oscillatory nature. In fact MP is the only method which determines as a parameter the duration of signal structure, which allows for explicit application of R&K criteria concerning the percentage of specific activity in the given epoch. One of the first applications of MP to biological signals concerned sleep spindles [Blinowska and Durka, 1994]. High accuracy identification and parametrization of sleep spindles by means of MP allowed for distinction of two classes of spindles of different frequency ranges and topographic localization [Zygierewicz et al., 1999]. In the following work the time evolution of SWA and spindle activities was investigated and an inverse relation in their magnitudes was found [Durka et al., 2005a]. Macro and microstructure of sleep EEG was considered in [Malinowska et al., 2006] and [Malinowska et al., 2007], where identification of K-complexes, detection of deep sleep stages (3 and 4) based directly upon the classical R&K criteria, continuous description of slow wave sleep, fully compatible with the R&K criteria, and detection of arousals were presented in the framework of the same unifying MP approach (Figure 4.13).

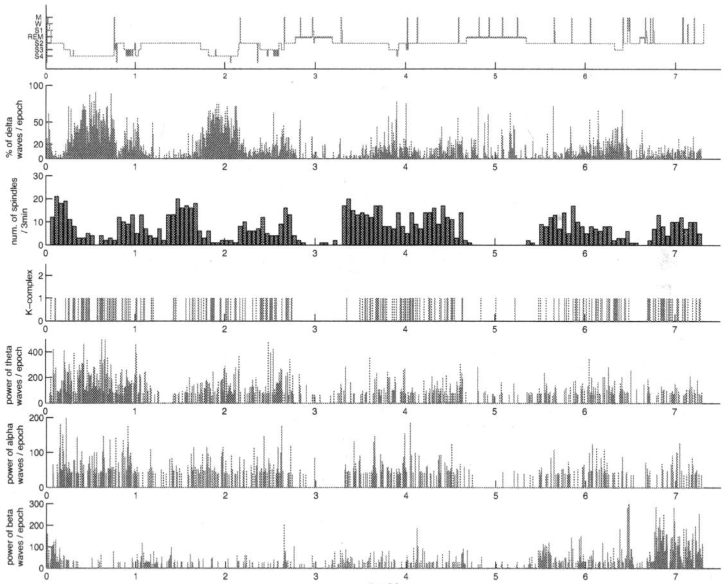

FIGURE 4.13: Time evolution during overnight sleep of the signal structures quantified by means of MP. From top to bottom: hypnogram; percentage of 20 s epoch occupied by waveforms classified as SWA; number of sleep spindles in 3 minutes periods; K-complex occurrence; power of theta waves per 20 s; power of alpha waves per 20 s; power of beta waves per 20 s epoch. By courtesy of U. Malinowska.

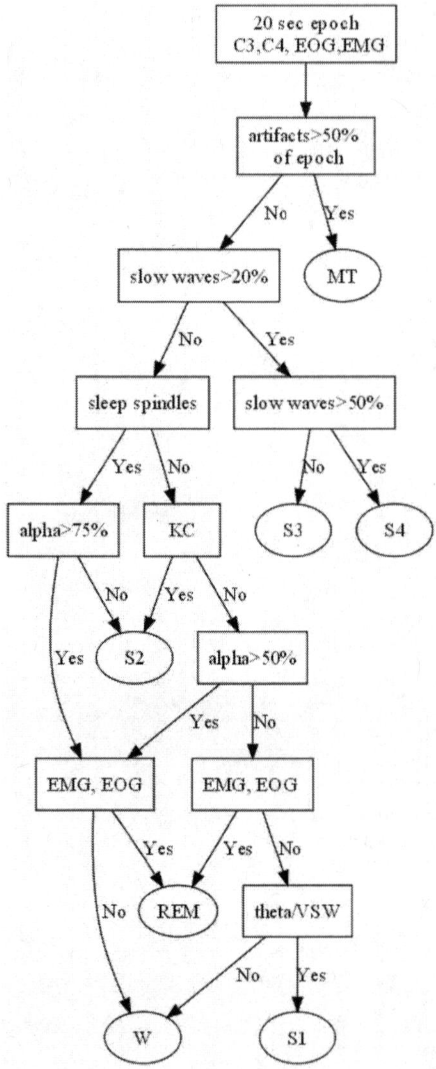

FIGURE 4.14: Block diagram of the sleep staging algorithm. KC—K-complex, MT—movement time, S—sleep stage. From [Malinowska et al., 2009].

Finally the results of the above mentioned works were gathered in an open system for sleep staging, based explicitly on the R&K criteria [Malinowska et al., 2009]. The system started with detection and parametrization of relevant waveforms and then combining these results (together with information from EMG and EOG signals) into a final decision assigning epochs to sleep stages. Automatic sleep staging was performed in a hierarchical way presented in Figure 4.14. In the first step, each 20-s epoch is tested for muscle artifacts in EEG or EMG derivation, which indicate the movement time (MT). If at least one of the analyzed derivations (C3, C4, or EMG) exceeded a corresponding threshold in more than 50% of the 20 s epoch, the epoch is scored as MT. In the second step, the algorithm detects slow wave sleep stages 3 and 4 by applying fixed 20% and 50% thresholds to the amount of epochs time occupied by slow waves. In the following step, the algorithm detects sleep spindles and K-complexes, which are related to stage 2. If at least one sleep spindle or K-complex occurs in the 20 s epoch, and less than 75% of the epoch is occupied by alpha activity, the epoch is scored as stage 2. If alpha activity occupies above 75%, EOG and EMG signals of this epoch are examined to distinguish stage REM from wake. Performance of the system was evaluated on 20 polysomnographic recordings scored by experienced encephalographers. The system gave 73% concordance with visual staging—close to the inter-expert concordance. Similar results were obtained for other systems reported in the literature [Penzel et al., 2007], however these expert systems were tuned explicitly for maximizing the concordance with visual scoring and were usually based on a black-box approach, so their parameters were difficult to relate to the properties of EEG signals observed in visual analysis. Also, most of these systems are closed-source commercial solutions. On the contrary, a bottom-up approach presented in [Malinowska et al., 2009] together with an open implementation (including complete source code) freely available from the internet (`http://eeg.pl/stager/`) is flexible and directly related to the visual analysis. The parameters of the system can be easily adapted to the other criteria concerning sleep stage classification.

4.1.6.4 Analysis of EEG in epilepsy

Epilepsy is one of the most common neurological disorders second only to stroke; it affects about 0.8% of the world's population. The most important treatment is pharmacological, however in 25% of patients seizures are drug resistant. Epilepsy is manifested by a sudden and recurrent brain malfunction which has its origin in excessive and hypersynchronous activity of neurons. The seizures occur at random and impair the normal function of the brain. During the seizure the electroencephalogram changes dramatically: its amplitude increases by an order of magnitude and characteristic patterns varying in time and frequency appear. The mechanisms of the seizure generation are still not known. It is presumed that seizure occurrence is connected with the decrease of the inhibition in brain. There are several models of seizure generation and some of them are compatible with the observed electroencephalographic activity [Lopes da Silva et al., 2003a].

Electroencephalography is the most useful and cost-effective modality for the study of epilepsy. For the purpose of the localization of epileptic foci imaging methods such as single photon emission computer tomography (SPECT), and high resolution MRI are useful. However, there are epileptic foci elusive to these imaging methods. In such cases EEG or MEG are indispensable.

The characteristics of epileptic seizure are determined based on the region of brain involved and the underlying epileptic condition. Two main classes of seizures are distinguished:

1. Partial (focal) seizures. They arise from an electric discharge of one or more localized areas of brain and then possibly they are secondarily generalized. They may or may not impair consciousness.

2. Generalized seizures. In this case electrical discharge involves the whole brain. The symptoms could be absence of consciousness (petit mal) or loss of consciousness connected with muscle contraction and stiffness (grand mal).

Digital analysis of epileptic activity serves different purposes:

- Quantification of the seizure
- Seizure detection and prediction
- Localization of an epileptic focus

The above problems will be addressed in the following paragraphs.

4.1.6.4.1 Quantification of seizures Investigations of seizure dynamics are conducted with the aim to answer some of the most fundamental and yet debatable questions in epileptology like how the propagating ictal discharges affect the ongoing electrical activity of the brain or why seizures terminate. Seizure dynamics have been investigated using many different mathematical methods, both linear and non-linear. Time course of seizure is a non-stationary process typically evolving from higher to lower frequencies; therefore the methods of time-frequency analysis are appropriate to estimate the seizure dynamics.

The most advanced method of time-frequency analysis—matching pursuit (Sect. 2.4.2.2.7) was used by [Franaszczuk and Bergey, 1999] to evaluate seizures originating from the mesial temporal lobe. Periods of seizure initiation, transitional rhythmic bursting activity, organized bursting activity, and intermittent bursting activity were identified (Figure 4.15). The authors showed in a subsequent study [Bergey and Franaszczuk, 2001] that the complexity of the EEG signal registered from the location closest to the epileptic focus increases during a seizure.

The non-linear method based on the computation of the correlation dimension D_2 was applied by [Pijn et al., 1991] for analysis of the time course of EEG recorded from different sites of the limbic cortex of rats. A comparison of the evolution of a seizure analyzed by means of correlation integral [Pijn et al., 1997] and matching pursuit[1] is shown in Figure 4.16 [Blinowska, 2002].

[1]Matching pursuit is a method in which no assumption about linearity or non-linearity of the signals is required.

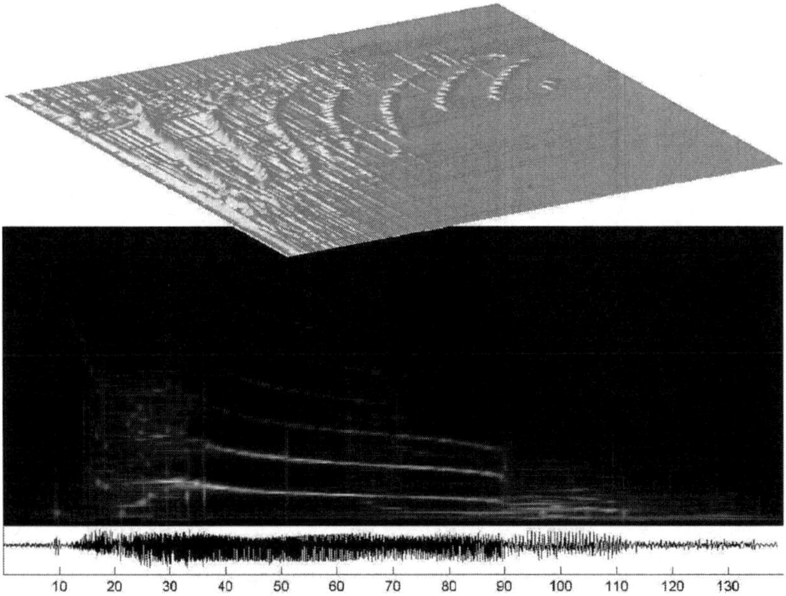

FIGURE 4.15: (SEE COLOR INSERT) From the bottom: signal, 2D time-frequency energy distribution obtained by means of MP, the same distribution in 3D. Courtesy of P. Durka, from [Durka, 2003].

At the beginning of the seizure (20–40 s) the occurrence of epileptic spikes resulted in a low value of D_2; in time-frequency energy distribution it was reflected by very broad-band structures of short duration. During the period of chaotic behavior (60–80 s) characterized by flatness of the plot we can see random distribution of time-frequency structures. Interesting is the period (150–170s) when the system tends toward limit cycle, which is accompanied by a low value of D_2. We can conclude that time-frequency distribution obtained by MP algorithm reveals the dynamics of the signal and explains the behavior of D_2 preventing its misinterpretation.

The problem of spatial synchronization in the pre-ictal and ictal periods has attracted a lot of attention. It is usually approached by assessing the correlation structure of multichannel EEG. Schindler [Schindler et al., 2007] found that the zero-lag correlations of multichannel EEG either remain approximately unchanged or, especially in the case of secondary generalization, decrease during the first half of the seizure, then gradually increase before seizure termination. However, zero-lag correlation doesn't give the full information on synchronization phenomena since there are phase differences between channels connected with the propagation of epileptic activity.

In [Schiff et al., 2005] canonical multivariate discrimination analysis based on singular value decomposition was applied to search for dynamically distinct stages of epileptic seizures. The input values were: total power, total correlation at both zero

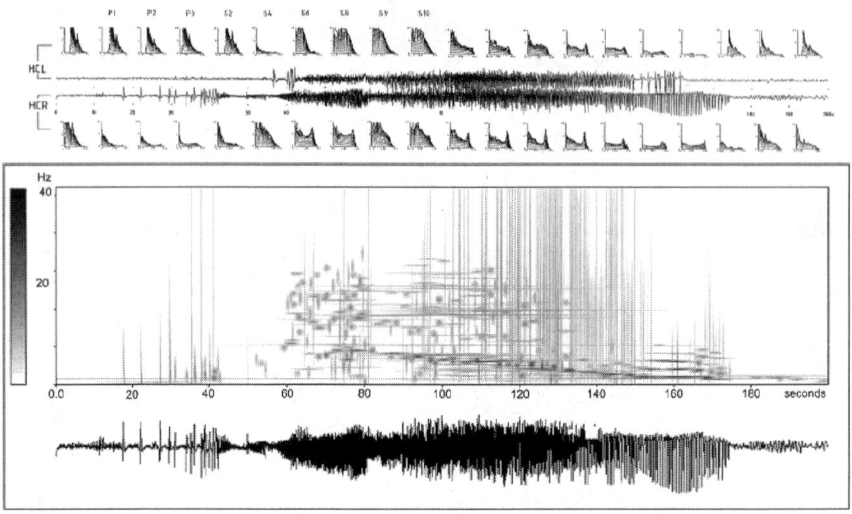

FIGURE 4.16: A comparison of the evolution of a seizure analyzed by means of correlation integral and matching pursuit. Top panel: the time sequence of plots of the correlation integral $\log C(\varepsilon,m)$ vs $log(\varepsilon)$ (Sect. 2.5.2) obtained from 10-s epochs of the signal from rat hippocampus in the peri-ictal time. Adapted from [Pijn et al., 1997]. Bottom panel: Time frequency representation obtained by means of MP from the HCR signal shown in top panel and repeated at the bottom of the figure. From [Blinowska, 2002].

and arbitrary time lag, average phase amplitude (calculated by Hilbert transform), phase angle, and amplitude dispersion. The authors distinguished rhythmic partial onset, tonic middle and clonic terminal activity. In respect to the postulated hyper-synchronous character of seizures they reported that synchronization was a prominent feature only once the seizure had passed through its initiation phase and was a variable feature of seizure termination depending on the subject. We have to distinguish here local synchronous activity of neurons which leads to the high amplitude of epileptic activity and hyper-synchronization in the larger spatial scale.

An interesting methodological approach to the assessment of synchronization of interictal and ictal EEG signals was presented by [Franaszczuk and Bergey, 1999]. They have used the multichannel autoregressive method of analysis that can be interpreted in stochastic and deterministic framework. As a measure of synchronization they have used a value connected with the goodness of fit of MVAR model:

$$SY = \log\left(\det(\hat{\mathbf{V}})\right) \tag{4.3}$$

where $\det(\hat{\mathbf{V}}))$ is a determinant of the residual matrix $\hat{\mathbf{V}}$ of MVAR (Sect. 3.2). For a purely uncorrelated Gaussian normalized white noise, $\hat{\mathbf{V}}$ is a diagonal identity matrix[2], and $SY = 1$ setting the upper bound value for SY. For a purely deterministic linear system or a dynamical system of a periodic or quasi-periodic trajectory the matrix $\hat{\mathbf{V}}$ represents a covariance matrix of measurement errors, setting the lower bound value of SY. For chaotic or stochastic colored-noise systems the value of SY will be between these bounds. The quantity SY can be interpreted as a measure of order in the system. There is a close relationship between Shannon entropy and residual variance of an autoregressive process [Serio, 1992]. In case of multichannel process, high correlation between channels increases predictability. If channels are highly correlated, one channel can be predicted using other channels, the number of variables necessary to describe dynamics of the system is lower, and the MVAR is better fitted, resulting in smaller values of SY. The changes to lower values of SY reflect higher spatiotemporal synchronization. The method was tested on a limited number of patients, however the results were very coherent. The relatively high and stationary level of SY in the interictal EEG remote from seizure onset reflected much less synchronization. The minimum of SY always occurred shortly after the onset of a seizure, reflecting high regional synchrony. The level of synchronization remained high after a seizure for prolonged periods which may explain in part the phenomena of seizure temporal clustering often observed in patients. It seems that the method has a high potential for explaining the evolution of seizures and can be helpful in seizure detection [Jouny et al., 2005].

4.1.6.4.2 Seizure detection and prediction The study of epileptic EEG dynamics can potentially provide insights into seizure onset and help in construction of systems for seizure detection and prediction. A method capable of predicting the occurrence of seizures from EEG would open new therapeutic possibilities. Namely long-term medication with anti-epileptic drugs could move toward EEG-triggered on demand therapy (e.g., excretion of fast-acting anticonvulsant substances or electrical stimulation preventing or stopping the seizure). The problem of forecasting seizures is not yet satisfactorily resolved which raises the question of whether the prediction is feasible at all.

There are two different scenarios of seizure generation [Lopes da Silva et al., 2003b]. One implies sudden and abrupt transition, in which case a seizure would not be preceded by detectable changes in EEG. The model describing this scenario assumes existence of a bi- or multistable system where the jump between coexisting attractors takes place, caused by stochastic fluctuations [Suffczynski et al., 2004]. This model is believed to be responsible for primary generalized epilepsy. Alternatively the transition to epileptic state may be a gradual process by which an increasing number of critical interactions between neurons in a focal region unfolds over time. This scenario, likely responsible for focal epilepsy, was modeled by [Wendling et al., 2002].

Since the 1970s many methods have been devised with the purpose of forecasting an epileptic seizure. They are critically reviewed in [Mormann et al., 2007]. The

[2]Signal must be normalized by subtracting the mean value and devision by standard deviation.

early idea was to use the rate of occurrence of interictal epileptic spikes as a measure indicating the time of seizure onset. There were some positive reports in this respect [Lange et al., 1983], but in the systematic study based on large experimental material it was demonstrated that spiking rates do not change markedly before seizures, but may increase after them [Gotman and Marciani, 1985, Katz et al., 1991]. Other attempts involved application of autoregressive modeling which allowed for detecting pre-ictal changes within up to 6 s before seizure onset [Rogowski et al., 1981, Salant et al., 1998].

In the 1980s, the non-linear methods based on chaos theory were introduced to epilepsy research. They involved calculation of attractor dimension, e.g., [Elger and Lehnertz, 1998], Lyapunov exponent [Iasemidis et al., 2003], correlation density [Martinerie et al., 1998]. However, more recently the optimistic results obtained by non-linear measures have been questioned. Starting from early 2000 a number of papers based on rigorous methodology and large experimental material showed that the results reported earlier were difficult to reproduce and too optimistic [Aschenbrenner-Scheibe et al., 2003, Lai et al., 2003, Harrison et al., 2005, McSharry et al., 2003]. In a study by [Wendling et al., 2009], concerning coupling of EEG signals in epileptic brain, nonlinear regression, phase synchronization, generalized synchronization were compared with linear regression. The results showed that regression methods outperformed the other tested measures. The re-evaluation of the earlier works led to the conclusion that the presence of non-linearity in the signal does not justify the use of non-linear complicated measures, which might be outperformed by simpler linear ones.

The detection of seizures is especially important in case of patients with frequent seizure episodes and for neonates. The methods applied for this purpose usually consist of two stages. In the first stage signal features such as spike occurrence, spike shape parameters, increase of amplitude or time-frequency characteristics found by means of wavelet transform were extracted. In the second step artificial neural networks (ANN) were used for identification of seizure onset. The review of the works based on this approach may be found in [Sanei and Chambers, 2009].

Saab and Gotman [Saab and Gotman, 2005] proposed to use for patient nonspecific seizure detection a wavelet transform (Daubechies 4 wavelet). From WT the characteristic measures were derived, namely: relative average amplitude (the ratio of the mean peak-to-peak amplitudes in the given epoch to the mean of amplitude of the background signal), relative scale energy (the ratio of the energy of the coefficients in the given scale to the energy of the coefficients in all scales), coefficients of the variation of the amplitude (square of the ratio of the standard deviation to the mean of the peak-to-peak amplitude). The classification was performed based on Bayes conditional formula which allows to describe behavior of the system based on how it behaved in the past. The a priori probabilities were obtained using the training data and the a posteriori probabilities served as the indicator of the seizure activity. The reported sensitivity was 76% and median value of delay in detecting seizure onset was 10 s.

The patient specific method for the detection of epileptic seizure onset was presented by Shoeb at al. [Shoeb et al., 2004]. They applied wavelet decomposition to construct a feature vector capturing the morphology and topography of the EEG epoch. By means of a support vector machine classifier the vector representative of seizure and non-seizure epochs was constructed individually for each patient. The method was validated on 36 subjects. The authors reported 94% sensitivity, and an average latency in detecting seizure onset of 8s.

The interesting method of seizure detection based on the decomposition of signal by means of MP algorithm was proposed by [Jouny et al., 2003]. The method relies on the observation that at the seizure onset the number of atoms needed to describe the signal rapidly increases. The authors introduced the measure called GAD (Gabor atom density). GAD was defined as the number of atoms obtained during the MP decomposition divided by the size of the reconstructed time-frequency space. The criterion for termination of iterative decomposition was based on the energy threshold found empirically for all studied subjects. GAD calculated for moving window together with the time-frequency plot is shown in Figure 4.17. GAD measures complexity of a signal as the number of elementary components needed to represent it. The method appeared to be very sensitive and specific for detecting intracranial ictal activity. The earliest changes during a seizure have been observed in the channels closest to the region of seizure onset.

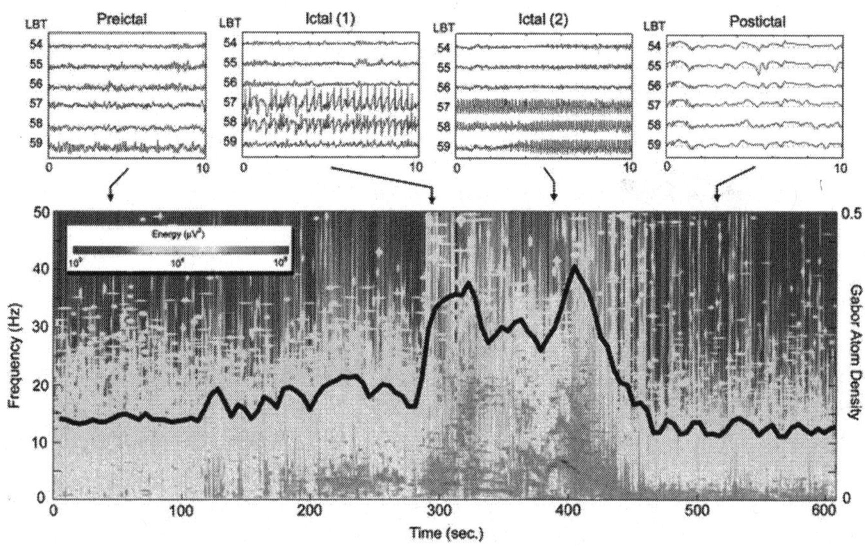

FIGURE 4.17: (SEE COLOR INSERT) GAD analysis of a complex partial seizure. Upper panels: EEG traces for 4 different epochs. Lower panel: color-coded time-frequency energy density of the signal from contact LBT 58. Superimposed over this plot is the corresponding GAD (black trace). From [Jouny et al., 2003].

New approaches in anticipating seizures involve the so-called active algorithms in contrast to passive algorithms described above. The active probe may be intermittent photic stimulation. It was found [Kalitzin et al., 2002, Parra et al., 2003] that in EEG and MEG signals, before the seizure onset, the phenomenon of phase clustering of harmonically related frequency components occurs. It doesn't mean that the stimulus provokes the seizure in any way. Based on this observation phase clustering index (PCI), which reflects the ability of neuronal systems to generate epileptic discharges, was introduced [Kalitzin et al., 2002]. PCI is zero if the phases are random, and tends toward high values if the phases are strongly clustered. The development of seizures was correlated with the value of PCI in the gamma band, therefore rPCI, PCI in gamma band relative to PCIs in the other frequency bands, was introduced. Based on rPCI the time horizon of the occurrence of seizures may be estimated [Kalitzin et al., 2005], which opens possibilities of counteracting the seizures. Using computational model it has been shown that active paradigms yield more information regarding the transition to a seizure than a passive analysis [Suffczynski et al., 2008]. The method seems to be promising, however it requires further experimental and model studies.

The evolution of the approach to seizure prediction leads to the conclusion that before addressing the question whether a developed method might be sufficient for clinical applications, it has to be tested if the performance of a new method is better than chance. To this aim methods for statistical validation are needed. They may rely on comparison with analytical results (derived from random or periodic prediction schemes) [Winterhalder et al., 2003] or on simulations by Monte Carlo method, e.g., [Andrzejak et al., 2003b, Kreuz et al., 2004, Jerger et al., 2005]. The statistical significance of results may be tested, e.g., using the concept of seizure time surrogates [Andrzejak et al., 2003a]. To this aim the seizure-onset times of the original EEG recordings are replaced by the seizure onset times generated by random shuffling of the original values. If a measure's predictive performance for the original data is higher than the one for a certain quantile of the shuffled ones, then the performance of this measure can be considered as significantly better than the random prediction. In the study of Mormann et al. [Mormann et al., 2005] comprising linear and non-linear methods the concept of surrogates introduced in [Andrzejak et al., 2003a] was used. The authors found significant predictive performance for the estimators of spatial synchronization, contrary to the univariate methods such as correlation dimension, Lyapunov exponent, and the signal energy which were unable to discriminate between pre-ictal and ictal periods. It was also found that non-linear measures didn't have higher predictive performance than linear ones.

To assure the methodological quality of future studies on seizure prediction the following guidelines were formulated [Mormann et al., 2007]:

- Prediction should be tested on unselected continuous long-term recordings
- Studies should assess both sensitivity and specificity
- Results should be tested using statistical validation methods based on Monte Carlo simulations. If prediction algorithms are optimized using training data, they should be tested on independent testing data.

These rules in fact should be mandatory for any kind of method aiming at prediction or classification based on the analysis of EEG/MEG signals.

4.1.6.4.3 Localization of an epileptic focus In case of drug resistant epilepsy surgery is a treatment option. Its goal is to remove a minimum volume of brain tissue responsible for seizure generation; therefore a seizure focus should be determined with maximal precision. To this aim presurgical evaluations are performed involving EEG, video-EEG monitoring, high resolution MRI, SPECT. In the last years magnetoencephalography has also contributed to the presurgical techniques applied for focus localization. The advantage of MEG technique over EEG is that the magnetic field is very weakly influenced by the tissues of the head. On the other hand EEG combined with video monitoring provides long-term recordings including pre-, post-, and seizure periods.

The application of to the MEG technique for presurgical evaluation of epilepsy activity was reported by [Fischer et al., 2005]. In this study authors used single equivalent dipole model for source reconstruction. The localizations obtained for epileptic structures were subjected to a hierarchical cluster analysis and ICA was used to construct an ellipsoid representing the epileptic focus. The ellipsoid volume was compared voxelwise with the resection volume generated from pre- and post-operative MRI scans. A low distance between the mass centers of both volumes and high coverage of the MEG ellipsoid by the resection volume correlated with the positive outcome of the surgical treatment.

MEG was compared with long-term EEG by [Paulini et al., 2007], who reported that in a studied group of 105 patients MEG was inferior to long-term EEG in recording interictal spikes, but yielded more circumscribed localizations if spikes were present. In another study [Wheless et al., 1999] the site of surgery was correctly predicted by MEG in 52%, in interictal scalp video-EEG in 45%, in ictal scalp video-EEG in 33%, and in the invasive EEG in 70%.

The surgical treatment is usually preceded by application of subdural and/or depth electrodes in the brain area delineated as responsible for seizures by the EEG, MEG, and imaging techniques. Intracranial EEG allows for more precise localization of epileptic focus. However some seizures are difficult to localize even with intracranial probes. It was reported by [Matsuoka and Spencer, 1993] that in patients with partial seizures originating from lateral temporal or extratemporal regions only 15% of seizures could be localized by visual analysis, because of the rapid regional spread of epileptic activity. The method which helps in this respect, providing insights into sources of EEG signal, is directed transfer function. DTF is a measure of the propagation of activity as a function of frequency. The method has been successfully applied by [Franaszczuk et al., 1994] to the localization of mesial temporal seizures foci. The authors demonstrated that the DTF method can determine whether during seizure development the initial focus continues to be the source of a given activity or whether the other more remote areas become secondary generators.

4.1.6.5 EEG in monitoring and anesthesia

4.1.6.5.1 Monitoring brain injury by quantitative EEG The monitoring technique of brain injury (which may be, e.g., a result of cardiac arrest) is based on the hypothesis that the injury causes unpredictable changes in the statistical distribution of EEG, which reduce the information content of the signal. The impairment of brain function decreases its ability to generate complex electrophysiological activity, leading to a reduction of the EEG entropy. Therefore the measure which is commonly used in monitoring brain injury is entropy (equation 2.28, Sect. 2.3.1.2).

Since the EEG of an injured brain is usually non-stationary, the time dependent entropy based on sliding window technique, was introduced, e.g., [Bezerianos et al., 2003]. In this technique the entropy is calculated within each window and its time course is followed.

The alternative measure called information quantity (IQ) is based on discrete wavelet transform DWT. In this approach the DWT is performed within each moving window and IQ is calculated from probability distribution of wavelet coefficients. In this way subband information quantity (SIQ) may be found [Tong et al., 2004].

The probability $p_n^k(m)$ is obtained from the probability distribution of the wavelet coefficients in the kth subband; m is the bin number in the probability histogram. Subband information quantity in k^{th} subband and in the epoch n may be expressed as [Thakor et al., 2009]:

$$SIQ^k(n) = -\sum_{m=1}^{M} p_n^k(m) \log_2 p_n^k(m) \tag{4.4}$$

In this way the time evolution in the subbands may be followed. It was found that SIQ depends on frequency band [Thakor et al., 2009]. Entropy or SIQ are early markers of brain injury and measure of recovery after ischemic incident.

4.1.6.5.2 Monitoring of EEG during anesthesia EEG monitoring is a routine aid to patient care in the operating room. It is used to reliably assess the anesthetic response of patients undergoing surgery and predict whether they respond to verbal commands and whether they are forming memories, which may cause unintentional unpleasant recall of intraoperative events. The algorithms working in both domains—time and frequency—were developed for evaluation of EEG during anesthesia. In this state patients may develop a pattern of EEG activity characterized by alternating periods of normal to high or very low voltage activity. The phenomenological parameter in time domain called burst suppression ratio (BSR) was introduced to quantify these changes [Rampil, 2009]. BSR is calculated as the fraction of the epoch length where EEG meets the criterion: occurrence of epoch longer than 0.5 s with the amplitude not exceeding $\pm 5\mu$V.

Spectral analysis is used in anesthesia monitoring as well. Specific parameters based on calculation of spectral density function were introduced, namely: median power frequency (MPF)—frequency which bisects the spectrum, with half the power above and the other half below—and spectral edge frequency (SEF) the highest frequency in the power spectrum. Spectral power as a function of time is used for tracking the changes of EEG during anesthesia. The methods described in Section 2.4.2 may be applied for this purpose. Another measure used in anesthesia monitoring is bicoherence (Sect. 2.3.2.1.4).

A measure based on the concept of entropy, so called spectral entropy, is applied during anesthesia and critical care. Spectral entropy *SE* for a given frequency band is given by the formula [Rampil, 2009]:

$$SE(f_1, f_2) = -\sum_{f_i=f_1}^{f_2} P_n(f_i) \log P_n(f_i) \tag{4.5}$$

where $P_n(f_i)$ is a normalized power value at frequency f_i. This measure may be normalized in the range $(0,1)$ taking into account the number of data points $n(f_1, f_2)$ in the band (f_1, f_2). Normalized spectral entropy SE_N is given by the formula:

$$SE_N(f_1, f_2) = \frac{SE(f_1, f_2)}{\log n(f_1, f_2)} \tag{4.6}$$

4.1.7 Analysis of epoched EEG signals

In the previous section we discussed methods of signal analysis that are suitable for getting insights into ongoing, spontaneous brain activity.

If we consider the brain as a system, we can learn a lot about it, observing and analyzing reactions of the system to specific perturbation of the spontaneous activity. The perturbations are usually caused by different stimuli. Studies of reaction of the brain to the stimuli require focusing the analysis on a period of time surrounding the moment of stimulation. The methods developed for this type of analysis try to quantify the changes in the EEG/MEG signal that are provoked by the external or internal stimuli. These changes can be globally referred to as event-related potentials (ERP), a subset of them related to sensory (visual, auditory, somatosensory) stimuli are commonly called evoked potentials (EP). In case of MEG signals, the respective names are: event-related fields (ERF) and evoked fields (EF). For the sake of simplicity we shall refer further in this section to the EEG signals: ERP and EP, keeping in mind that the described methods can be applied equally well to the analysis of MEG signals.

The voltage deflections observed in the ERP reflect the reception and processing of sensory information, as well as higher level processing that involves selective attention, memory updating, semantic comprehension, and other types of cognitive activity. In clinical practice ERPs are viewed as a sequence of components which are defined by their positive or negative polarity (in respect to the reference electrode potential), their latencies, their scalp distribution, and the relation to experimental variables. The clinical and research interest in ERPs relies on the fact that they are linked in time with physical or mental events [Duncan et al., 2009].

In clinical neurophysiology the ERP or EF is described as the response averaged across tenths or hundreds of repetitions. The diagnostic strength of analysis of ERP comes from the observations, accumulated over years, of correlations between the latency and amplitude of positive (P) and negative (N) deflections of the average ERP and the behavioral or clinical state of the subject. According to the published guidelines ([Picton et al., 2000], p. 141), "the simplest approach is to consider the ERP waveform as a set of waves, to pick the peaks (and troughs) of these waves, and to measure the amplitude and latency at these deflections." These peak amplitude measurements are not representing absolute values of electric brain activity, but are obtained either relative to a pre-stimulus baseline (so called baseline to peak analysis) or sometimes to an immediately preceding or following peak (peak-to-peak analysis). Example of an evoked field is shown in Figure 4.18.

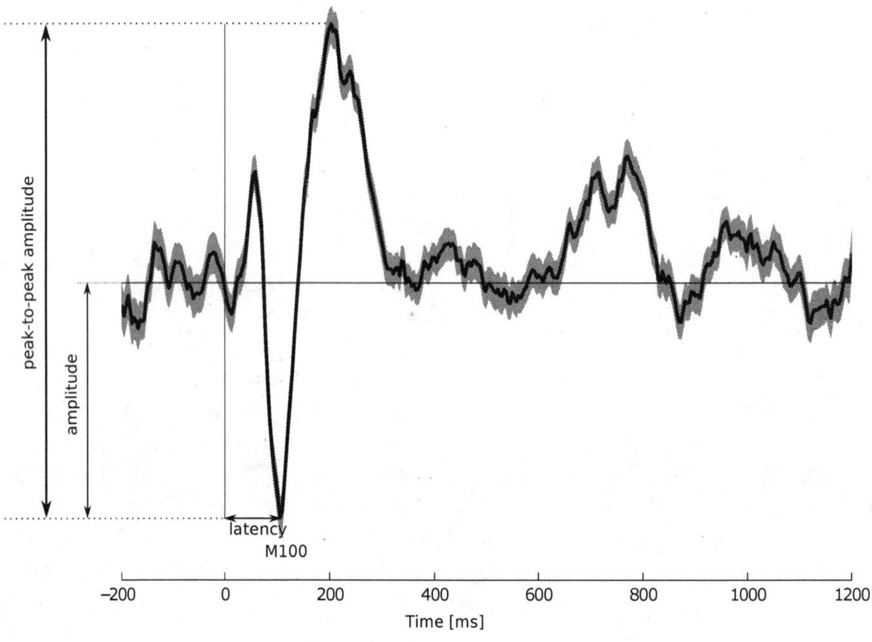

FIGURE 4.18: An example of auditory evoked field. The black line is the average of 200 trials, the gray outline is the standard deviation of the mean.

The deflections appearing in the first tenths of a second are the easiest to interpret. They reflect the first stages of sensory processing. For instance, the brainstem auditory evoked potentials, (BAEP) consists of 7 waves with latencies between 1 and 12 ms generated in well defined locations in the auditory pathways. BAEPs can be used as markers of the integrity of the auditory pathway.

The deflections appearing after the first hundred of ms are related to activity of specific brain systems, which, in some cases, have been delineated (e.g., [Giard et al., 1995, Pineda, 1995, Nieuwenhuis et al., 2005]). Some of these deflections are related to cognitive operations. As such, their presence depends on the experimental

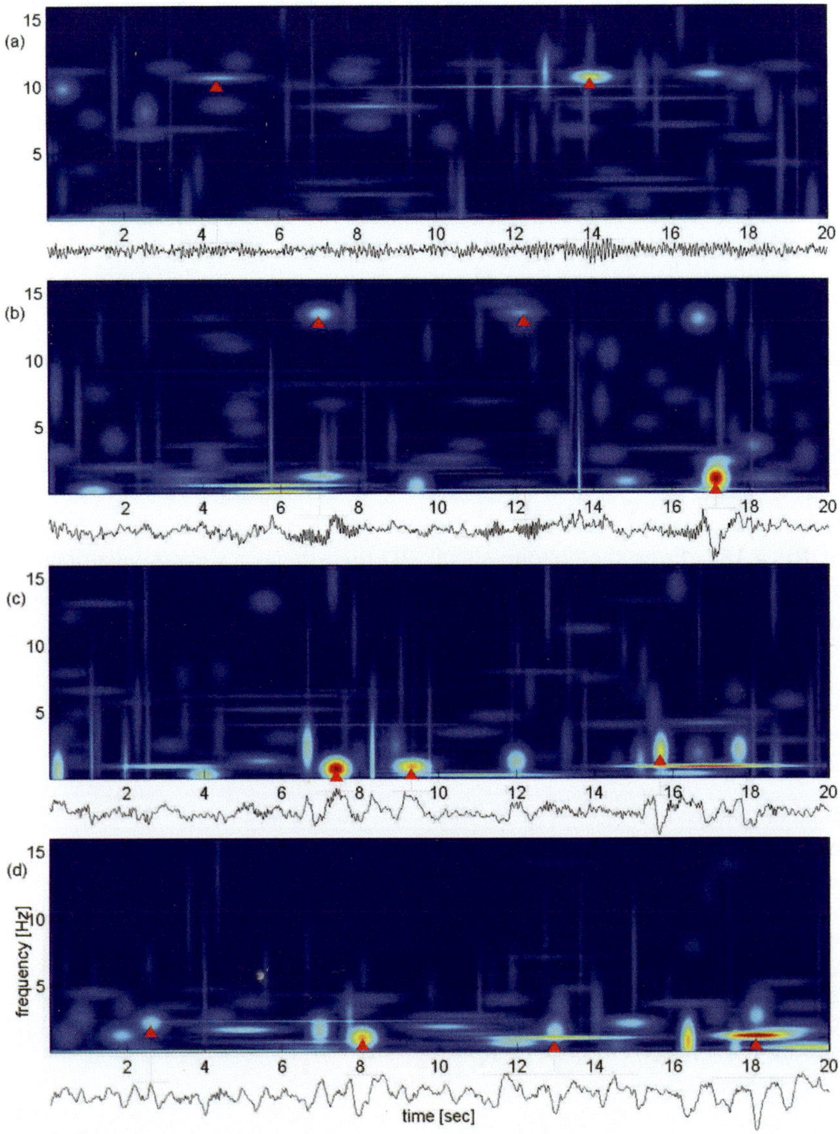

FIGURE 4.12: Time frequency maps of EEG signal energy for 20-second epochs from different sleep stages shown above the corresponding signals. a) Wakefulness, alpha waves marked by arrows; b) stage 2, two spindles and K-complex marked; c) stage 3, two slow waves and a K-complex marked; d) stage 4, slow waves marked. From [Malinowska et al., 2006].

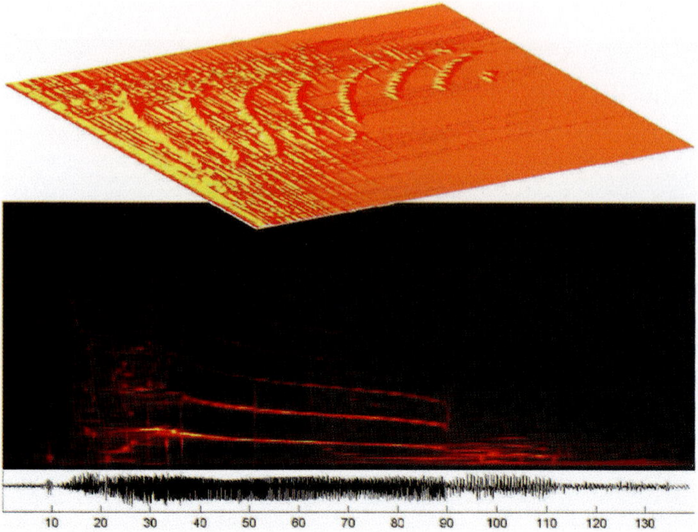

FIGURE 4.15: From the bottom: signal, 2D time-frequency energy distribution obtained by means of MP, the same distribution in 3D. Courtesy of P. Durka, from [Durka, 2003].

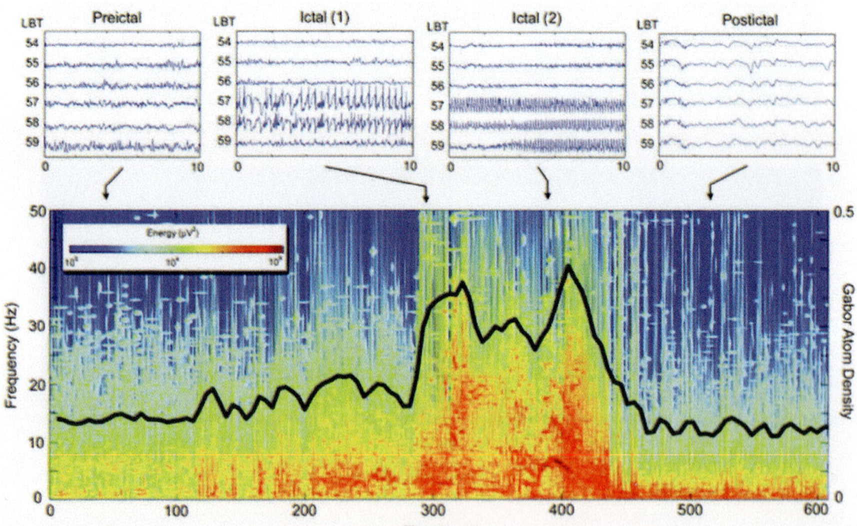

FIGURE 4.17: GAD analysis of a complex partial seizure. Upper panels: EEG traces for 4 different epochs. Lower panel: color-coded time-frequency energy density of the signal from contact LBT 58. Superimposed over this plot is the corresponding GAD (black trace). From [Jouny et al., 2003].

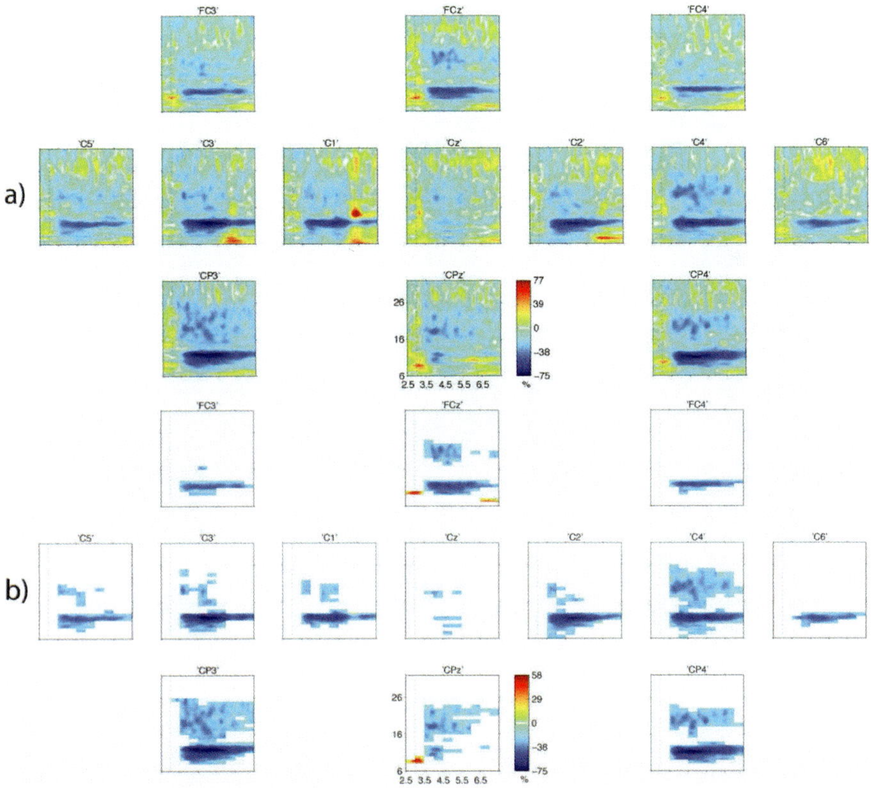

FIGURE 4.20: An example of time-frequency analysis of ERD/ERS (left hand movement imagination). The t-f maps of selected electrodes are located topographically. Horizontal axis—time [s]; vertical—frequency [Hz]. Onset of imagination marked by a dotted line. a) ERD/ERS map, b) significant effects (Sect. 4.1.7.3.3).

FIGURE 4.21: Propagation of EEG activity in the left hemisphere during right finger movement. In each panel SDTF as a function of time (horizontal axis) and frequency (vertical axis). The flow of activity is from electrode marked under the column to the electrode marked at the relevant row. Red—the highest intensity, blue—the lowest. From [Ginter et al., 2001].

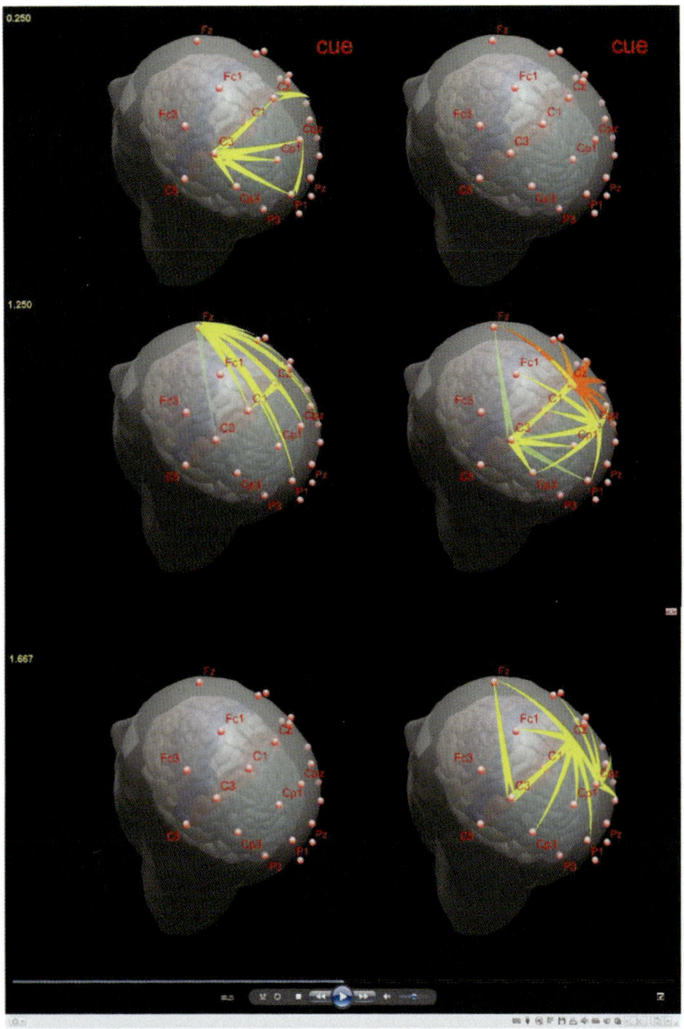

FIGURE 4.22: Snapshots from the animation representing transmission in the gamma band during right finger movement (left column) and its imagination (right column) obtained by SDTF. The intensities of flows represented as colors of arrows (red the strongest). Numbers indicate the time in seconds after the cue presentation.

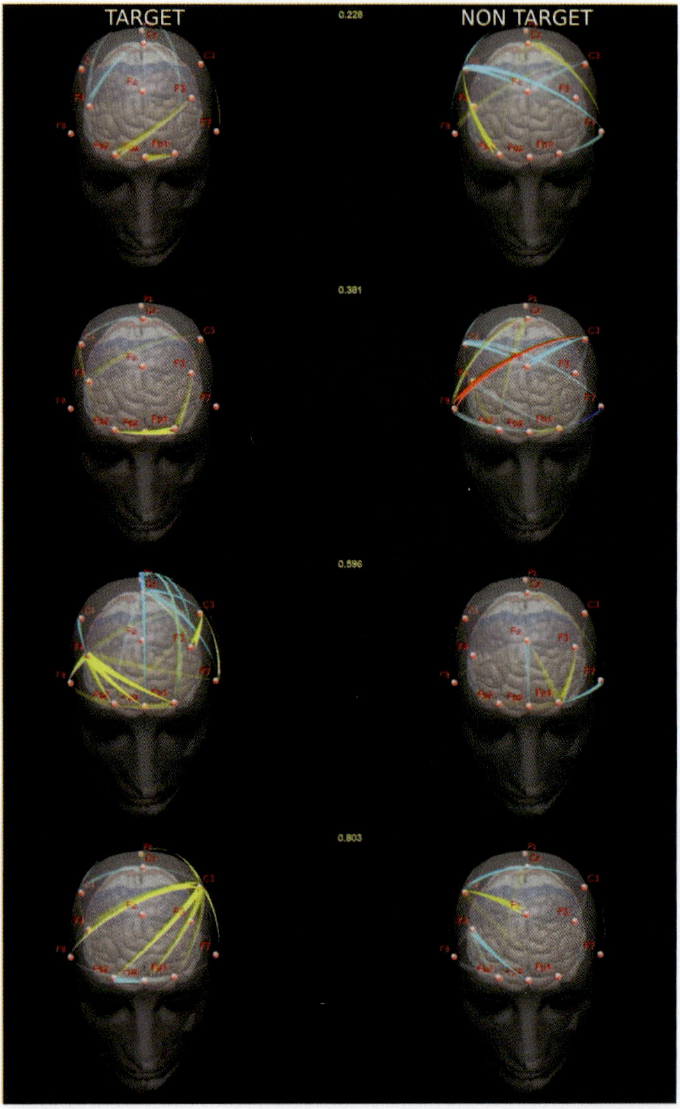

FIGURE 4.23: The snapshots from the animation representing the significant changes of propagation during the CAT test obtained by SDTF. Red and yellow arrows—increase of propagation, blue—decrease of propagation. At the right—target; at the left—non-target. In the middle—time after the stimulus in seconds. Please note in the non-target case at 0.381 s strong increase of propagation F8→ C3 connected with active inhibition of movement; and in the case of target, at 0.803 s, propagation from C3 connected with command to move the finger.

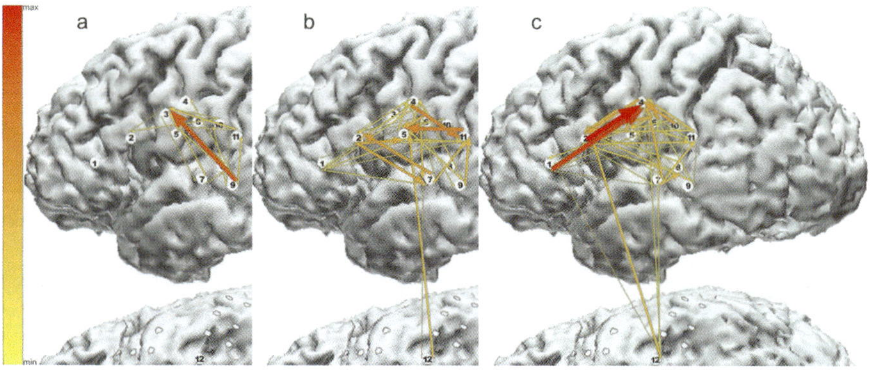

FIGURE 4.24: Integrals of ERC for frequency range 82–100 Hz calculated for three stages of an auditory word repetition task. (a) Auditory perception stage, (b) response preparation, (c) verbal response. Arrows indicate directionality of ERC. Width and color of each arrow represent the value of the ERC integral. Color scale at the left. For clarity, only integrals for event-related flow increases are shown. From [Korzeniewska et al., 2008].

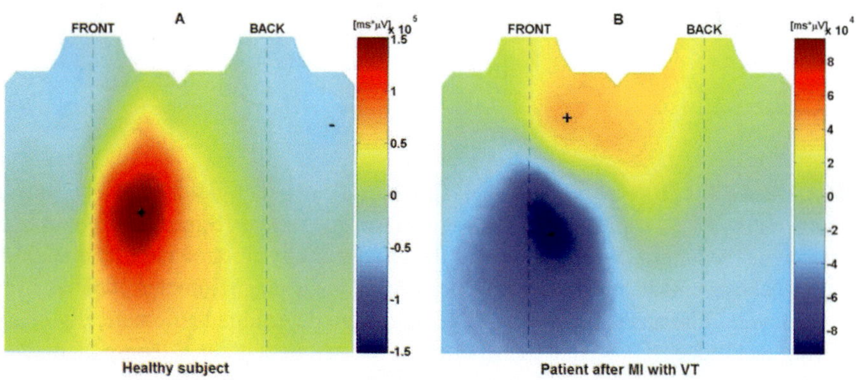

FIGURE 4.30: Body surface maps of QRST integral: A) healthy volunteer, B) patient after myocardial infarction (MI) with documented episode of ventricular tachycardia (VT). From [Fereniec et al., 2008].

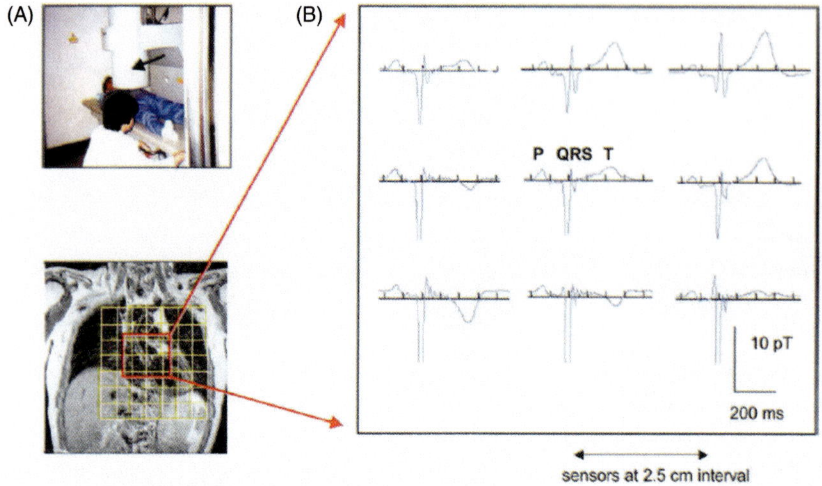

FIGURE 4.43: A 64-channel magnetocardiographic system. Overview of the system (A), of 8 by 8 matrix of sensors (sensor interval: 2.5 cm, a measuring area: 17.5 by 17.5 cm) superimposed on magnetic resonance image (B), signals at 9 out of 64 channels. From [Yamada and Yamaguchi, 2005].

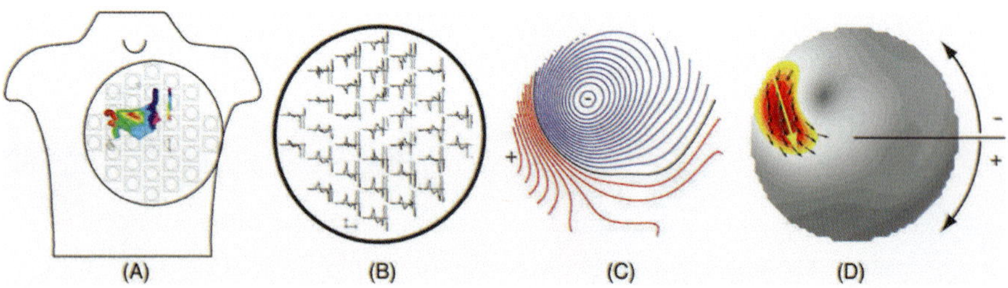

FIGURE 4.45: Recording and analysis of atrial magnetic fields. (A) The sensor arrangement of a 33 triple sensor (99-channel) magnetometer. Superimposed on sensor array is the antero-posterior view of the left atrium by electroanatomic mapping. (B) Signal—averaged magnetic field density on each magnetometer channel over cardiac cycle. The onset and end of atrial signal are determined automatically using filtering technique. (C) Spatial distribution of the magnetic field B_z component over the middle part of atrial complex interpolated from the measurement using multipole expansion. The blue color indicates flux out of the chest (-) and red color flux into the chest (+). The step between two consecutive lines is 200 fT. (D) Pseudocurrent map derived by rotating magnetic field gradients by 90°. The red-yellow color indicates the area of the top 30% of strongest currents, and the large arrow indicates their mean direction. Zero angle direction is pointing from subject's right to left and positive clockwise. From [Jurkko et al., 2009].

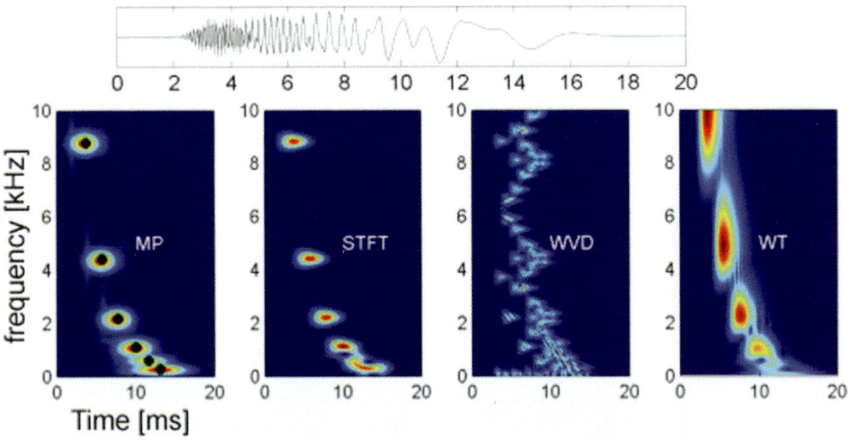

FIGURE 4.61: Time-frequency energy distributions obtained by means of different methods (named on pictures) for simulated signal (shown above) consisting of six gamma tones of frequencies 280, 550, 1100, 2200, 4400, 8800 Hz.

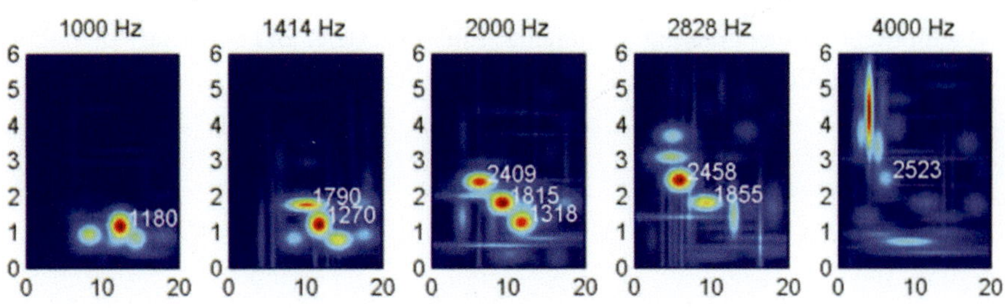

FIGURE 4.62: Time-frequency energy distribution of OAE obtained for tonal stimulation of frequencies shown above the pictures. The same resonance modes are excited by tonal stimuli of different frequencies.

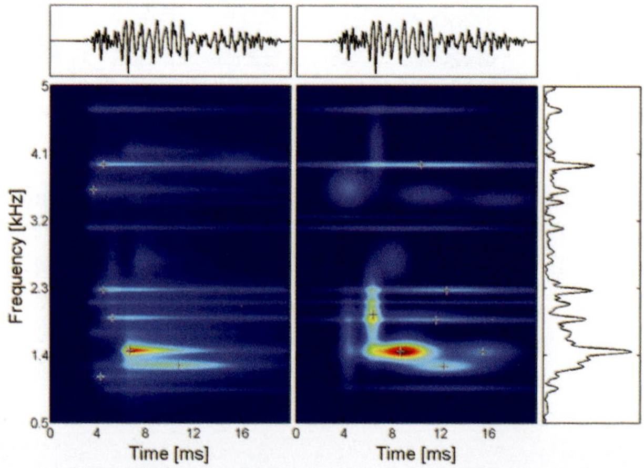

FIGURE 4.63: Time-frequency distributions of OAE signal obtained by means of MP with dictionary containing asymmetric functions (left) and dictionary of Gabor atoms only (right). At the far right, Fourier spectrum of the signal is shown.

paradigm. Deviations in amplitude and topopgraphy of components can lead to inferences about the nature and locus of brain dysfunction.

One of the most commonly considered ERP component is the P300. In fact, the "300" is a naming convention rather than the latency value since this positive deflection may appear between 300 and 600 ms after the stimuli of different modalities (e.g., visual, auditory, sensory). P300 is thought to reflect processes involved in stimulus evaluation or categorization. It is usually elicited using the oddball paradigm in which low-probability target items are inter-mixed with high-probability non-target (or "standard") items. The topography of the P300 recorded by scalp electrodes reveals a maximum over the midline centro-parietal regions. The most direct way to localize functional activation may be obtained by means of iEEG signal analysis. In particular the P300 components were extensively studied with the aim of identification of distributed networks responsible for their generation [Sinai et al., 2009]. A number of studies addressed in the above reference explored the neural substrates of a variety of ERPs elicited by functional activation of motor and sensory cortices. The literature on experimental paradigms for evoking and the interpretation of amplitudes, latencies, and topographies of late ERPs such as P300 (and its subdivision into P3a and P3b), N200, N2pc, N170, P200, N400, P600 is very rich and reviewing it is far beyond the scope of this book. However, from the signal analysis point of view the problems and techniques for dealing with the ERPs are common to all of them and are described in the following paragraphs.

The main difficulty in analysis of ERPs is the fact that, in comparison to spontaneous activity of the brain, the changes provoked by the stimuli are often very subtle; they are by an order of magnitude smaller than the ongoing EEG. The aim of the analysis is separation of the activity of brain not related to the stimuli (called the ongoing or background activity) which is treated as noise, and the activity related to the stimulus which is the signal of interest. Thus in technical terms the problem in quantification of ERPs is the poor signal to noise ratio. The choice of methods to improve the signal to noise ratio depends on the assumptions that one makes about the relation between the signal, the stimulus, and noise. In the following sections we discuss methods appropriate for analysis of responses that are phase-locked to the stimulus (Sect. 4.1.7.1) and these that are non-phase-locked (Sect. 4.1.7.3).

4.1.7.1 Analysis of phase locked responses

4.1.7.1.1 Time averaging The basic technique used in the analysis of EP is the time averaging of multiple responses time locked to the stimulus. The first approach of this type, which resulted in visualization of the EPs was reported by Dawson [Dawson, 1947]. In his solution a number of time locked responses were photographically superimposed. Today, the averaging of the digitized epochs of EEG, aligned to the stimulus is easily done by means of a computer.

The time averaging method is based on three assumptions:

- The brain response is time locked (invariably delayed) in respect to the stimulus

- The brain response to the consecutive stimuli is of the same morphology
- The ongoing activity is a stationary, zero mean noise

In commonly used analysis a simple additive model is assumed where the measured stochastic signal $x(t)$ is a sum of two terms: the deterministic EP signal $s(t)$ and the uncorrelated zero mean noise $n(t)$. According to this model the ith realization of the signal is:

$$x_i(t) = s(t) + n_i(t) \tag{4.7}$$

Averaging across N repetitions gives:

$$\bar{x}(t) = \frac{1}{N}\sum_{i=1}^{N} x_i(t) = \frac{1}{N}\left(Ns(t) + \sum_{i=1}^{N} n_i(t)\right) \tag{4.8}$$

The expected value of the averaged signal is:

$$E[\bar{x}(t)] = s(t) \tag{4.9}$$

since for zero mean noise we have $E[\frac{1}{N}\sum_{i=1}^{N} n_i(t)] = 0$. The variance of $\bar{x}(t)$ is:

$$\sigma^2_{\bar{x}(t)} = E\left[\left(\frac{1}{N}\sum_{i=1}^{N} n_i(t)\right)^2\right] = \frac{1}{N}\sigma^2_{n(t)} \tag{4.10}$$

since $s(t)$ is deterministic. Therefore the signal to noise ratio in terms of amplitude improves proportionaly to $\sqrt{N}$.

For the potentials appearing in the first tenths of a second after the sensory stimulation, such as BAEP, the simple additive model (equation 4.7) can be regarded as valid [John et al., 1978]. The validity of the model for longer latency potentials is doubtful. Nevertheless, due to its simplicity, its applications in the practice of EP analysis is overwhelming. A model closer to reality can be formulated assuming that the response is a stochastic process $s_i(t)$. Then the model (equation 4.7) becomes

$$x_i(t) = s_i(t) + n_i(t) \tag{4.11}$$

This model implies that the response may not be completely described by the mean value $\bar{x}(t)$ but also by higher statistical moments (Sect. 1.1) as, e.g., variance.

Indeed, it should be stressed that the averaging procedure is not as simple as it may appear at first glance. The simple additive model in which an EP is a sum of deterministic signal and uncorrelated background noise holds only in a few ideal cases. One can expect that there are interferences between the neural activity evoked by the stimulus and the ongoing activity resulting in the reorganization of the latter [Sayers et al., 1974]. In spite of this fact, one may still accept the simple additive EP model, but he must be aware of the violation of the assumptions. It should be realized that the signal in reality is not deterministic, but stochastic. Thus it is recommended to evaluate the variance of the EP along with its mean value. The assumption of "noise" as an uncorrelated signal is also a great simplification and it is necessary to take into account the statistical characteristics of the ongoing activity as will be shown in the next paragraph. In any case, it is always suggested to include some pre-stimulus time in the EEG epochs to be averaged to obtain some estimate of the mean and variability of the ongoing activity.

4.1.7.1.2 Influence of noise correlation In quantification of EP, the main issue is to increase the signal-to-noise ratio so as to minimize the contamination of EP by the background EEG activity. Equation 4.10 shows that the ratio of signal variance to the noise variance for N repetitions of the stimulus is $\frac{1}{N}$. This result holds for the noise samples that are uncorrelated. This however, often is not true, e.g., in the presence of strong rhythmic background activity like for example alpha rhythm. In such a case subsequent samples of $n(t)$ are not independent. The degree of dependence of noise samples is given by its autocorrelation function, which in this case has the following form [Niedermayer and Lopes da Silva, 2004, chap. 50]:

$$R_{xx}(\tau) = \sigma^2 \exp(-\beta|\tau|)\cos(2\pi f_0 \tau) \tag{4.12}$$

where σ^2 is the variance of noise, f_0 is the mean frequency of rhythmic activity, β/π is the corresponding bandwidth, and τ is the delay parameter. In this case the EP variance takes another form [Spekreijse et al., 1976]:

$$\sigma^2_{\bar{x}(t)} = \frac{\sigma^2}{N}\left[\frac{1-\exp(-2\beta T)}{1-2\exp(-\beta T)\cos(2\pi f_0 T)+\exp(-2\beta T)}\right] \tag{4.13}$$

where T is the inter-stimulus period. It is important to note that rhythmicity in the background activity will influence the signal-to-noise ratio. It can be seen from the above equation that the variance ratio tends to $\frac{\sigma^2}{N}$ as βT becomes large. A related problem is whether the stimulus should be periodic or aperiodic. Aperiodic stimulation can lead to reduced EP variance [Ten Hoopen, 1975]. Periodic stimulation can result in generation of a rhythmic activity with frequency corresponding to the repetition rate of the stimulus.

4.1.7.1.3 Variations in latency One of the forms of randomness in single trial response is the variation in latency. The latency jitter may lead to a distortion of the averaged EP. The distortion may result in an underestimation of the averaged EP peak amplitude and its greater spread over time. In more severe cases it may render the EP undetectable. A possible consequence of this fact is that a significant difference in amplitude of an average EP recorded in two experimental conditions can come from a difference in latency jitter between conditions. Moreover, if the distribution of latency jitter is skewed, the difference in latency jitter between conditions will also lead to a difference in the latency of the deflections of the average ERP. The extent to which latency jitter affects ERPs averaged in the time domain varies as a function of the duration of the ERP deflection. Indeed, for the same latency variance, a long-lasting deflection will be relatively more repeatable across trials than a short-lasting deflection and, consequently, less distorted by time-domain averaging. For longer latency of brain response, latency jitter is greater, so short-lasting components with long latency can hardly by observed [Mouraux and Iannetti, 2008].

We can assume that in equation 4.11 the variability of the signal is due only to the variable latency and for each trial i the shape of the response $s(t)$ is invariant. It implies that there exists a time translation Δt_i such that:

$$s_i(t) = s(t + \Delta t_i). \tag{4.14}$$

The first approach to correction for the latency variation was proposed by Woody [Woody, 1967]. In his method the Δt_i was estimated by means of the cross-correlation function of the response i with a template (i.e., the initial estimate for the average EP). The required translation was evaluated as the lag of maximum of cross-correlation. The individual trials are then readjusted by Δt_i and averaged, giving the new estimate of average EP. This estimate was used as a template in the next step of the process. This iterative procedure leads to a better estimate of the average EP. Woody showed also that the magnitude of the cross-correlation function can be used to sort the single trial EP to different sub-classes. The refinement of Woody's method relied on taking into account the statistical properties of the ongoing EEG activity [McGillem and Aunon, 1977].

A method of latency correction, involving additionally a classification step, was proposed by [Pfurtscheller and Cooper, 1975]. According to this technique single trial EPs are cross-correlated with different templates. The maxima of cross-correlation functions are used to estimate the amplitude and latency of particular components. Next, these trials that fall within the certain amplitude or latency limits are averaged together. In this way, different classes of selected averages are obtained that may correspond to different subsystems or physiological conditions. Averaging of all trials as one class may lead to a considerable information loss.

4.1.7.1.4 Habituation The observed trial-to-trial variability in amplitude and latency of the components of an evoked potential may arise from two major sources. One is systematic and depends on the changes in reaction to the repetitive stimulation, and the other is stochastic. Systematic changes in response to repeated stimulation are a general property of a nervous system. New stimuli first elicit an arousing response, named the orientation response [Sokolov, 1960]; it consists of changes in a large number of autonomic variables, such as the skin conductance reaction, phasic heart rate changes, and it also involves a reaction of the cerebral cortex. Habituation is, in a wide sense, defined as decrease of response as a function of stimulus repetition and it is a constant finding in almost any behavioral response. The simplest mathematical model of habituation is an exponential decay in case of regularly repeated stimulus [Sokolov, 1960]. The rate of decay depends on physical properties of the stimulus, its relevance, the interstimulus interval, and the properties of the neural systems involved in the reaction. Habituation is not the only possible reaction to a sequence of stimuli. The increases in the responses during the first stimuli related to a sensitization process have been described in the literature [Thompson and Spencer,

1966, Groves and Thompson, 1970]. Moreover, these authors proposed that response to repeated stimulation is the interaction of sensitization and habituation.

In order to study these systematic changes over time, there are in general two possible approaches. One is the adjustment of the experimental paradigm, and this is described further in this paragraph. Second is the development of methods able to extract single-trial responses; these are described in the following paragraph.

In the classical works on habituation/sensitization we can find two possible paradigms: sub-ensemble averages and block averages. In case of sub-ensemble averages recording sessions were subdivided into consecutive sub sets consisting of a few consecutive trials. In each of the sub sets the ensemble average is computed. The time evolution of these sub-ensemble averages is used to assess the habituation/sensitization process. This approach is successful only if changes between trials are much slower than the time span of the number of trials included in each sub-average. Another approach to study the trial-to-trial changes is to repeat a sequence of trials as a block, each sequence starting after an assumed recovery time. Then, corresponding responses from each block are averaged. This second method, assumes that consecutive blocks are equivalent and the habitation/sensitization processes are exactly the same in each block. Moreover, both sub-ensemble and block-averaging require long recording sessions, and thus general arousal changes are likely to occur during the experiment [Quian Quiroga and van Luijtelaar, 2002].

4.1.7.2 In pursuit of single trial evoked responses

The averaging techniques do not allow for direct access to variations of ERP latencies and amplitudes. In particular, short lasting changes, which may provide relevant insights into cognitive operations, are blurred or even masked by averaging processes, whereas single trial analysis would allow a reliable extraction of signal characteristics from a single ERP sequence. From this point of view, single-trial analysis of ERP is the ultimate goal of almost all related research, crucial for understanding subtle changes in individual brain responses to a repeated stimulus.

Several techniques have been proposed in order to improve access to individual ERPs. In each case specific assumptions must be made about the nature of the response.

4.1.7.2.1 Wiener filters Many approaches involve filtering of the single-trial traces, in particular using techniques based on the Wiener formalism (Sect. 2.1.4), which provides an optimal filtering in the mean square error sense [Woody, 1967, Walter, 1969, Doyle, 1975]. However, these approaches have the common drawback of considering the signal as a stationary process and, since the ERPs are compositions of transient responses with different time and frequency signatures, they are not likely to give optimal results.

4.1.7.2.2 Model based approach A model based approach was proposed in [Cerutti et al., 1987]. The background EEG activity was modeled as an AR process and event-related activity as autoregressive moving average process ARMA. In this way it was possible to find a filter extracting single ERP. However, the method relied on average ERP in the identification of ARMA model, and as such was not free from the assumptions pointed out in Sect. 4.1.7.1.1. A similar approach was presented in [von Spreckelsen and Bromm, 1988]. The method relies on the assumption that the measured activity can be separated into its evoked and spontaneous parts. A compound state-space model trying to incorporate the observable properties of both parts is formulated on the basis of additivity of the two components. Within this model, spontaneous activity is described as an autoregressive process, while the EP is modeled by an impulse response of a parametrically described system. Based on the state-space representation of the model, a Kalman filter for the observation of the system's state can be designed which yields estimates for both activities.

4.1.7.2.3 Time-frequency parametric methods The discrete wavelet decomposition (Sect. 2.4.2.2.3) is particularly suitable to parameterize the transients which are time locked to the stimuli. The procedures aiming at extraction of single-trial ERP based on discrete wavelet decomposition were proposed by [Bartnik et al., 1992]. In their approach the individual trials were first split into two parts: one which was assumed to contain components related to the stimulus (ERP epoch) and the second, which was assumed to contain only the spontaneous EEG activity (reference epoch). Both datasets were decomposed by dyadic wavelet transform. The coefficients of the basis functions that comprise the detail functions were then analyzed by regression and discriminate analysis to identify the coefficient best suited for distinguishing ERP from spontaneous EEG. Finally, the coefficients that most significantly differentiated the ERP epochs from the reference epochs were selected and used for reconstruction of the underlying single-trial ERPs. These reconstructed ERPs had significantly improved signal to noise ratio. The authors verified the method on a set of 40 auditory potentials. The correct classification to the group of ERP epochs was obtained for 34 cases. It was demonstrated that by means of DWT and discriminant analysis it is possible to construct a method capable of extracting components strongly related to the ERP with minimal assumptions. Nothing has to be assumed about the particular form of the components. A similar method based on wavelet denoising, although less advanced in the selection of relevant wavelet coefficients, was successfully applied to the study of the habituation effect of rat auditory evoked potentials [Quian Quiroga and van Luijtelaar, 2002].

A further development of the idea of parametric decomposition application for extracting single-trial ERPs, was proposed in the framework of multivariate matching pursuit (Sect. 2.4.2.2.7 and 3.6.3). The flexible parametric approach offered by MMP allows for explicit implementation of assumptions made about the ERPs. For instance, one may assume that the ERP morphology is constant and that only the amplitude varies across trials [Sielużycki et al., 2009a]:

$$x_i(t) = a_i s(t) + n_i(t) \tag{4.15}$$

where a_i is the amplitude of ERP in trial i. This assumption is equivalent to imposing constraints on the parameters $\gamma = \{u, f, \sigma, \phi\}$ (eq. 2.114) of the Gabor functions such that all parameters, except the amplitude weight $a_i = \langle R^i x, g_{\gamma_i} \rangle$, are kept identical across all trials, yet not set a priori to any particular values. This approach guarantees that only these features, which consistently occurred across all trials, emerge from the multi-trial decomposition. Relaxing the constraint of a constant phase allows for accounting for the jitter and some limited shape variability of s_i:

$$x_i(t) = a_i s_i(t) + n_i(t) \tag{4.16}$$

Both approaches were successfully validated in the task of analysis of habituation influence on amplitude and in the latter case also on the latency of auditory M100 evoked magnetic field. At first glance, the relaxation of the phase constraint may be seen as a favorable reduction of the bias of the estimation; however, the algorithm becomes more prone to individual noise patterns in each trial [Sielużycki et al., 2009b]. Therefore, neither of the two approaches is superior to the other one in an absolute sense. They should be used according to the particular conditions of the paradigm, taking into account the inherent assumptions and limitations of the two methods.

Recently, [Bénar et al., 2009] introduced a so-called consensus matching pursuit algorithm, with which the authors tried to explain the trial-to-trial variability in the parameter space by means of a certain restricted variation of the Gabor functions approximating each trial independently within a certain neighborhood of the so-called consensus atom. This consensus atom is selected by means of a specific voting procedure, and is the most representative atom for all trials.This approach constitutes a further step in releasing constraints on the parameters of the Gabor functions, since it allows amplitude, scale, frequency, and phase variations, at least to a certain degree, which is controlled by a Gaussian kernel.

4.1.7.2.4 ERP topography So far we discussed the evaluation of the time course of the ERP. Much valuable information is obtained by considering ERP topography since the distribution of the current sources in the brain is reflected in the spatial pattern of evoked potential, which can be visualized in the form of an ERP map.

The first problem one has to face when analyzing multichannel EEG data, is the problem of reference. EEG can only be measured as a difference of potentials, and as such the measured waveform differs when the reference electrode changes.This may influence the interpretation of the results [Murray et al., 2008].

One popular solution to this problem is the use of global field power (GFP) introduced by Lehmann and Skrandies [Lehmann and Skrandies, 1980]. GFP is computed at any moment in time as the standard deviation of voltage across the set of channels. The GFP tells how steep on average are gradients of potentials across the electrode montage. In this sense the GFP is a measure of the strength of activation at a given time point. It can be evaluated with the standard ERP parameters like: latencies, area, peak amplitude, etc.[3] The GFP extrema can be used in order to identify brain activity components in a reference invariant way.

[3]It is important to note that because GFP is a non-linear transformation, the GFP of the group-average ERP is not equivalent to the mean GFP of the single-subject ERPs.

We should note that the spatial distribution of potentials on the head surface is reference invariant[4], since if the reference potential shifts up or down the whole "potential landscape" shifts by the same amount. It is important to keep in mind that the absolute locations of the potential maxima or minima on the scalp do not necessarily reflect the location of the underlying generators. This is due to the propagation of the intracranial signals by volume conduction. One has to realize this fact in order to avoid confusion in the interpretation of EEG data.

One possibility to obtain reference-free waveforms, which exploits the reference invariance of topography, is the computation of the so-called current source density (CSD) waveforms. The CSD or Laplacian derivation is obtained by estimation of the 2nd spatial derivative across the electrode montage (Sect. 4.1.3). In this way, CSD derivations are intrinsically based on spatial gradients in the electric field at the scalp. This procedure eliminates to some extent the problem of reference-dependance inherent to voltage waveforms and it also minimizes contributions of volume conduction within the plane of the scalp, thus effectively working as spatial "sharpening" (high-pass) filters. However this kind of reference is not proper for methods relying on estimation of correlation matrix between the channels of the process, i.e., for the MVAR model. The Laplacian operator disturbs the correlation structure of the multivariate set of signals, and hence the phase dependencies.

It is important to note that the changes in the topography of the electric field at the scalp can only be caused by changes in the configuration of the underlying intracranial sources (excluding artifacts such as eye movements, muscle activity, etc.).[5] Mapping of electrical brain activity in itself does not constitute an analysis of the recorded data but it is a prerequisite to extracting quantitative features of the scalp recorded electrical activity. In the second step, in order to find the differences between experimental conditions or between groups of subjects, the derived topographical measures must be subjected to statistical tests.

Due to volume conduction, potentials measured on the scalp are spatially blurred. The linear superposition of electric fields allows to assume that each electrode records a signal which is a linear mixture of all the underlying sources of electrical activity (not only from the brain, but also from muscle, cardiac, eye movement etc.). Successful separation of the signals from different sources gives possibilities of more correct functional interpretation of the different components of recorded brain responses. In general there are three approaches to the problem of separating the sources. Two of them operate in the sensor space and will be discussed in this paragraph, the third approach is the solution of the inverse problem. Methods of solutions to the inverse problem are beyond the scope of signal processing, therefore they will not be considered here. The methods operating in the sensor space do not require an explicit head model. They include: principal component analysis (PCA, see Sect. 3.6.1) and blind source separation (BSS), most often implemented as in-

[4]Please note that this does not hold for maps of power.

[5]The opposite need not be the case, since there is an infinite number of current source configurations inside the volume conductor that give the same distribution of potentials on the surface.

dependent component analysis (ICA, see Sect. 3.6.2). Linear data decompositions, such as PCA or ICA, separate multichannel data into a sum of activities of components each comprised of a time course of its activity in every trial and a single scalp map giving the strength of the volume conducted component activity at each scalp electrode. Note that polarity seen in the ICA and PCA maps of topographical component distribution has no real meaning. The actual potential distribution is the product of the map and the value of the component in a given moment in time.

The goals of PCA and ICA are different. Let us consider for a k channel signal that the k values of potential measured at a certain moment in time are coordinates of a point in the k-dimensional space. The PCA extracts temporally orthogonal directions (axes) in that space.[6] The components are the projections of the original data onto these axes, with the first principal component representing the maximum variation in the data; the second principal component is orthogonal to the first and represents the maximum variation of the residual data. This process of component extraction is repeated several times, and since the original variables are correlated only a small number of principal components accounts for a large proportion of the variance in the original data [Skrandies, 1989, Skrandies, 2005].

The goal of ICA, on the other hand, is to find components in the data whose activities are statistically as distinct from one another as possible, meaning that their signals have the least possible mutual information. Minimizing mutual information implies not only decorrelating the component signals, but also eliminating or reducing their higher-order joint statistics. This stronger statistical independence allows ICA to relax the orthogonality of the component scalp maps, which is physiologically more plausible, and to separate phase-uncoupled activities generated in spatially fixed cortical domains (or non-brain artifact sources).

If the scalp maps associated with different activities are not orthogonal, PCA will mix portions of them into one or more principal components, rather than separating them into different components, as is the case in ICA. Thus, if the recorded data are in fact the sum of (nearly) independent signals from spatially confined and distinguishable sources, PCA will lump, and ICA will split the source activities across resultant signal components.

Data reduction by PCA may be efficient for compression, but the higher flexibility of ICA decomposition, with spatially overlapping scalp maps allowed, may result in components which are easier to interpret in the physiological context. ICA can be applied to EEG epochs averaged in the time domain [Makeig et al., 1997] or to non-averaged concatenated EEG epochs [Makeig et al., 2004]. In both cited studies ICA allowed separating late cognitive ERPs into distinct, independent constituents.

Applying ICA one has to be aware that an obtained independent component (IC) does not necessarily represent an anatomically defined source of activity. Indeed, an IC is defined by its temporal independence relative to the other sources of activity. If the activities within two distinct regions of the brain strongly covary, they will be represented within a single component. In case of unconstrained ICA, the total num-

[6] Corresponding to necessarily orthogonal scalp maps.

ber of estimated ICs equals the total number of recording electrodes. If that number is considerably greater than the true number of independent sources contributing to the recorded signals, ICs containing spurious activity will appear because of overfitting. Contrarily, if the total number of ICs is considerably smaller than the true number of underlying sources, valuable information will be lost because of underfitting. This important limitation (that the number of independent components is arbitrarily defined) could explain why, until now, ICA has not been able to clearly and unequivocally disentangle event-related electrical brain responses into physiologically meaningful independent constituents [Mouraux and Iannetti, 2008]. Another problem that has to be considered when applying ICA is the amount of available data needed to correctly estimate the unmixing matrix (see Sect. 3.6.2).

4.1.7.3 Analysis of non-phase locked responses

A given sensory or internally generated stimulus or a motor response can provoke not only the time-locked event-related response, but can also induce changes in the ongoing EEG/MEG activities. These effects can be quantified by observing the relative changes: (i) in the power of specific frequency bands—known as event-related synchronization ERS or desynchronization ERD [Pfurtscheller and Lopes da Silva, 1999], (ii) the changes in the phase distribution—known as phase resetting [Sayers et al., 1974], (iii) phase synchrony [Engel and Singer, 2001], and (iv) phase clustering [Kalitzin et al., 2002].

4.1.7.3.1 Event-related synchronization and desynchronization To introduce the ERD and ERS measures we have to realize that in the spontaneous ongoing brain activity (recorded as EEG) rhythmic components are present (see Sect. 4.1.2). The measure of ERD is defined as the relative decrease and ERS as the relative increase of power in a given frequency band relative to some baseline level. Thus mathematically both effects can be described by the same formula:

$$ERD/ERS = \frac{S_f - R_f}{R_f} \tag{4.17}$$

with ERD relating to negative values and ERS to the positive values; S_f is the band power in the analysis time; and R_f is the band power in the baseline time. The baseline time should be selected in such a way that the signal can be considered as not affected by the processes under investigation. The baseline epoch should contain only the spontaneous activity.

The physiological interpretation of ERD/ERS is explicitly expressed in their names. The EEG signal can be observed mainly due to synchronous activity of neurons under the electrode which corresponds to the area of cortex of the order of a few squared centimeters. The amplitude of the rhythmic oscillation is proportional to the number of synchronously active neurons (Sect. 4.1.1). In this sense the ERD and ERS observed in EEG measure the level of synchronization of huge ensembles of cortical neurons. However, one should be aware of the spatial scale of the physiological processes reflected in the ERD/ERS. Especially in the case of ERD one can imagine two scenarios leading to decrease of power measured on the EEG level. First, the simplest, is that in the whole patch of cortex the oscillations in a given frequency band disappeared; perhaps the frequency of oscillation had shifted to some other frequency band. The second scenario is that, during the baseline time, an ensemble of neural populations oscillates synchronously in a given rhythm. Then during the information processing or mental activity the ensemble splits into many smaller ensembles, each oscillating with its own frequency and phase. In the superposition of their activity we would observe a decrease of power of the rhythm in respect to that which was present during baseline time.

In the following paragraphs we shall review methods developed for quantification of ERD/ERS. In order to determine the statistically significant ERD/ERS values one needs an appropriate number of repetitions of the experiment.

4.1.7.3.2 Classical frequency band methods The standard method for evaluation of ERD/ERS was proposed by Pfurtscheller and Aranibar [Pfurtscheller and Aranibar, 1979]. It can be described by the following algorithm:

- Let $x_f(t;i)$ be the band-pass filtered signal in trial $i \in \{1,\ldots,N\}$, and the baseline time $t_b \in \{t_1,\ldots,t_k\}$
- Square the filtered signal samples to obtain instantaneous band-power

$$S_f(t;i) = x_f^2(t;i) \tag{4.18}$$

sometimes the mean of the data across trials is subtracted before squaring to remove the ERP component, giving so called inter-trial variance:

$$S_f(t;i) = (x_f(t;i) - \bar{x}(t,f))^2 \tag{4.19}$$

- Compute averaged band-power by averaging instantaneous band-power across trials:

$$\bar{S}_f(t) = \frac{1}{N}\sum_{i=1}^{N} S_f(t;i) \tag{4.20}$$

- Average the power in the baseline time

$$R_f = \frac{1}{N}\sum_{t\in t_b} \bar{S}_f(t) \tag{4.21}$$

- Compute the relative power change

$$ERD/ERS(t) = \frac{\bar{S}_f(t) - R_f}{R_f} \cdot 100\% \tag{4.22}$$

- Smooth the relative power change by moving average or low-pass filtering or complex demodulation using the Hilbert transform.

Illustration of the above algorithm is shown in Figure 4.19

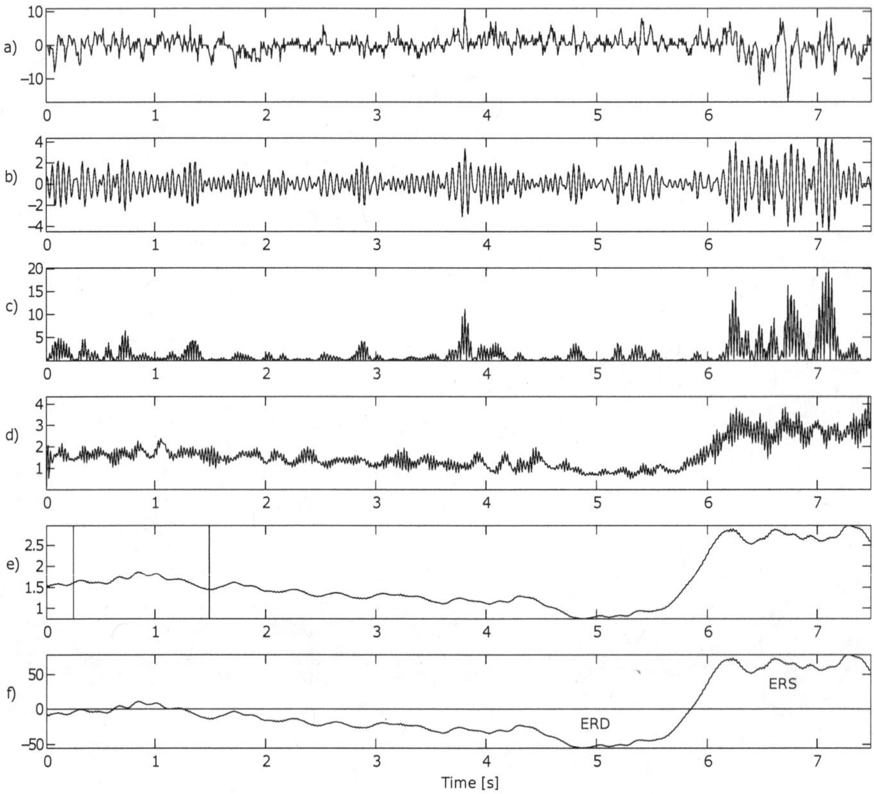

FIGURE 4.19: Illustration of the classical ERD/ERS estimation for the beta band. a) One of the N EEG epochs; b) the above signal filtered in the 15– 25 Hz band; c) instantaneous power of the filtered signal; d) instantaneous power averaged over N epochs; e) smoothed version of the signal in (d), the reference time marked by two vertical lines; f) ERD/ERS time course—average instantaneous power relative to the mean value in the reference time.

4.1.7.3.3 Time-frequency methods The frequency bands at which the ERD/ERS effects are observed and ERD/ERS unique temporal and topographic patterns related to the functional brain activation vary considerably between subjects. In the classical approach one has to try out a number of possible frequency bands in the search for the ones that display the most pronounced effects—so called reactive frequency bands. As was pointed out by [Pfurtscheller, 1999] it is important to examine event-related changes in signal energy from the broader perspective of the entire time-frequency plane.

In order to accomplish this, one needs to estimate the energy density of the signal in the time-frequency space. In the literature many methods were proposed for this purpose, e.g.:

- Spectrogram [Makeig, 1993],
- Bandpass filtering in overlapping bands [Graimann et al., 2002],
- Scalogram [Tallon-Baudry et al., 1996],
- Smoothed pseudo Wigner-Ville transform [Lachaux et al., 2000]),
- Estimate of energy density derived from the matching pursuit (MP) parameterization [Durka et al., 2001b],

The mathematical basis, properties, and MATLAB routines to compute the above mentioned estimators of the energy are described in Sect. 2.4.2. Different estimates of the time-frequency distribution of signal energy offer different trade-offs between temporal and spectral resolution. For example, the scalogram has high temporal resolution and low frequency resolution at high frequencies, low temporal resolution and high frequency resolution at low frequencies. The presence of cross-terms in some of the time-frequency estimators must be taken into account when interpreting the ERD/ERS in the time-frequency space. In contrast, the matching pursuit (MP) parameterization theoretically provides optimal time-frequency resolution throughout the time-frequency plane. Different estimators of energy density were compared in [Zygierewicz et al., 2005]. The results yielded by different estimators gave compatible results, however those obtained by MP procedure provided more detailed information of the time-frequency structure of the ERD/ERS.

The time frequency estimator of energy density $E(t,f)$ can be used to evaluate the ERD/ERS in the time-frequency plane in accordance with the general definition (4.17):

$$ERD/ERS(t,f) = \frac{\langle E(t,f)\rangle_{tr} - B(f)}{B(f)} \tag{4.23}$$

where $\langle E(t,f)\rangle_{tr}$ is the energy density at (t,f) averaged across trials, and $B(f)$ is the mean energy of baseline at frequency f averaged across trials.

The important question to be answered is: which of the ERD/ERS effects are significantly above the fluctuation level? The statistical problem related to that question is the assessment of significance in the statistical maps (see Sect. 1.5.3.2). In [Durka et al., 2004, Zygierewicz et al., 2005] a massive univariate approach with the multiple comparisons problem controlled with FDR (Sect.1.5.3.3) was proposed as an efficient solution to that question, and will be described below.

For the null hypothesis of no changes at the given time-frequency coordinates, we have to reduce the resolution to time-frequency boxes ($\Delta t \times \Delta f$). There are two reasons for decreasing the time-frequency resolution:

- The time and frequency resolution are bounded by the uncertainty principle, which for the frequencies defined as the inverse of the period (Hz) gives: $\Delta t \times \Delta f \geq \frac{1}{4\pi}$. The lower bound is only reached by the Gabor functions
- In real data there are big variations of energy estimates due to different kinds of noise. Increasing the product $\Delta t \times \Delta f$ reduces the variance of energy estimate

Results from [Durka et al., 2004], suggest that $\Delta t \times \Delta f = 1/2$ gives robust results. This introduces a discretization of the time-frequency plane into resels (resolution elements) $r(i,j)$. In order to obtain $E_n(i,j)$ — energy in resels $r_n(i,j)$ (subscript n denotes the trial) — we integrate[7] energy density $E_n(t,f)$:

$$E_n(i,j) = \int_{i\cdot\Delta t}^{(i+1)\cdot\Delta t} \int_{j\cdot\Delta f}^{(j+1)\cdot\Delta f} E_n(t,f)dtdf \tag{4.24}$$

At this point, we may proceed to testing the null hypothesis of no significant changes in $E_n(i,j)$. The distribution of energy density for a given frequency is not normal. However, in many practical cases it can be transformed to an approximately normal one using an appropriate Box-Cox transformation [Box and Cox, 1964]. The Box-Cox transformations are the family of power transformations:

$$BC(x,\lambda) = \begin{cases} \frac{x^\lambda - 1}{\lambda} & if\ \lambda \neq 0 \\ \log(x) & if\ \lambda = 0 \end{cases} \tag{4.25}$$

For each frequency j the λ parameter is optimized by maximization of the log-likelihood function (LLF) [Hyde, 1999] in the reference period:

$$\lambda_{\text{opt}}^j = \max_\lambda \{LLF(\lambda)\} = \max_\lambda \left\{ -\frac{m}{2} \log \sigma^2_{BC(x,\lambda)} + (\lambda - 1) \sum_{k=1}^{m} \log x \right\} \tag{4.26}$$

where m is the length of data x, $x \in \{E_n(i,j) : i \in t_b, n = 1,\dots,N\}$. The optimal λ_{opt}^j is then used to transform all the resels in frequency j. Standard parametric tests can be applied to the normalized data. However, we cannot a priori assume equal variances in the two tested groups. The known solution to the problem of variance

[7] In the spectrogram equation (2.84) and scalogram equation (2.95), the energy is primarily computed on the finest possible grid and then the integration is approximated by a discrete summation. In case of MP the integration of equation (2.117) can be strict. The procedure was described in detail in [Durka et al., 2004].

heterogeneity in the t-test is Welch's [Welch, 1938] correction of the number of degrees of freedom. The test can be formulated as follows for each of the resels (i, j). The null hypotheses $H_0^{i,j}$ and alternative hypotheses $H_1^{i,j}$:

$$H_0^{i,j} : \langle X(j)\rangle_{\mathrm{b,tr}} = \langle X(i,j)\rangle_{\mathrm{tr}} \tag{4.27}$$

$$H_1^{i,j} : \langle X(j)\rangle_{\mathrm{b,tr}} \neq \langle X(i,j)\rangle_{\mathrm{tr}} \tag{4.28}$$

where: $\langle X(j)\rangle_{\mathrm{b,tr}}$ is the normalized energy in the baseline time averaged across baseline time and trials, and $\langle X(i,j)\rangle_{\mathrm{tr}}$ is the normalized energy in resel (i, j) averaged across trials. The statistics is:

$$t(i,j) = \frac{\langle X(j)\rangle_{\mathrm{b,tr}} - \langle X(i,j)\rangle_{\mathrm{tr}}}{s_\Delta} \tag{4.29}$$

where s_Δ is the pooled variance of the reference epoch and the investigated resel. The corrected number of degrees of freedom ν is:

$$\nu = \frac{\left(\frac{s_1^2}{n_1} + \frac{s_2^2}{n_2}\right)^2}{\frac{\left(\frac{s_1^2}{n_1}\right)^2}{n_1 - 1} + \frac{\left(\frac{s_2^2}{n_2}\right)^2}{n_2 - 1}} \tag{4.30}$$

where s_1 is the standard deviation in the group of resels from the reference period, $n_1 = N \cdot N_\mathrm{b}$ is the size of that group, s_2 is the standard deviation in the group of resels from the event-related period, and $n_2 = N$ is the size of that group.

If we cannot assume $X_n(i, j)$ to be distributed normally, we estimate the distribution of the statistic t from the data using (4.29), separately for each frequency j:

1. From the $X(i, j)$, $i \in t_\mathrm{b}$ draw with replacement two samples: A of size N and B of size $N \cdot N_\mathrm{b}$
2. Compute t as in (4.29): $t_\mathrm{r}(j) = \frac{\langle X_A\rangle - \langle X_B\rangle}{s_\Delta}$, where s_Δ is pooled variance of samples A and B.

...and repeat steps 1 and 2 N_{rep} times. The set of values $t_\mathrm{r}(j)$ approximates the distribution $T_\mathrm{r}(j)$ at frequency j. Then for each resel the actual value of (4.29) is compared to this distribution:

$$p(i,j) = 2\min\{P(T_\mathrm{r}(j) \geq t(i,j)), 1 - P(T_\mathrm{r}(j) \geq t(i,j))\} \tag{4.31}$$

yielding two-sided $p(i, j)$ for the null hypothesis H_0^{ij}. The relative error of p is (c.f. [Efron and Tibshirani, 1993])

$$err = \frac{\sigma_p}{p} = \sqrt{\frac{(1-p)}{pN_{\mathrm{rep}}}} \tag{4.32}$$

Although the above presented method is computationally intensive, at present it causes no problems in most of the standard applications. However, corrections for multiple comparisons imply very low effective critical values of probabilities needed to reject the null hypothesis. For the analysis presented in [Durka et al., 2004] critical values of the order of 10^{-4} were routinely obtained. If we set $p = 10^{-4}$ in (4.32), we obtain a minimum $N_{\text{rep}} = 10^6$ resampling repetitions to achieve 10% relative error for the values $p(i, j)$.

In [Durka et al., 2004] either parametric or resampled statistical tests were applied to energies in each resel separately. However, the very notion of the test's confidence level reflects the possibility of falsely rejecting the null hypothesis. For example, a confidence level of 5% means that it may happen in approximately one in 20 cases. If we evaluate many such tests we are very likely to obtain many such false rejections. This issue is known in statistics as the issue of multiple comparisons, and there are several ways to deal with it properly.

To get a valid overall map of statistically significant changes we suggest the approach chosen in [Durka et al., 2004] that is a procedure assessing the false discovery rate (FDR, proposed in [Benjamini and Hochberg, 1995]). The FDR is the ratio of the number of falsely rejected null hypotheses (m_0) to the number of all rejected null hypotheses (m). In our case, if we control the FDR at a level $q = 0.05$, we know that among resels declared as revealing a significant change of energy, at most 5% of them are declared so falsely. [Benjamini and Yekutieli, 2001] proves that the following procedure controls the FDR at the level q under positive regression dependency, which can be assumed for the time-frequency energy density maps:

1. Order the achieved significance levels p_i, approximated in the previous section for each of the resels separately, in an ascending series: $p_1 \leq p_2 \leq \cdots \leq p_m$

2. Find
$$k = \max_i \left\{ p_i \leq \frac{i}{m} q \right\} \tag{4.33}$$

3. p_k is the effective significance level, so reject all hypotheses for which $p \leq p_k$.

Resels $r(i, j)$ are marked significant if the null hypothesis $H_0^{i,j}$ can be rejected using the significance level p_k for the probabilities $p(i, j)$ of the null hypothesis (4.27). An example of ERD/ERS time-frequency maps together with the assessment of their significance obtained with the above described procedure is illustrated in Figure 4.20.

4.1.7.3.4 ERD/ERS in the study of iEEG The ERD/ERS methodology was also applied to iEEG signals. The most advanced method providing the highest and most adaptive time-frequency resolution—matching pursuit was used, e.g., by [Zygierewicz et al., 2005] for evaluation of ERD/ERS responses from cortex during hand movement and for iEEG analysis during speech perception [Ray et al., 2003].

The ERD/ERS methodology applied to the iEEG made it possible to study high gamma responses (HGR), which are difficult to record in standard EEG. They were

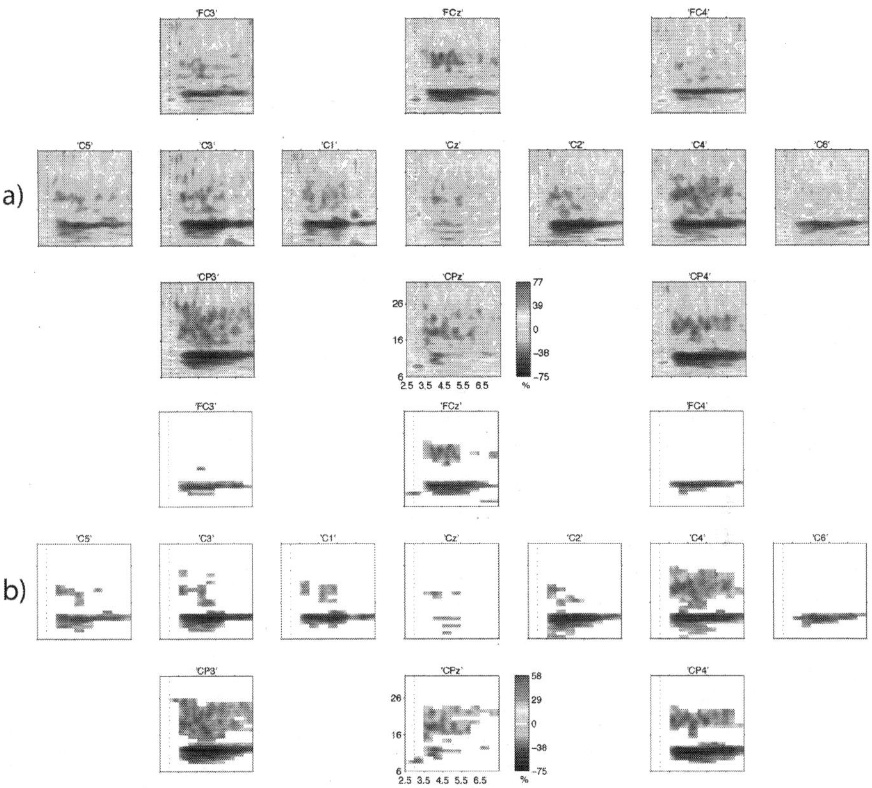

FIGURE 4.20: (SEE COLOR INSERT) An example of time-frequency analysis of ERD/ERS (left hand movement imagination). The t-f maps of selected electrodes are located topographically. Horizontal axis—time [s]; vertical—frequency [Hz]. Onset of imagination marked by a dotted line. a) ERD/ERS map, b) significant effects (Sect. 4.1.7.3.3).

first described in detail in human ECoG recorded in sensorimotor cortex [Crone et al., 1998]. HGRs have a broadband spectral profile, which range from 60 Hz to 200 Hz with the majority of event-related energy in the band 80–150 Hz. The ubiquitous occurrence of HGRs which have been found in motor, somatosensory, auditory, and visual cortex [Sinai et al., 2009] during functional activation suggests that it is a general electrophysiological index of cortical processing, since they are correlated with the degree of functional activation. Detailed time-frequency analysis of HGRs from LFP of somatosensory cortex revealed that they are temporally tightly linked to neuronal spikes [Ray et al., 2008].

Recently high frequency gamma components have been measured also in scalp EEG [Ball et al., 2008, Darvas et al., 2010] and in MEG [Dalal et al., 2008, Gunji et al., 2007], but iEEG and LFP give the best opportunity to study HGR and especially to correlate their characteristics with the spike occurrence.

4.1.7.3.5 Event-related time-varying functional connectivity The methods described in previous paragraphs concentrated on the analysis of ERD/ERS that is on the variations of power related to a certain event. The other phenomenon, which is equally interesting, is related to the event variation of couplings between the different distant populations. In this respect the methods derived from multivariate autoregressive model (Sect. 3.2) can be very useful.

Phenomenon involving the phase information, namely a time varying directional connectivity between neural population is recently the focus of interest. In the past the bivariate methods usually based on coherences were used, however the demonstration that they give misleading results (Sect. 3.5) turned attention to the methods based on extension of Granger causality defined in the framework of MVAR.

When applying MVAR one has to keep in mind that the number of model parameters should be preferably smaller by an order of magnitude than the number of samples in the data window. Number of MVAR parameters is pk^2 (where p—model order, k—number of channels), number of data points is kN (where N—number of points in the window). Effectively we get a condition: $pk/N < 0.1$.

In order to get a time-varying estimate we need to use a short time window, which is in contradiction with the above relation. The number of data points may be effectively increased by means of ensemble averaging over realizations when they are available from repeated trials of the experiment. To solve the problem of the time-varying estimate two approaches may be distinguished: sliding window or time-continuous fit of MVAR model, which may be performed adaptively, e.g., by means of a Kalman filter. Usually in both approaches ensemble averaging is applied.

The Kalman filter approach was extended for multivariate non-stationary processes by Arnold et al. [Arnold et al., 1998]. However, the computation effort of the technique is high as was shown, e.g., in [Kaminski et al., 2010]. In the case of the Kalman filter the computation time rises fast with the number of channels and number of realizations. Taking into account that the estimation of the errors is usually based on bootstrap methods, which involves repetition of computations hundreds of times, the computation time in the Kalman filter approach (much longer than in case of SDTF) can be a serious drawback, which hampers its wide application. In [Kaminski et al., 2010] it was also reported that the Kalman method has difficulties in adaptation for a large number of channels.

Another adaptive method of MVAR fitting is the least mean square (LMS) algorithm. The adaptation capability was found to be better for its modification—recursive least-square (RLS) algorithm with forgetting factor [Moller et al., 2001]. This algorithm takes into consideration a set of EEG epochs and the RLS estimation is controlled (similarly to Kalman filter) by the value of adaptation factor. The algorithm was initially used for estimation of multivariate coherences [Moller et al., 2001]. It was also applied for estimation of bivariate Granger causality in the Stoop test[8] [Hesse et al., 2003] and by [Astolfi et al., 2008].

An example of application of time-varying formulation of PDC was a study of a foot movement task [De Vico Fallani et al., 2008]. Time-dependent MVAR parameter matrices were estimated by means of the RLS. The MVAR model was fitted to the signals representing the current density on cortex found by solving the linear inverse problem [Grave de Peralta Mendez and Gonzales Andino, 1999]. The dense connections obtained between 16 regions of interest formed a complicated pattern and to find the meaningful transmissions between the regions of interest the theoretical graph indexes [Newman, 2003] have to be applied. The directed connectivity measures found by means of the projection of EEG activity on cortex represent some causality measures, but they don't reflect the actual propagation of EEG waves, since in the procedure of inverse projection the information on the phase difference between channels is lost. Taking into account the fact that DTF and PDC estimators are practically not influenced by a volume conduction (which is zero phase propagation) the operation of finding the current density on cortex seems to be unnecessary.

The determination of the dynamic propagation during performance of finger movement and its imagination [Ginter et al., 2001, Ginter et al., 2005, Kus et al., 2006] may serve as an example of the application of short-time DTF to the scalp EEG. The established evidence [Pfurtscheller and Lopes da Silva, 1999] is that during the movement the decrease of activity in alpha and beta bands (ERD) in the primary motor area corresponding to the given part of the body is observed, and increase of beta activity called beta rebound follows. In the gamma band, brief increase during movement was reported [Pfurtscheller and Lopes da Silva, 1999]. These findings corresponded very well with the results obtained by means of SDTF. Figure 4.21 shows the SDTFs as functions of time and frequency for a right finger movement experiment. The gaps in the propagation of alpha and beta activity for electrodes overlying left motor cortex (most pronounced for electrode C3) and subsequent fast increase of propagation in beta band more frontally are compatible with spectrally, temporally, and topographically specific ERD/ERS phenomena.

By virtue of selectivity of SDTF to phase difference (neglecting zero phase or random phase dependencies) the evolution in the gamma band may be studied (after filtering of prevailing alpha activity). In Figure 4.22 the EEG propagation in the gamma band during right finger movement and its imagination, obtained by SDTF, is shown. In case of movement the short burst of gamma propagation from C3 followed by propagation from frontal electrodes was observed. In case of movement imagination this propagation started later and a cross-talk between different sites overlying motor area and other sensorimotor areas may be noticed. (The dynamics of propagation may be observed in animations available at: `http://brain.fuw.edu.pl/~kjbli/DTF_MOV.htm`). This kind of transmission is compatible with the notion that a more difficult task requires involvement of the several sensorimotor areas and it is in agreement with neurophysiological hypotheses

[8] In the Stroop test congruent and non congruent stimuli are compared, e.g., word blue written in blue with the word blue written in red.

FIGURE 4.21: **(SEE COLOR INSERT)** Propagation of EEG activity in the left hemisphere during right finger movement. In each panel SDTF as a function of time (horizontal axis) and frequency (vertical axis). The flow of activity is from electrode marked under the column to the electrode marked at the relevant row. Red—the highest intensity, blue—the lowest. From [Ginter et al., 2001].

concerning the interactions of brain structures during simple and complicated motor tasks.

Another application of SDTF concerned evaluation of transmission during cognitive experiments. In the continuous attention test (CAT) geometrical patterns were displayed and the subject had to react to two subsequent identical pictures by pressing the switch (situation target) and do nothing when the patterns were not identical (situation non-target). The results confirmed the engagement of pre-frontal and frontal structures in the task and elucidated the mechanisms of active inhibition [Blinowska et al., 2010a]. Figure 4.23 shows the snapshots from the movie illustrating transmissions. Only significant flows showing increase or decrease of propagation in respect to reference period are shown. They were calculated according to the procedure devised in [Korzeniewska et al., 2008], which will be described in Sect. 4.1.7.3.7. In Figure 4.23 one can observe for the non-target situation the flow from electrode F8 (overlying right inferior cortex) to electrode C3. This kind of transmission was present in six out of nine studied subjects. In three subjects the propagation for non-target was from the Fz electrode (overlying pre-supplementary motor area).

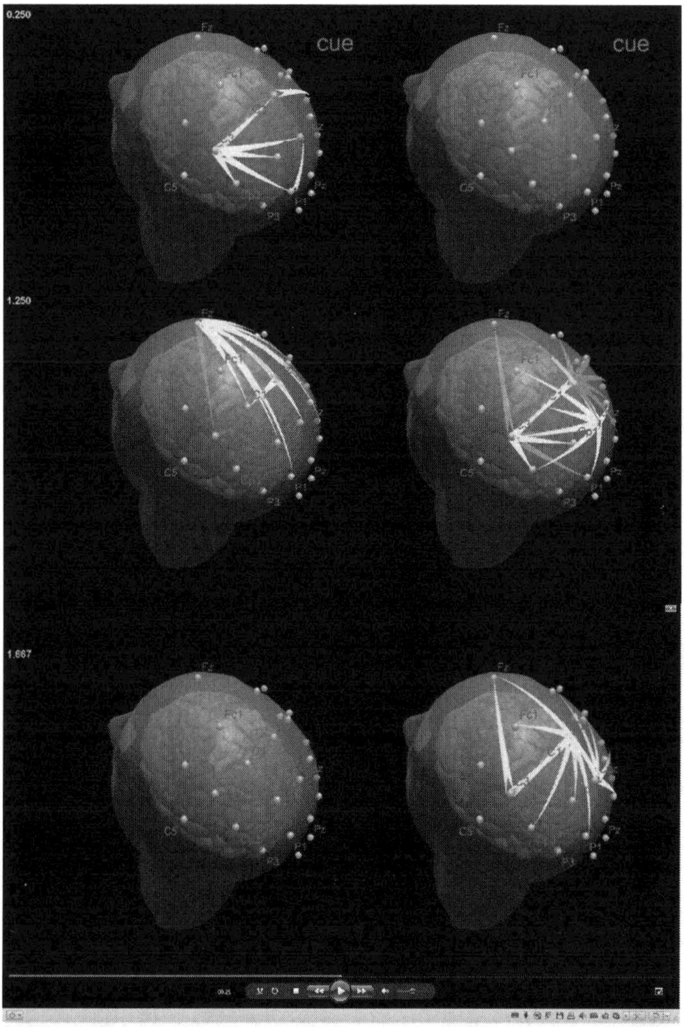

FIGURE 4.22: (SEE COLOR INSERT) Snapshots from the animation representing transmission in the gamma band during right finger movement (left column) and its imagination (right column) obtained by SDTF. The intensities of flows represented as colors of arrows (red the strongest). Numbers indicate the time in seconds after the cue presentation.

Both above mentioned structures are connected with movement inhibition. One can conjecture that the observed propagation was the expression of withdrawal from the action of pressing the switch. In case of target the propagation from the C3 electrode initiating the motor action was observed. The obtained results supported the hypoth-

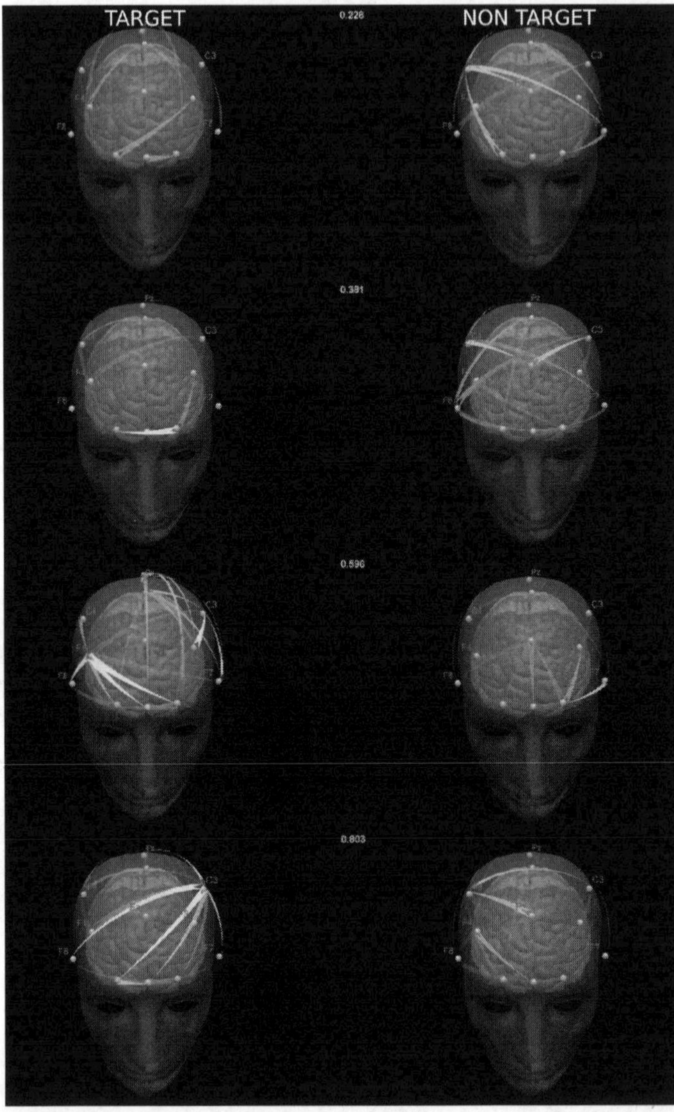

FIGURE 4.23: **(SEE COLOR INSERT)** The snapshots from the animation representing the significant changes of propagation during the CAT test obtained by SDTF. Red and yellow arrows—increase of propagation, blue—decrease of propagation. At the right—target; at the left—non-target. In the middle—time after the stimulus in seconds. Please note in the non-target case at 0.381 s strong increase of propagation F8→ C3 connected with active inhibition of movement; and in the case of target, at 0.803 s, propagation from C3 connected with command to move the finger.

esis of an active inhibition role in motor tasks played by right inferior frontal cortex and pre-supplementary motor area. The animations of propagation during a CAT test are available at URL `http://brain.fuw.edu.pl/~kjbli/CAT_mov.html`.

The results obtained by means of SDTF in experiments involving working memory were compatible with fMRI studies on the localization of the active sites and supplied the information concerning the temporal interaction between them [Brzezicka et al., 2011, Blinowska et al., 2010b].

The above described experiments point out that by means of SDTF estimated from scalp recorded EEG a very clear-cut and consistent evidence concerning the activity propagation may be found. The findings are compatible with the known facts and at the same time they have supplied new evidence concerning the information processing in the brain.

4.1.7.3.6 Functional connectivity estimation from intracranial electrical activity The study of intracranial EEG (iEEG) gives the opportunity to record the electrical activity, practically free of artifacts, directly from the brain structures. In the past, this research was mostly concentrated on the analysis of spike trains; the role of slow cortical activity was mostly neglected. Nowadays, the role of oscillatory activity of neural networks has become more and more recognized. There is a consensus that the higher cortical functions depend on dynamic interplay between spatially distributed multiple cortical regions coupled by the transmission of oscillatory activity, e.g., [Singer, 1993, Buzsaki and Draguhn, 2004].

The most popular method to study the interaction between cortical areas is coherence (in spite of the limitations of bivariate coherence). To investigate the dynamics of interactions between brain structures, temporal fluctuations in coherence were studied, e.g., by [Bullock et al., 1995], and oscillatory synchrony was considered by [Tallon-Baudry, 2003]. Recently attention has focused on the directionality of interactions between brain regions. Among the methods used for the study of directionality, those based on Granger causality seem to be the most appropriate.

For the analysis of signals recorded from specific brain structures it is important to determine the direct interaction. Direct directed transfer function (Sect. 3.3.2.3.1) was introduced by [Korzeniewska et al., 2003] for the purpose of investigating the propagation of LFP recorded from electrodes chronically implanted in the brain of a behaving animal. The patterns of propagation were recorded from four brain structures involved in processing of emotions. The patterns of transmission were studied for an animal walking on a runway and for walking on a runway accompanied by a stressful stimulus (bell ringing). More flows between the brain structures appeared in the case of the stressful stimulus and most of them were reciprocal.

The investigation of the intracranial human EEG is limited to cases of patients who are being prepared for surgical intervention. The limitations in the iEEG research are caused by the fact that the electrode placement is dictated solely by clinical concerns. However, the area covered by electrodes is usually not confined to the location of the diseased tissue because this location is not precisely determined before implantation and also the neighboring areas have to be checked to be sure surgery will not seriously disturb sensory, motor, or cognitive functions. iEEG provides unprecedented opportunity for studying indices of cortical activation, since it is characterized by high temporal resolution and mesoscopic spatial resolution that is intermediate between the macroscopic scale of EEG/MEG and multiunit recording of neuronal activity.

iEEG allows for investigation of high-gamma activity, which is hard to observe using scalp electrodes. Causal interaction between signals in high-gamma range (above 60 Hz) for the word repeating task were studied by [Korzeniewska et al., 2008]. The authors introduced a new measure—short-time direct directed transfer function (SdDTF), which combined the benefits of directionality, directedness, and short-time windowing. This function appeared to be an effective tool for analyzing non-stationary signals such as EEG accompanying cognitive processes. The performance of the function was tested by means of simulations, which demonstrated that the SdDTF properly estimates directions, spectral content, intensity, direct causal interactions between signals, and their time evolution. To evaluate event-related changes in SdDTF, that is, event-related causality (ERC), a new statistical methodology was developed to compare prestimulus and poststimulus SdDTF values. This procedure is described in Sect. 4.1.7.3.7.

In order to quantitatively describe the transmissions, ERC for high gamma activity was integrated in the 82–100 Hz range, which was empirically derived based on the mean ERC over all time points and all pairs of analyzed channels. Figure 4.24 shows the magnitudes of the interaction in the form of arrows of different widths. The most prominent identified connections involved: in the first phase (listening) flows from the auditory associative cortex to mouth/tongue motor cortex, and in the second phase (repeating of the word) propagation from the Brocas area (responsible for speech) to mouth/tongue motor cortex.

One of the problems important for neuroscience is the relation between the spike trains and local field potentials. For the spike train evaluation the methods developed in the field of point processes analysis are customarily applied [Brown et al., 2004]. Nevertheless there is a possibility of also using the broad repertoire of stochastic continuous signal analysis methods, described in this book, to spike trains.

In the approach proposed by [Kocsis and Kaminski, 2006] the spike trains were processed in the following way: the spikes were low-pass filtered by an order 1 Butterworth filter with cutoff frequency at 10% of Nyquist frequency (the filtering procedure was applied as zero phase filter; Sect. 2.1); then 10% of stochastic noise uncorrelated with the signal was added in order to make the spike train better match the stochastic character of the AR model. The procedure is illustrated in Figure 4.25.

The described approach was used in the experiment where LFP was recorded from hippocampus and spike trains from the supramammilliary nucleus (SUM) of a rat with the aim of finding the dynamic coupling between the structures in a situation when the sensory stimulus was applied. The MVAR model was fitted to the spike signal from SUM (approximated in the above described way) and the hippocampal LFP; then the SDTF functions were estimated. The temporal dynamics of the direction of influence revealed sharp reverses in the direction of the theta drive in association with sensory-elicited theta rhythm. It was found that in this situation the

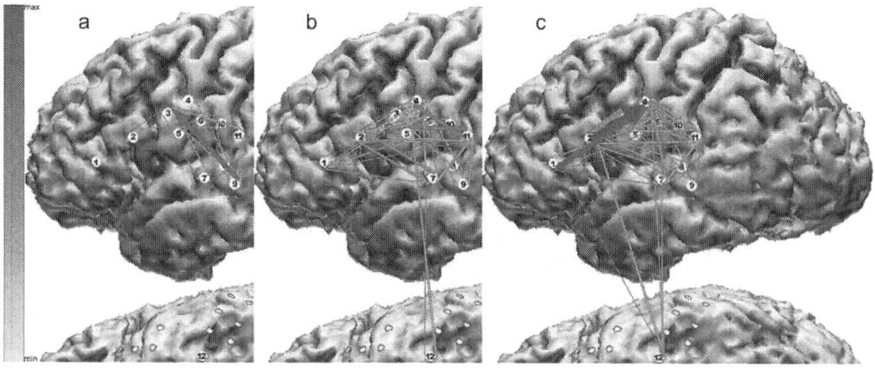

FIGURE 4.24: **(SEE COLOR INSERT)** Integrals of ERC for frequency range 82–100 Hz calculated for three stages of an auditory word repetition task. (a) Auditory perception stage, (b) response preparation, (c) verbal response. Arrows indicate directionality of ERC. Width and color of each arrow represent the value of the ERC integral. Color scale at the left. For clarity, only integrals for event-related flow increases are shown. From [Korzeniewska et al., 2008].

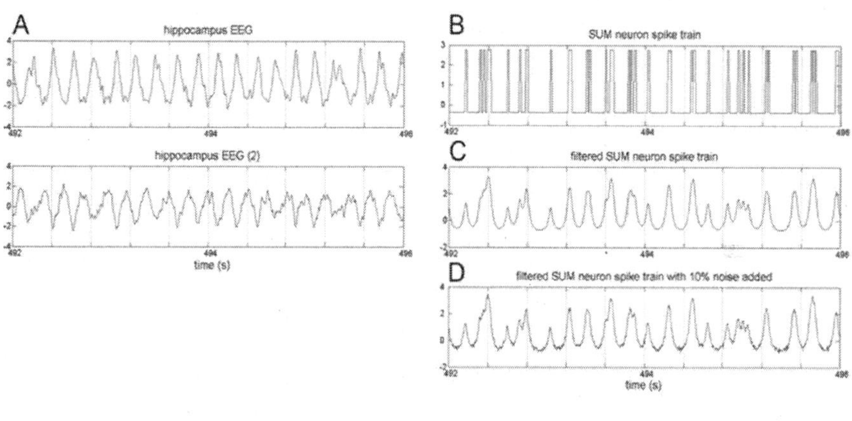

FIGURE 4.25: Transforming spike train into continuous signal. A, LFP recorded from hippocampus, B, standardized spike train, C, low-pass filtered spike train, D, low-pass filtered spike train with 10% noise added. From [Kocsis and Kaminski, 2006].

subpopulation of SUM neurons contains information predicting future variations in the LFP rhythm in hippocampus. In contrast, during slow spontaneous theta rhythm it was the SUM spike train that can be predicted from the hippocampal activity.

The described procedure of turning the spike trains to the continuous signals opens new perspectives of application of the methods of signal processing to point processes. In particular this approach may be useful for spike train analysis and for the investigation of the relations between point processes and continuous signals.

4.1.7.3.7 Statistical assessment of time-varying connectivity Functional connectivity is expected to undergo rapid changes in the living brain. This fact should be taken into account when constructing the statistical tool to test the significance of the changes in the connectivity related to an event. Especially, it is difficult to keep the assumption that during the longer baseline epoch the connectivity is stationary. ERC involves statistical methods for comparing estimates of causal interactions during pre-stimulus baseline epochs and during post-stimulus activated epochs that do not require the assumption of stationarity of the signal in either of the epochs. Formally the method relies on the bivariate smoothing in time and frequency defined as:

$$Y_{f,t} = g(f,t) + \varepsilon_{f,t} \tag{4.34}$$

where $g(f,t)$ is modeled as penalized thin-plate spline [Ruppert et al., 2003] representing the actual SdDTF function and $\varepsilon_{f,t}$ are independent normal random ($N(0,\sigma_\varepsilon^2)$) variables. The thin-plate spline function can be viewed as spatial plates joined at a number of knots. The number of knots minus the number of spline parameters gives the number of the degrees of freedom.

To introduce the testing framework proposed in [Korzeniewska et al., 2008], we first introduce some notations. Denote by $f_1,\ldots,f_m$ the frequencies where f_m is the number of analyzed frequencies. Also, denote by $t = t_1,\ldots,t_n$ the time index corresponding to one window in the baseline, where t_n is the total number of baseline windows. Similarly, denote by $T = T_1,\ldots,T_n$ the time index corresponding to one window in the post-stimulus period, where T_n is the total number of post-stimulus windows.

The goal is to test for every frequency f, and for every baseline/stimulus pair of time windows (t,T), whether $g(f,t) = g(f,T)$. More precisely, the implicit null hypothesis for a given post-stimulus time window T at frequency f is that:

$$H_{0,f,T} : g(f,t_1) = g(f,T) \text{ or } g(f,t_2) = g(f,T) \text{ or } ,\ldots, g(f,t_n) = g(f,T) \tag{4.35}$$

with the corresponding alternative

$$H_{1,f,T} : g(f,t_1) \neq g(f,T) \text{ and } g(f,t_2) \neq g(f,T) \text{ and } ,\ldots, g(f,t_n) \neq g(f,T) \tag{4.36}$$

To test these hypotheses a joint 95% confidence interval for the differences $g(f,t) - g(f,T)$ for $t = t_1,\ldots,t_n$ is constructed. Let $\hat{g}(f,t), \hat{\sigma}_g^2(f,t)$ be the penalized spline estimator of $g(f,t)$ and its associated estimated standard error in each baseline time window. Similarly, let $\hat{g}(f,T), \hat{\sigma}_g^2(f,T)$ be the penalized spline estimator of $g(f,T)$ and its associated estimated standard error in each post-stimulus time window. Since residuals are independent at points well separated in time, the central limit theorem applies and we can assume that for every baseline/stimulus pair of time windows (t,T)

$$\frac{(\hat{g}(f,t) - \hat{g}(f,T)) - (g(f,t) - g(f,T))}{\sqrt{\hat{\sigma}_g^2(f,t) + \hat{\sigma}_g^2(f,T)}} \sim N(0,1) \tag{4.37}$$

approximates a standard normal distribution. A joint confidence interval with at least 95% coverage probability for $g(f,t) - g(f,T)$ is:

$$\hat{g}(f,t) - \hat{g}(f,T) \pm m_{95}\sqrt{\hat{\sigma}_g^2(f,t) + \hat{\sigma}_g^2(f,T)} \tag{4.38}$$

where m_{95} is the 97.5% quantile of the distribution

$$MAX(t_n, T_n) = max_{t \in \{t_1,\ldots,t_n\}, T \in \{T_1,\ldots,T_n\}, f \in \{f_1,\ldots,f_m\}} |N_{t,T,f}| \tag{4.39}$$

where $N_{t,T,f}$ are independent $N(0,1)$ random variables. This is equivalent to applying the Bonferonni correction for $t_n T_n f_m$ tests to control the family-wise error rate (FWER).

The utility of the ERC approach was demonstrated through its application to human electrocorticographic recordings (ECoG) of a simple language task [Korzeniewska et al., 2008]. ERC analyses of these ECoG recordings revealed frequency-dependent interactions, particularly in high gamma (>60 Hz) frequencies, between brain regions known to participate in the recorded language task, and the temporal evolution of these interactions was consistent with the putative processing stages of this task.

4.1.8 Multimodal integration of EEG and fMRI signals

In recent years there has been a growing awareness in scientific society that for a better understanding of information processing in the brain there is a need to integrate the modalities offering different spatial and temporal resolution. EEG/ERP provide high temporal resolution, but much poorer spatial resolution; in contrast fMRI (functional magnetic resonance imaging) offers high spatial resolution, but poor time resolution. The blood oxygenation response (BOLD) measured by means of fMRI depends on the oxygen uptake which is connected with neuronal activity. However in contrast to EEG, BOLD response is an indirect and delayed metabolic correlate of neuronal process. The challenges to scientists are: to better understand the link between BOLD and neural activity and to capitalize on complementary properties of EEG and BOLD in the investigation of brain processes. To this aim multimodal analyses are undertaken based on concurrent measurement of both modalities. In [Blinowska et al., 2009] the connection between neuronal activity and BOLD response is delineated and four approaches to the analysis of simultaneously acquired EEG/ERP-fMRI are described, namely: i) fMRI informed EEG for constrained source localization, ii) EEG or ERP-informed fMRI analysis, iii) parallel independent component analysis (ICA), and iv) joint ICA (jICA) application for matching temporal sources from the EEG with spatial sources from fMRI.

fMRI informed EEG uses hemodynamic priors for cortical activity localization. In [Urbano et al., 1998] and [Babiloni et al., 2000], a statistically significant percentage increase of the BOLD signal during the task compared to the rest state was used to define the priors in the solution of the linear inverse problem. The information was derived from the BOLD responses of the relevant cortical areas, taking into account strength of activated voxels. The hemodynamic information offered by fMRI provided the norm in the source space for solution of the inverse problem.

Studies concerning EEG informed fMRI aim to identify BOLD correlates of EEG events. In case of epilepsy the early applications involved a spike-triggered acquisition mode whereby each fMRI scan was acquired following visualization of spike [Symms et al., 1999]. In a more advanced approach [Liston et al., 2006] the semi-automatic system was proposed based on spatio-temporal clustering of interictal EEG events. The clusters of interictal epileptic discharges were correlated to BOLD activations. In a second step signal space was used to project scalp EEG onto dipoles corresponding to each cluster. This allowed for identification of previously unrecognized epileptic events, the inclusion of which increased the experimental efficiency as reflected by significant BOLD activation.

The general idea behind ERP-informed fMRI analysis is to correlate active brain regions identified by fMRI with amplitude modulation of individual ERP. In the experiment concerning auditory oddball paradigm [Eichele et al., 2005] in the first step the data were de-noised by means of application of wavelets and ICA. Then single-trial N1, P2, and P3 amplitudes of ERP were found. These amplitude vectors were convolved with the function of hemodynamic response and used as regressors to find the BOLD time course.

In the simultaneous measurements of ERP and fMRI the signals are recorded which are spatially and temporally mixed across the brain, since they are volume-conducted and temporally extended by hemodynamic response. The data in both modalities are generated by multiple, simultaneously active overlapping neural populations. In solving problems, where specific hypotheses regarding spatial and temporal relationships are lacking or are ill-specified, blind source separation methods are useful. In order to address the problem of mutual ERP-fMRI relation the ICA approach was proposed in [Eichele et al., 2008]. The ICA was used in parallel: to recover spatial maps from fMRI in terms of spatial independent components (sIC) and time-courses from ERP in terms of time related independent components (tIC). Then the components were matched across modalities by correlating their trial-to-trial modulation. Inspecting the results, only one tIC component was found which predicted selectively the time-course of one sIC component, there were no other covariances between components corresponding to both modalities.

An alternative approach is to fuse the ERP and fMRI signals in a common data space. In [Moosmann et al., 2008] joint independent component analysis was proposed for analysis of simultaneous single trial ERP-BOLD measurements from multiple subjects. The authors presented the results based on simulated data, which indicated a feasibility of the approach. It seems that there is a long way to go to obtain a satisfactory integration of EEG and fMRI modalities. Comparative experiments involving intracranial recordings would be helpful in this respect.

4.2 Heart signals

4.2.1 Electrocardiogram

Electrocardiogram (ECG) is a record of electrical activity of a heart. The contraction activity of a heart is driven by the electric pacemaker—the assembly of cardiac cells that possess the property of automatic generation of action potentials. Such cells are located in the sino-atrial node (SA), the atrio-ventricular node (AV), and in the specialized conduction systems within the atria and ventricles. Myocardial cells have the unique property of transmitting action potentials from one cell to an adjacent cell by means of direct current spread.

4.2.1.1 Measurement standards

Einthoven—the pioneer in the field of electrocardiology—as a model of electrical activity of the heart muscle, proposed a two-dimensional dipole placed in a homogenous conducting volume of thorax in the middle of a triangle with vertexes defined by the electrodes placed on the arms and left leg. Einthoven standard leads I, II, III measured the potential differences: I- between the left arm (LA) and the right arm (RA), II - (between the left leg (LL) and the RA), III- (between the LL and the LA). The propagation of the electrical activity throughout the cardiac cycle can be represented by an instantaneous electrical heart vector (Figure 4.26). In spite of the crude approximation of the body geometry, the projections of heart vector on the arms of the Einthoven triangle became a standard in electrocardiography.

The typical leads used in electrocardiography are shown in Figure 4.26. They include three standard ECG leads: I, II, III, and the so-called augmented limb leads: aVR, aVL, aVF corresponding to the potential recorded between a given limb (correspondingly: right arm, left arm, left foot) with respect to the average of the potentials of the other two limbs. In standard clinical applications signals from six leads placed on torso are also used. The potentials measured between electrodes placed on the limbs correspond to the projections on the frontal plane. The electrodes located on a torso supply the information about ECG vector projection on the horizontal plane (Figure 4.26).

In a three dimensional system proposed by Frank electrodes are placed according to Cartesian XYZ axes (Figure 4.27). In principle three orthogonal projections are sufficient to reconstruct the vector; however in case of ECG the redundancy in the information is beneficial since the projection axes are non-orthogonal; the body is not a homogenous sphere and the vector changes in time. Besides some particular features of ECG are more visible in specific derivations.

Sampling frequency of ECG has to be chosen according to the aim of the analysis. Commonly accepted range of ECG is 0.05–100 Hz. However, when ventricular late potentials are considered the upper frequency range reaches 500 Hz.

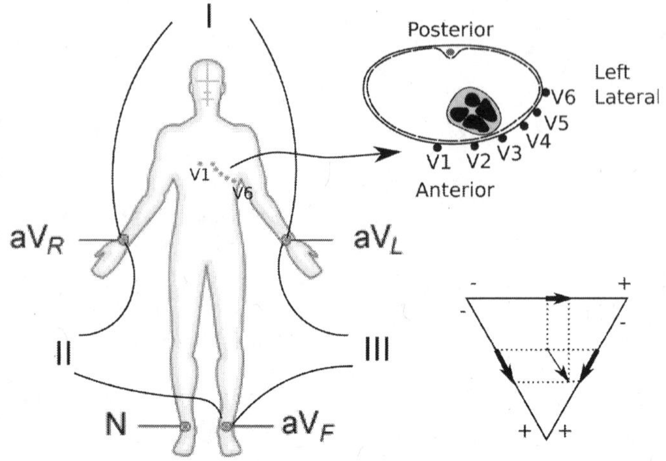

FIGURE 4.26: The typical leads used in electrocardiography. At left—limb leads; top right—chest leads. Bottom-right: the heart vector and its projection on the Einthoven triangle.

4.2.1.2 Physiological background and clinical applications

The transmission of electrical heart activity is reflected in the features of the ECG evolution. When depolarization propagates toward the given electrode the detected voltage is seen as positive and is represented by an upward deflection in the ECG. Let us consider the potential recorded by the lead II during the heart cycle. The activation starts in the SA node and propagates toward AV, the depolarization of atria follows with heart vector pointing down and left[9] which is reflected in the positive P wave (Figure 4.28). Then there is a delay at the AV node corresponding to the flat portion of the ECG. The septal depolarization with vector pointing to the right corresponds to the Q wave. The period between the P wave and the QRS complex corresponds to the time between the atrial and ventricular depolarization when ECG is on isoelectric level. This period serves as a reference for ECG amplitude estimation. The prominent QRS peak corresponds to the depolarization of ventricles with a heart vector pointing to the left and down. The width of the QRS complex depends on conduction speed through the ventricle. S-T segment corresponds to the pause between ventricular depolarization and repolarization. Next, a phase of repolarization follows, during which both the direction of propagation and the electrical polarity are changed which results in a positive T wave. U wave, which can be observed by means of high resolution electrocardiography, and which is also connected with repolarization, follows the T wave. In case of a short heart cycle it might be impossible to register the U wave.

[9]The left and right concerns subjects' left and right.

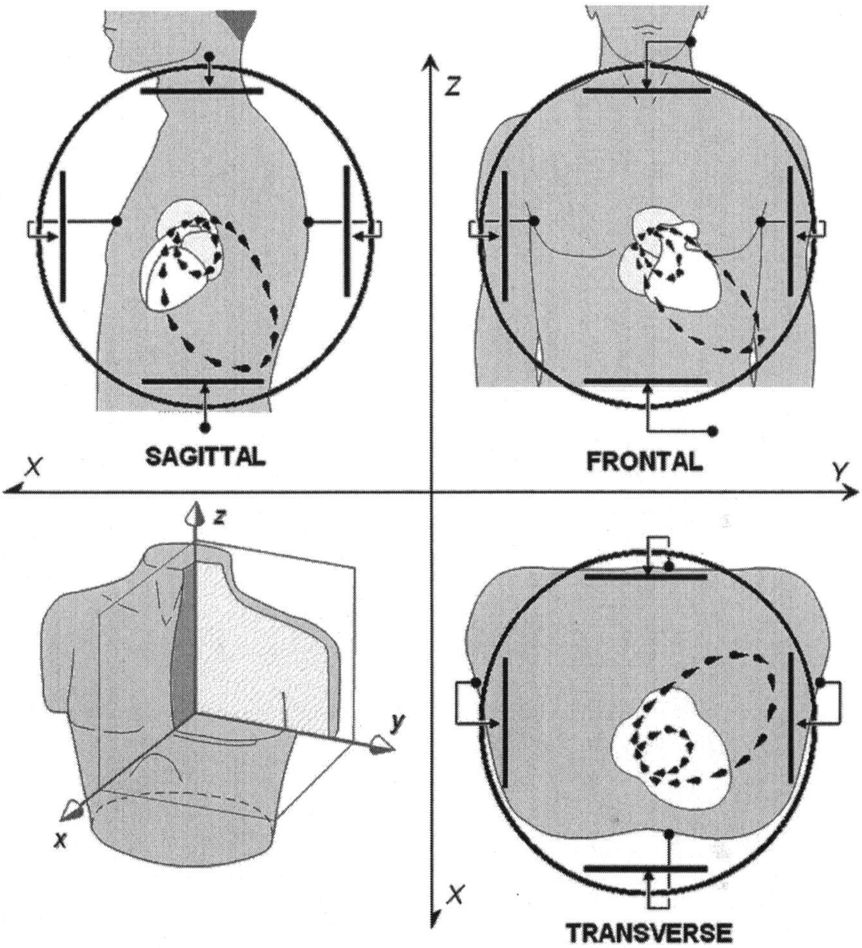

FIGURE 4.27: The 3-D Frank's electrode system. Black dots—the heart vector trajectory projections on sagittal, frontal, and transverse planes are marked. From [Malmivuo and Plonsey, 1995].

The heart pacemakers are influenced by the sympathetic and parasympathetic portions of the autonomic nervous system. The sympathetic system acts quickly to increase heart rate, blood pressure, and respiration; the parasympathetic system acts to decrease heart rate and slow down respiration and blood flow (e.g., during falling asleep).

The disturbances in the generation or propagation of electrical activity of the heart are reflected in the ECG, hence it is a widely used clinical test for heart diseases. However, this doesn't mean that all abnormal features of electrical heart activity will be observed in ECG, since according to the Helmholtz law, already mentioned in

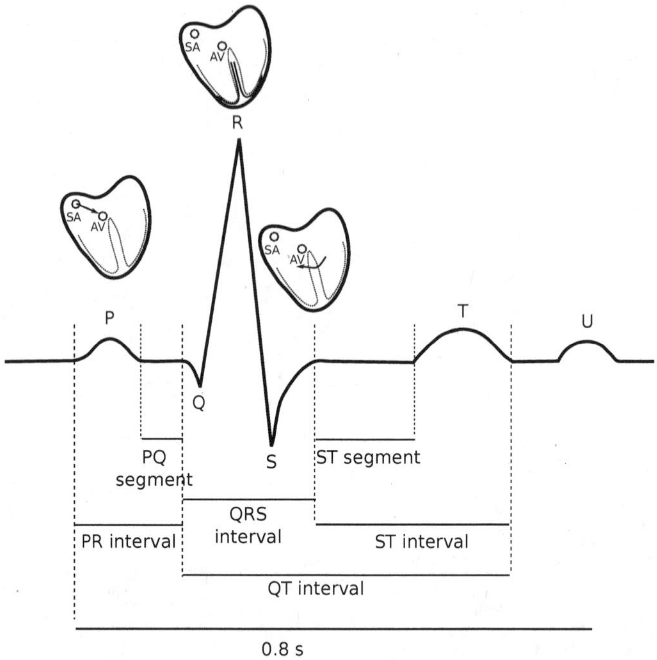

FIGURE 4.28: The potential recorded by lead II during the heart cycle. Main waves, intervals, and segments marked. Above schematic representations of electrical activity propagation during the cycle are marked.

the context of EEG, the same potential difference on the body surface can be due to different configurations of sources; hence abnormal activity may be masked. The information on the electromagnetic activity of the heart, its generation and measurement may be found in [Malmivuo and Plonsey, 1995].

In the clinical environment ECG recording is often accompanied by the measurement of respiration and blood flow, since these signals are mutually dependent, e.g., respiration rhythm is visible in the heart rate signal.

The ECG is a clinical test in most heart diseases. Other medical conditions such as diabetes and metabolic disturbances may also have an impact on ECG. In ECG the activity of autonomous nervous system is reflected, therefore ECG can be used as an indicator of activation of sympathetic and parasympathetic systems. Among the heart activity disturbances reflected in ECG are the arrhythmias, bradycardia (slowing of the rhythm), tachycardia (increase of heart rate), ectopy, fibrillation, myocardial infarction, premature atrial/ventricular contraction, atrial/ventricular flutter, conduction blocks, cardiac ischemia, hyper- or hypokalemia (excess/deficiency of potassium), hyper-, hypocalcemia (excess/ deficiency of calcium). The description of influence of these and other pathological heart conditions on ECG may be found in [Chee and Seow, 2007].

4.2.1.3 Processing of ECG

In ECG deterministic and stochastic features are manifested and the repertoire of the methods applied for its analysis is very broad. It includes: morphological features extraction, analysis in time and frequency domain, statistical methods, classification procedures. From ECG signal another time series, namely HRV (heart rate variability), is derived which in turn is a subject of statistical analysis.

In case of signal processing before applying a new method to the given type of data or for the differentiation of a particular pathological state it is recommended to compare the results with other methods and test it on standard data. In case of ECG fortunately there exists a Physionet database [Goldberger et al., 2000], which can be helpful in this respect. It is available at `http://www.physionet.org/physiobank/database/`. PhysioBank is a large and continuously updated archive of well-characterized digital recordings of physiologic signals and related data that may be downloaded at no cost. PhysioBank currently includes databases of multi-parameter cardiopulmonary, neural, and other biomedical signals from healthy subjects and patients with a variety of conditions with major public health implications, including sudden cardiac death, congestive heart failure, arrhythmia (MIT-BIH arrhythmia database). In PhysioNet the software for analysis of ECG and HRV may be found as well.

4.2.1.3.1 Artifact removal The first step in ECG analysis is elimination of the artifacts. ECG signals may be corrupted by the technical artifacts such as: power line interferences, artifacts due to bad electrode contacts, quantization or aliasing errors, noise generated by other medical equipment present in the patient care environment and biological artifacts: patient-electrode motion artifacts, muscular activity, baseline drift—usually due to respiration. Technical artifacts may be avoided by designing proper measurement procedures, however elimination of biological artifacts is much more difficult and requires special signal analysis techniques. The filtering techniques used for denoising ECG involve linear and non-linear methods, model based or model free approaches.

For rejecting high frequency noise FIR filters or Butterworth 4-pole or 6-pole low-pass digital filters appear to be suitable [Weaver, 1968]. The cut-off frequency is usually set around 40 Hz. For baseline wander high-pass linear filters of cut-off frequency up to 0.8 Hz may be used. Cut-off frequency above 0.8 Hz would distort the ECG waveform. Among the linear, model based approaches to denoising, Wiener filtering is an often used method. Unfortunately, one of the Wiener filter assumptions is that both signal and noise are stochastic. However, ECG has a prominent deterministic character, which diminishes the performance of the Wiener filter. One of its drawbacks is the reduction of the signal amplitude.

Wavelet transform is a model free method of ECG denoising. The discrete wavelet transform may be used to decompose ECG into time-frequency components (Sect. 2.4.2.2.4) and then the signal may be reconstructed only from presumably free of noise approximations. It is essential to choose the mother wavelet properly; it has to be maximally compatible with the ECG signal structures. In [Clifford, 2006] biorthogonal spline wavelets were recommended for ECG reconstruction, which was performed by keeping only the first approximation $A_2^d(x)$. The procedure is equivalent in a way to low-pass filtering followed by resampling. However, the wavelet transform approach seems to have advantages over conventional FIR filters, which may produce oscillating structures in filtered signal (so called Gibbs oscillations) [Clifford, 2006]. In Figure 4.29 denoising by means of wavelet transform is illustrated. The biorthogonal wavelets family is avaliable in MATLAB Wavelet Toolbox. The wavelets in the family are named: `biorJ.K`, where `J` and `K` correspond to the number of vanishing moments in the low-pass and high-pass filters, respectively. The wavelets with higher `J` and `K` seem to reproduce ECG signal structures better.

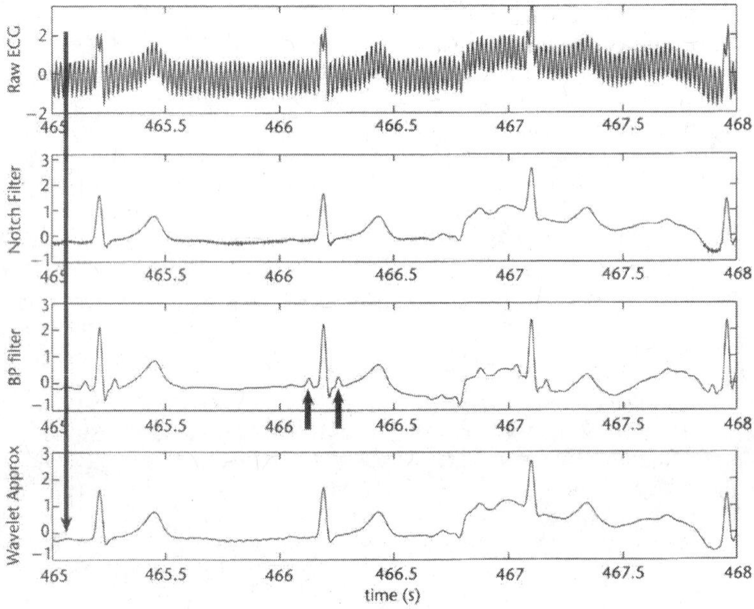

FIGURE 4.29: Raw ECG with 50 Hz mains noise, IIR 50 Hz notch filtered ECG, 0.1–45 Hz bandpass filtered ECG and bior3.3 wavelet filtered ECG. The left most arrow indicates the low amplitude P wave. Central arrows indicate Gibbs oscillation in the FIR filter causing a distortion larger than the P wave. Reproduced by permission from author (Clifford, G., *Advanced Methods and Tools for ECG Data Analysis*, Norwood, MA: Artech House, Inc., 2006).

Other methods useful in removing the noise components from ECG are PCA and ICA. It is an advantage of ICA that its subspace axes are not necessarily orthogonal and the projections on them are maximally independent. ICA is a blind source separation method. The independent components for a set of signals can be found if the mixing matrix **D** is known (Sect. 3.6.2, equation 3.52). We can expect that specific ICA components will be connected with the noise/artifact sources. Since ICA components are transformed combinations of leads, in order to recover the information about the ECG signals the back projection has to be performed (Sect. 3.6.2). During this procedure the noise sources may be removed. Namely the components corresponding to artifact/noise may be set to zero during reconstruction.

However, this procedure is not quite straightforward, since we have to determine which ICA components are noise. The system based upon kurtosis and variance have been devised to automatically distinguish noise components [He et al., 2006], but the quality of results may be dependent on the particular data set. The problem is complicated, since the ICA mixing/demixing matrix must be tracked over time and the filter response is constantly evolving. A robust system of separating artifacts from ECG is still to be found.

4.2.1.3.2 Morphological ECG features Clinical assessment of ECG mostly relies on evaluation of its time domain morphological features such as positions, durations, amplitudes, and slopes of its complexes and segments (Figure 4.28). The morphological feature may be estimated using a sequence of individual heartbeats or using averaged heartbeat. The first step in the analysis is usually detection of the QRS complex; it serves as a marker for averaging of heart cycles, for evaluation of heart rate, for finding the heart axis. Usually the analyzed parameters include amplitudes of Q,R,S,T peaks, depression or elevation of the ST segment, durations of the QRS complex, QT interval, dispersion of QT (difference between the longest and the shortest ST).

The algorithms for the determination of ECG morphological time features based on wave boundaries, positions, amplitudes, polarizations may be found, e.g., in [Pahlm and Sornmo, 1984, Pan and Tompkins, 1985, Laguna et al., 1994].

For feature extraction from ECG beats, with the aim of beat recognition, several approaches based on different formalisms have been proposed. Examples of such approaches include: Fourier transform [Minami et al., 1999], Hermite functions [Lagerholm et al., 2000], wavelet transform [Senhadji et al., 1995, Yu and Chen, 2007].

An interesting approach to ECG features quantification was proposed by [Christov et al., 2006]. Each training heartbeat was approximated with a small number of waveforms taken from a Wavelet Packet dictionary (Symlet 8). Matching pursuit algorithm was used as the approximation procedure. During the training procedure each of the five classes of heartbeats was approximated by 10 atoms, which were then used for the classification procedures. The results showed high classification accuracy. Matching pursuit was also used in the procedure of fitting the wavelets from the Daubechies family to the ECG structures, with the aim of obtaining sparse time-frequency representation of ECG [Pantelopoulos and Bourbakis, 2010].

At present there are also ready to use systems for ECG features extraction available on the Internet. The software for determination of the characteristic time points of ECG may be found in PhysioNet (`http://www.physionet.org/physiotools/`

ecgpuwave/) as a routine ecgpuwave.

A classification system for the electrocardiogram features called the Minnesota Code [Pineas et al., 1982] utilizes a defined set of measurement rules to assign specific numerical codes describing ECG morphology. It provides the ranges of normal values of ECG morphological features in the standard 12 leads. The system was developed in the late 1950s in response to the need for reporting ECG findings in uniform and objective terms and has been updated and improved since then. Minnesota Code incorporates ECG classification criteria that have been validated and accepted by clinicians [Kors et al., 1996].

ECG ST segment analysis is of special interest for detection of ischemia, heart rate changes, and other heart disturbances. The review of the ST segment analysis approaches, comparison of performance of ST analyzers, and algorithm for detection of transient ischemic heart rate related ST episodes may be found in [Jager, 2006].

Another feature of ECG important for clinical diagnosis is T-wave alternans (TWA). T wave is connected with the processes of repolarization of the heart muscle and its character changes from beat to beat and in particular its amplitude has alternating higher and lower values in consecutive cycles. TWA is a valuable ECG index that indicates increased susceptibility for ventricular arrhythmia and the risk of cardiac arrest, one of the most frequent causes of sudden death. The parameter commonly used in diagnostics is T wave amplitude, although its phase and shape are also considered. To find subtle TWA fluctuations spectral analysis is usually applied to the series constructed from T wave amplitude values, found in consecutive cycles. The alternation of amplitude from beat to beat is reflected in the power spectrum of these signals as a peak at 0.5 Hz. The peak amplitude normalized in respect to noise gives the information about the T wave variability; its higher value reflects greater repolarization dispersion and indicates risk of sudden cardiac arrest [Narayan, 2006].

An important TWA index is its phase, which may be detected by methods quantifying sign-change between successive pairs of beats, or by means of Fourier transform. TWA phase reversal is helpful in sudden cardiac arrest risk stratification.

4.2.1.3.3 Spatial representation of ECG activity; body surface potential mapping and vectorcardiography New diagnostic possibilities were opened by the methods involving mapping of the spatial distribution of TWA on a thorax [Klingenheben et al., 2005, Fereniec and Karpinski, 2004]. It is known that sometimes there can be no changes in the standard ECG despite the evidence of coronary artery disease or myocardial infarction. Analysis of high-resolution ECG recorded from multiple electrodes placed on the torso enables detection of electric activity of individual fragments of the cardiac muscle which opens up a possibility of developing a qualitatively new diagnostic method and a means to control the efficiency of treatment.

Figure 4.30 shows body surface maps of the QRST integral for a healthy volunteer and a patient after myocardial infarction (MI) with a documented episode of ventricular tachycardia (VT). The drastic change in the shape of isopotential boundaries may be easily observed.

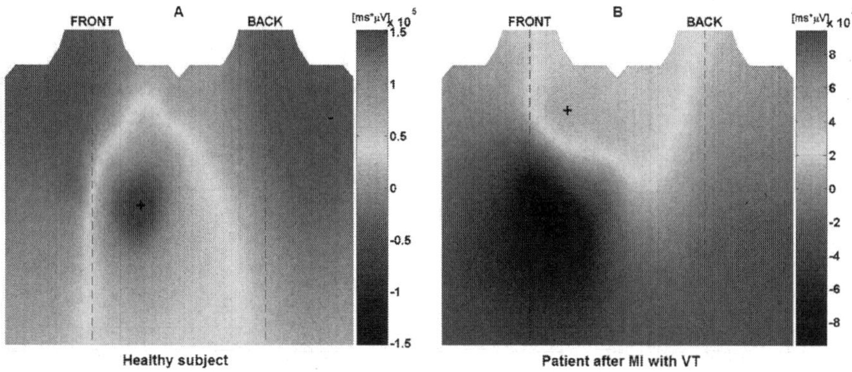

FIGURE 4.30: (SEE COLOR INSERT) Body surface maps of QRST integral: A) healthy volunteer, B) patient after myocardial infarction (MI) with documented episode of ventricular tachycardia (VT). From [Fereniec et al., 2008].

The pioneer of body surface potential mapping (BSPM) is B. Taccardi [Taccardi, 1962, Taccardi, 1963] who recorded the ECG from 200 electrodes. He described the time-varying potential distribution on the thorax with the aim of relating its evolution to the propagation of the activity in the heart muscle. At present the technique is used in many laboratories. The construction of maps involves triangulation of the body surface and interpolation, usually by means of biharmonic splines [Sandwell, 1987]. The BSPM technique was used for quantification of different phenomena occurring during the heart cycle, e.g., the spatial distribution of the late atrial potentials [Jokiniemi et al., 2003] and the distribution of electric field during the phase of repolarization [Khaddoumi et al., 2006]. In the above work the quantification of the electric field shape dispersion was proposed.

Diagnostic application of BSPM includes, e.g., detection of myocardial infarction, [De Ambroggi et al., 1988] , left ventricular hypertrophy [Corlan and De Ambroggi, 2000], and myocardial ischemia [Hanninen et al., 2001]. At present even portable devices for high resolution ECG mapping are available [Rosik et al., 2001].

Vectorcardiography is a method of recording the magnitude and direction of the electrical forces that are generated by the heart by means of a continuous series of vectors that form curving lines around a central point. In vectorcardiography the trajectory of the heart vector in 3D space is traced (Figure 4.27). The strength and direction of electric currents passing through the heart are represented as vector loops and the information about the directional evolution of the heart vector in time is found. This technique enables observation of the depolarization and repolarization in particular fragments of the cardiac muscle during its evolution which means that it permits detection of even small changes in electric activity of individual fragments of the cardiac muscle caused for example by ischemia or applied treatment. Usually evolution of electrical activity connected with QRS complex, P-wave, or ST complex is followed. High resolution vectorcardiography reveals changes related either to an untypical propagation of the depolarization vector or distinctly weaker amplitude of depolarization from particular fragments of the cardiac muscle. It was reported that high resolution vectorcardiography analysis permits a fast, non-invasive, and cheap confirmation of myocardial infarction and determination of its localization in persons whose standard and exercise ECG as well as echocardiography do not reveal changes. It also provides the recognition of changes in the depolarization amplitude related to, e.g., the effect of a drug on the cardiac muscle in ischemia [Krzyminiewski et al., 1999]. At present the on-line vectorcardiography is available and may be used for patient monitoring in hospital [Lundin et al., 1994].

4.2.1.3.4 Statistical methods and models for ECG analysis Statistical methods and models are applied for ECG segmentation, feature extraction, and quantification. For ECG analysis probabilistic models may be used, in particular hidden Markov models (HMM) (Sect. 2.2.1) were applied for the signal segmentation. The first step is to associate each state in the model with a particular feature of the ECG, namely to connect the individual hidden states with structures such as: P wave, QRS complex, etc. The next step involves training of HMM, usually by supervised learning. The important aspect is the choice of observation model to be used for capturing the statistical characteristics of the signal samples from each hidden state. As the observation models, Gaussian density, autoregressive model, or wavelets may be used. The latter two methods were reported as performing better [Hughes, 2006].

For ECG quantification and identification of the features helpful for distinction of pathological changes in ECG, the analysis based on statistical procedures such as PCA and models such as AR may be applied without making explicit reference to time-amplitude features of ECG structures.

AR model technique was proposed to classify different types of cardiac arrhythmias [Ge et al., 2007]. Preprocessing involved removal of the noise including respiration, baseline drift, and power line interference. The data window of 1.2 s encompassing R peak was used. AR model of order 4 was fitted to the data. Four model coefficients and two parameters characterizing noise level were used as input parameters to classification procedure. For classification, quadratic discriminant function was used. The performance of classification averaged over 20 runs (different training and testing data sets), showing specificity and sensitivity above 90% for each of six classes of arrhythmias. These results demonstrated the usefulness of the AR model for quantification of ECG. The model may also be used for ECG data compression and in the portable telemedicine ECG systems, since the algorithms are relatively simple and work fast.

Wavelet transform was used in analysis of high-resolution ECG for risk evaluation in tachycardia [Lewandowski et al., 2000]. The time-frequency maps were obtained by modified Morlet wavelet (Figure 4.31). For quantification of late potentials (appearing after QRS complex) the index called irregularity factor, quantifying the variation of energy, was proposed.

For ECG feature extraction and reduction of redundancy, the orthonormal function model may be used. It is based on the Karhunen-Loeve transform, which provides

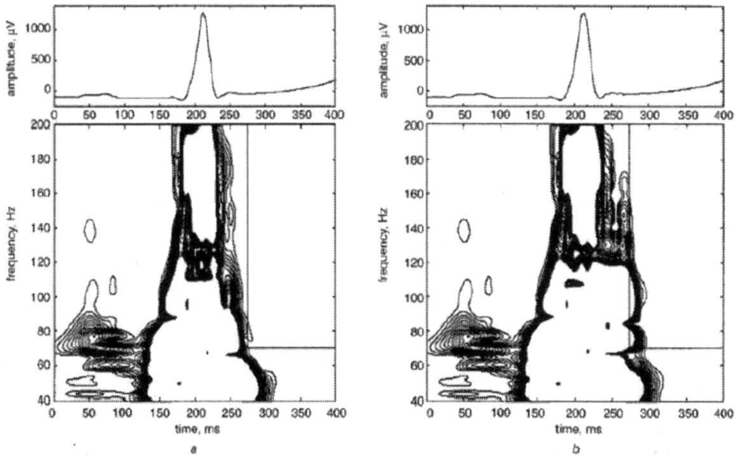

FIGURE 4.31: The time-frequency maps obtained by modified Morlet wavelet. a) ECG of healthy subject; b) the same signal with simulated late potentials. Time and frequency ranges for irregularity factor calculation are marked by straight lines. From [Lewandowski et al., 2000, Fig. 2].

signal representation by means of principal components (Sect.3.6.1). Representing ECG morphology by means of PCA provides robust estimates in terms of few descriptive parameters and allows for effective comparison of normal and deviating ECG pattern vectors exposing differences between them. In the procedure the noisy outliers may be eliminated on the basis of criteria concerning normalized residual error.

The application of PCA in ECG signal analysis included compression, filtering and QRS complex classification [Moody and Mark, 1990], shape representation of ECG morphology [Taddei et al., 1993], analysis of repolarization ST-T segment [Laguna et al., 1999], detection of transient ST-segment episodes during ambulatory ECG monitoring [Jager et al., 1998].

4.2.1.3.5 ECG patterns classification ECG feature extraction serves diagnostic purposes and is usually followed by the classification procedures. Artificial intelligence methods such as artificial neural networks are helpful in this respect. Different kinds of ANN including multilayer perceptrons and support vector machine classifiers have been used for ECG classification; training of networks may be performed by means of supervised or non-supervised learning. Introduction to supervised and unsupervised learning for ECG classification may be found in [Osowski et al., 2006, Wang and Azuaje, 2006].

4.2.2 Heart rate variability

The sequence of RR intervals (sometimes called normal-to-normal (NN) intervals)—that is, all intervals between adjacent QRS complexes resulting from sinus node depolarizations—forms the RR interval time series or RR tachogram. The sequence of successive times, t_i, $i \in \{1,2,\dots,n\}$, when the R wave appeared, e.g., identified by applying QRS detector to the ECG signal is:

$$\mathrm{RR}_i = t_i - t_{i-1} \tag{4.40}$$

A corresponding sequence of instantaneous heart rate is defined as:

$$\mathrm{ff}_i = \frac{1}{\mathrm{RR}_i} \tag{4.41}$$

Heart rate variability (HRV) is a physiological phenomenon of the variation of the time interval between heartbeats. For determination of HRV series ECG should be appropriately sampled. A low sampling rate may produce a jitter in the estimation of the R-wave fiducial points. The optimal recommended range is 250 to 500 Hz or perhaps even higher, while a lower sampling rate (in any case $\geq$ 100 Hz) may behave satisfactorily only if a properly chosen algorithm of interpolation is used to refine the R-wave fiducial point.

Since each RR interval is related to the previous one, the RR tachogram is unevenly sampled; therefore it has to be resampled to produce evenly sampled time series. Common resampling schemes involve linear or cubic spline interpolative sampling with resampling frequencies between 2 and 10 Hz. This procedure may introduce some bias in estimating power spectra. It was reported that the bias is lower for the procedure based on cubic splines than in case of linear resampling [Clifford and Tarassenko, 2004].

The Task Force of the European Society of Cardiology and the North American Society of Pacing Electrophysiology [Malik, 1996] provided an extensive overview of standards of measurement, physiological interpretation, and clinical use of HRV. It is believed that HRV is an indicator of sympathovagal interaction after acute myocardial infarction [Lombardi et al., 1987] and that it is helpful in diagnosis of congestive heart failure and predicting survival in premature babies. HRV is also relevant to the pathological conditions not necessarily connected with the heart action, e.g., diabetic neuropathy and susceptibility to SIDS. Since HRV is influenced by the autonomous nervous system the phenomena connected with its functioning have an impact on HRV. An example may be the sleep-wake cycle. HRV series may be analyzed in time or in the frequency domain.

4.2.2.1 Time-domain methods of HRV analysis

Time domain statistics are usually calculated on RR intervals without resampling. A review of the time domain methods of HRV analysis may be found in [Malik, 1996]. The basic indices involve: the standard deviation of RR intervals (which may be calculated over a 24-hour period, or short, usually 5 minutes, periods), the square root of the mean squared difference of successive RRs, the number of pairs of successive RRs that differ by more than 50 ms (so called NN50), the proportion of NN50 divided by total number of NNs (pNN50). These markers and other tools helpful in HRV quantification may be found in PhysioNet [Goldberger et al., 2000].

The series of RR intervals can also be represented by density distribution of RR interval durations or density distribution of differences between adjacent RR intervals. In construction of the density distribution a reasonable number of RR intervals is needed. In practice, recordings of at least 20 minutes (but preferably 24 hours) are recommended to ensure correct performance. The discrete scale of histogram should be appropriate—not too fine or too coarse, permitting the construction of smoothed histograms.

The deviations from normal distribution may be quantified by fitting to it some geometrical shapes or by means of higher order moments: skewness and kurtosis [Clifford, 2006]. However the higher the moment, the more sensitive it is to outliers and artifacts. Skewness—the indicator of asymmetry—is used to detect possible sudden acceleration followed by longer deceleration in the heart rate, the phenomenon which may be connected with clinical problems. Another measure quantifying RR distribution is the HRV triangular index measurement; it is the integral of the density distribution (that is, the number of all RR intervals) divided by the maximum of the density distribution [Malik and Camm, 1995].

4.2.2.2 Frequency-domain methods of HRV analysis

In HRV several rhythms can be distinguished, which are conventionally divided in four frequency bands:

- Ultra low frequency (ULF)): $0.0001\,\mathrm{Hz} \leq \mathrm{ULF} < 0.003\,\mathrm{Hz}$
- Very low frequency (VLF): $0.003\,\mathrm{Hz} \leq \mathrm{VLF} < 0.04\,\mathrm{Hz}$
- Low frequency (LF): $0.04\,\mathrm{Hz} \leq \mathrm{LF} < 0.15\,\mathrm{Hz}$
- High frequency (HF): $0.15\,\mathrm{Hz} \leq \mathrm{HF} < 0.4\,\mathrm{Hz}$

Fluctuations in VLF and ULF bands are thought to be due to long-term regulatory mechanisms such as thermoregulatory system, systems related to blood pressure, and chemical regulatory factors [Cerutti et al., 1995]. VLF appear to depend primarily on the parasympathetic system. HF is a measure of respiratory sinus arrythmias and can be considered an index of vagal modulation [Malliani et al., 1991]. Some studies suggest that LF, when expressed in normalized units, is a quantitative marker of sympathetic modulations; other studies view LF as reflecting both sympathetic activity and vagal activity. Consequently, the LF/HF ratio is considered by some investigators to mirror sympathovagal balance or to reflect the sympathetic modulations [Malliani et al., 1991]. It is important to note that HRV measures fluctuations in autonomic inputs to the heart, rather than the mean level of autonomic inputs.

In a spectrum calculated from short-term recordings of two to five minutes three main spectral components can be distinguished: VLF, LF and HF. The ULF component may be found from long-term recording, usually 24 hours. In case of long-term recordings the problem of stationarity arises. If the modulations of frequencies are not stable, the interpretation of the results of frequency analysis is less well defined. In particular, physiological mechanisms of heart period modulations responsible for LF and HF power components cannot be considered stationary during the 24-hour period, thus they should be estimated on the basis of shorter time epochs. To attribute individual spectral components to well-defined physiological mechanisms, such mechanisms modulating the heart rate should not change during the recording. Transient physiological phenomena should be analyzed by time-frequency methods. To check the stability of the signal in terms of certain spectral components, traditional statistical tests may be used.

Both, non-parametric and parametric methods are used for frequency analysis of HRV. The advantages of the non-parametric methods are: the simplicity of the algorithm used (fast Fourier transform) and high processing speed, while the advantages of parametric methods are: smoother spectral components that can be distinguished independent of pre-selected frequency bands, straightforward post-processing of the spectrum with an automatic calculation of low- and high-frequency power components (easy identification of the central frequency of each component), and an accurate estimation of PSD even on a small number of samples which is important for the quasi-stationary signal. A disadvantage of parametric methods may be the need to determine the model order, but this can be achieved by means of known criteria (Sect. 2.3.2.2.2).

The HRV signal with a period of artery occlusion is shown in Figure 4.32 a). The HRV spectral variability is illustrated in Figure 4.32 b). The running power spectra were obtained by means of the adaptive AR model with recursively fitted parameters [Mainardi et al., 1995]. In the above work the quantification of changes in HRV was performed by means of tracking the poles of transfer function (Sect. 2.3.2.2.4).

For determination of ECG derived respiratory information spectral analysis is usually applied. The respiratory activity is estimated as the HF component of HRV signal. The respiratory rhythm may be determined as the central frequency of the HF peak of the power spectrum or in case of AR model directly as a frequency determined by means of FAD (Sect. 2.3.2.2.4). The time-varying AR model was used for analysis of coupling of respiratory and cardiac processes [Meste et al., 2002]. Respiratory activity may also be found on the basis of beat morphology, namely from amplitude fluctuation of main complexes of ECG [Bailon et al., 2006], or possibly both methods can be combined [Leanderson et al., 2003].

The time-varying features of HRV may be studied by means of wavelet analysis. Both, continuous, e.g., [Toledo et al., 2003] and discrete wavelet analysis, e.g., [Roche et al., 2003] were applied in HRV analysis. Wavelet analysis of HRV of a patient with myocardial infarction is shown in Figure 4.33. Alternation of low (LF) and high (HF) frequency power is visible after the time of reperfusion. The advantage of discrete wavelet transform is the parametric description of the signal, which is useful for further statistical analysis and classification.

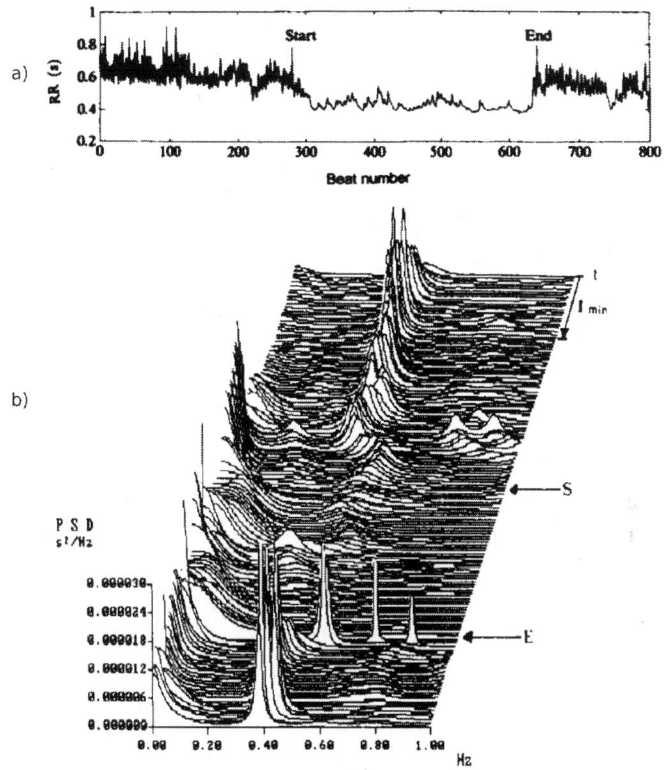

FIGURE 4.32: a) Tachogram during coronary artery occlusion. b) Spectra calculated from signal shown. Frequency—horizontal axis, time—oblique axis. S and E mark the beginning and end of occlusion respectively. From [Mainardi et al., 1995].

4.2.2.3 Relation of HRV to other signals

Blood pressure, respiration, and HRV are mutually related and the study of the relationships between these signals is a rich source of information on the function of the autonomous nervous system and its pathologies. In particular, cross-spectral analysis is a helpful tool in simultaneous analysis of these signals. The methodology and clinical applications of mutual analysis of HRV and blood pressure are described in [Baselli et al., 1988] and [Rienzo et al., 1993].

The possibility of performing a sleep evaluation on the basis of HRV and respiratory signal was indicated in [Bianchi et al., 2010]. The HRV and the respiration were analyzed in the frequency domain, and the statistics on the spectral and cross-spectral parameters put into evidence the possibility of sleep staging, arousals detection, and apnea recognition on their basis. This method may be especially useful for home monitoring of sleep.

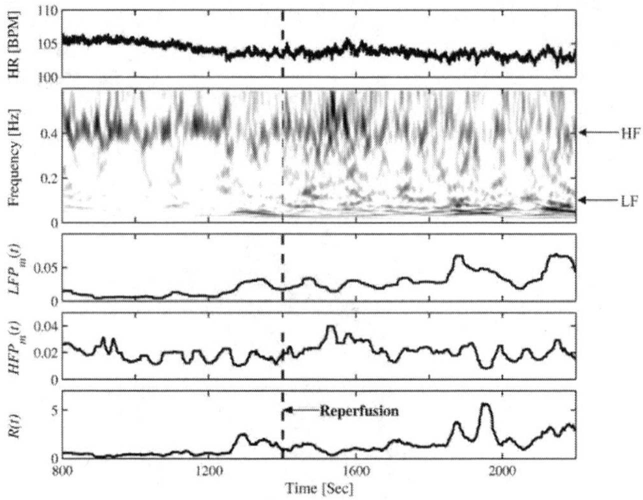

FIGURE 4.33: Wavelet analysis of HRV of patient with myocardial infarction. From top: signal, time-frequency energy distribution obtained by wavelet transform, low frequency power (LFP), high frequency power (HFP), ratio LFP/HFP. From [Toledo et al., 2003].

4.2.2.4 Non-linear methods of HRV analysis

Given the complexity of the mechanisms regulating heart rate, it is reasonable to assume that applying HRV analysis based on methods of non-linear dynamics will yield valuable information. The study concerning the properties of heart rate variability based on 24 hour recordings of the 70 subjects was conducted by [Urbanowicz et al., 2007]. It was reported that variability due to the linearly correlated processes was dominant (in normal group 85%), but with the development of the disease and risk of cardiac arrest the non-linear component increased. The contribution of the random noise variance was found at 5–15% of the overall signal variance. The exception was the case of atrial fibrillation where the contribution of random noise achieved 60%. One of the first applications of the non-linear methods was computation of the correlation dimension of the HRV [Babloyantz and Destexhe, 1988]. The most frequently used parameters that have been applied to measure non-linear properties of HRV include the correlation dimension, Lyapunov exponents, and Kolmogorov entropy. In spite of multiple attempts conducted with the use of the above methods to demonstrate the prevalence of chaos in heart rate series, more rigorous testing based primarily on surrogate data techniques did not confirm this hypothesis and revealed the weakness of classical non-linear methods connected with lack of sensitivity, specificity, and robustness to noise [Poon, 2000].

More recently scientists have turned to the methods which are less biased by noise and restricted length of the available data, namely: empirical mode decomposition, detrended fluctuation analysis, modified measures of entropy, Poincaré plots.

4.2.2.4.1 Empirical mode decomposition Empirical mode decomposition (EMD) is based on the decomposition of the original signal into components called instantaneous mode functions (Sect. 2.4.2.2.9). The method may be compared to band-pass filtering, but the sub-bands are not predetermined. The difference relies on the fact that high versus low frequency discrimination applies only locally, so selection of modes corresponds to adaptive, signal dependent, time-variant filtering. EMD found application in separation and tracking of rhythms present in HRV [Echeverria et al., 2001, Souza Neto et al., 2004].

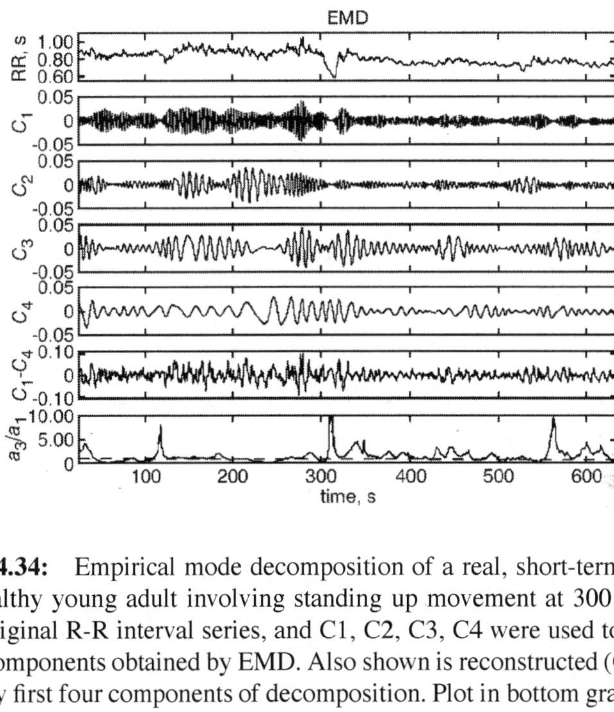

FIGURE 4.34: Empirical mode decomposition of a real, short-term HRV signal from a healthy young adult involving standing up movement at 300 s. Top graph presents original R-R interval series, and C1, C2, C3, C4 were used to describe the first four components obtained by EMD. Also shown is reconstructed (C1-C4) series obtained by first four components of decomposition. Plot in bottom graph represents Hilbert amplitude ratio of third and first components. From [Echeverria et al., 2001].

In [Echeverria et al., 2001] a series of simulations was performed illustrating the behavior of the method for stationary, non-stationary, and chirp signals. The calculations were also performed for experimental data involving measurements during rhythmic breathing and for HRV connected with the change of posture from seated to standing. First, the decomposition into instantaneous mode functions was performed, then the Hilbert transform was calculated for the consecutive components. Next, the instantaneous frequency was estimated from the phases of component time series

(Sect. 2.4.1). In Figure 4.34, the first four components, difference between component one and four, and Hilbert amplitude ratio of third and first component (a3/a1) as a function of time are illustrated. The value of component ratio may be considered as a convenient measure of the HRV rhythms evolution. One can observe the increase of the ratio a3/a1 after changing posture.

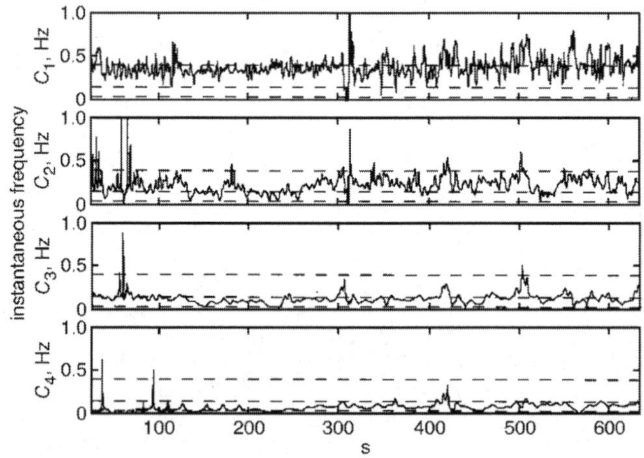

FIGURE 4.35: Instantaneous frequency as a function of time for the first four components presented in Figure 4.34. From [Echeverria et al., 2001].

In Figure 4.35 instantaneous frequency as a function of time for first components presented in Figure 4.34 is shown. The instantaneous frequency (IF) seems to be a good index for tracing the changes in HRV rhythms, but in the procedure of IF determination appropriate boundary requirements have to applied, which is not a trivial problem [Huang et al., 1998]. Another problem in application of EMD is the reproducibility of results. In the study of [Maestri et al., 2007] (described in more detail in Sect. 4.2.2.4.5) the reproducibility of EMD was found to be very poor. It seems that the EMD method has to be used with caution, especially in respect of fulfilling the assumptions of the method and in respect of statistical significance and reproducibility of results.

4.2.2.4.2 Entropy measures The indexes connected with information theory (Sect. 2.5.6): approximate entropy (ApEn) [Pinkus and Singer, 1996], conditional entropy [Porta et al., 1999], and sample entropy (SaEn) [Richman and Moorman, 2000] were introduced and found applications in recognition of certain cardiovascular pathologies. For example in [Perkiomaki et al., 2005] a progressive decrease in ApEn indicating reduced complexity of heart rate variability was observed before the onset of atrial fibrillation. Conditional entropy was used for evaluation of regularity, synchronization, and coordination in cardiovascular beat-to-beat variability [Porta et al., 2000].

SaEn was applied for a study of heart rate variability during obstructive sleep apnea episodes by [Al-Angari and Sahakian, 2007]. When compared with spectral analysis in a minute-by-minute classification, sample entropy had an accuracy, sensitivity and specificity slightly worse than the results supplied by the spectral analysis. The combination of the two methods improved the results, but the improvement was not substantial. Nevertheless, SaEn has begun to be applied in practice, namely in wearable devices. It was reported that SaEn was used to assess respiratory biofeedback effect from HRV [Liu et al., 2010].

4.2.2.4.3 Detrended fluctuation analysis Another measure of complexity is provided by detrended fluctuation analysis (DFA). In its framework the fractal index of self-similarity is determined (Sect. 2.5.3).

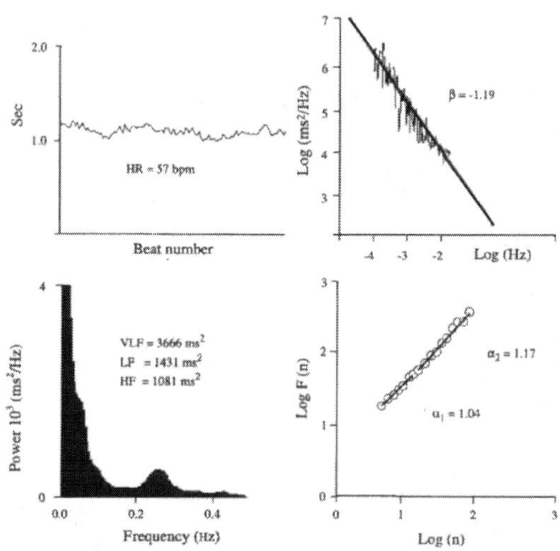

FIGURE 4.36: Scaling properties of 24 hour tachogram of a healthy subject. RR interval tachogram (upper left), power spectra (lower left), a power-law scaling slope of long-term fluctuations of heartbeats (upper right), and DFA results (lower right). β—long-term scaling slope, α_1 short-term fractal scaling exponent (4–11 beats), α_2 intermediate scaling exponent (> 11 bits). From [Perkiomaki et al., 2005].

The HRV scaling exponent value depends on the window length. In Figure 4.36 the examples of DFA results illustrating dependence of scaling exponents on data length are shown. In the lower right part of the figure it may be observed that the slope of the log-log plot changes with increase of the window size > 11 heartbeats. The short-time fractal scaling exponent has a relatively good correlation with the ratio of the low- to high spectral component in controlled recording conditions [Perkiomaki et al., 2005]. It was reported that short-term scaling exponents derived from HRV series can be predictors of vulnerability to ventricular tachycardia, ventricular fibrillation, arrhythmic death, and other heart conditions (review in [Perkiomaki et al., 2005]). Power law relationship of heart rate variability was used as a predictor of mortality in the elderly [Huikuri et al., 1998]. However the mono-fractal (scale invariant) behavior need not testify for the chaotic character of the data. It was reported that surrogate data test revealed the same α and β values for original HRV series as for the ones with randomized phases [Li et al., 2008].

In [Peng et al., 1995] by means of DFA the scaling exponent was determined for 24 h recordings of HRV of a healthy group and patients with congestive heart failure. Short term and long term scaling exponents α_1 and α_2 differed significantly for normal and pathological groups, but the separation was not complete—there was substantial overlap between both groups in parameter space (8 out of 12 healthy subjects and 11 out of 15 pathological subjects exhibited a "cross-over" between groups).

The same data (from PhysioNet database, Beth Israel Hospital) as those used by [Peng et al., 1995] were subjected to wavelet analysis and DFA by [Thurner et al., 1998]. For classification, standard deviations of wavelet coefficients at scales 4–5 (16–32 heart beats) was used, which yielded 100% classification (sensitivity and specificity) between the groups. In the above work, from the wavelet coefficients scaling exponents α were also determined. The discrimination between groups based on their values resulted in 87% sensitivity and 100% specificity, which was better than in [Peng et al., 1995] but worse than for parametrization by wavelet standard deviation. Thurner et al. conclude that their multiresolution approach succeeded not only because it discriminates trends, but also because it reveals a range of scales over which heart failure patients differ from normals.

4.2.2.4.4 Poincaré and recurrence plots Other non-linear methods of analyzing heart rate variability are the Poincaré map (PM) and recurrence plot (RP). The advantage of these methods is that they can be applied to short and non-stationary data. Recurrence plots were applied to heart rate variability data by [Marwan et al., 2002]. The examples of recurrence plots for HRV epochs before ventricular tachycardia (VT) in comparison to control epochs HRV are shown in Figure 4.37. The RP for epoch before VT onset shows big rectangular structures. Usually RP plots are characterized by specific shape parameters, among them diagonal line lengths and vertical line lengths, which are important for characterizing different conditions.

The study of HRV from 24 h ECG by means of entropy measures and Poincaré maps was conducted by [Żebrowski et al., 1994]. The examples of three-dimensional RR interval return maps (Poincaré maps) are shown in Figure 4.38. One can observe different patterns for cases: before cardiac arrest and one year later after medication. The authors point out that although pathological patterns differ from the normal ones, it is difficult to distinguish between different pathological cases (Figure 4.38 c and d).

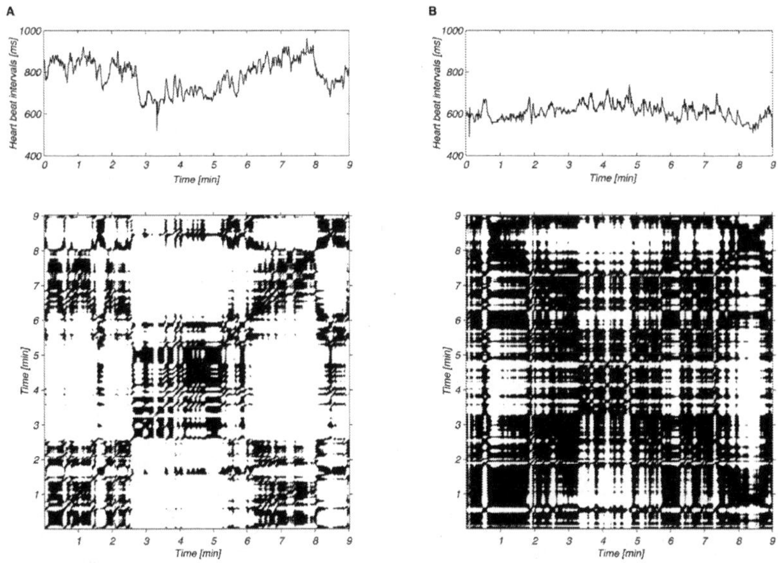

FIGURE 4.37: Recurrence plots of the heartbeat interval time series at a control time (a) and before a VT (b) with an embedding dimension of 6. The RP before a life-threatening arrhythmia is characterized by big black rectangles, whereas RP from the control series shows only small rectangles. From [Marwan et al., 2002].

4.2.2.4.5 Effectiveness of non-linear methods Different non-linear methods were compared in [Maestri et al., 2007]. The authors computed 11 non-linear HRV indices found from: detrended fluctuation analysis, entropy estimators including SaEn, empirical mode decomposition, and Poincaré plots. The performance of these measures was evaluated by comparison of results of the two 5-minute sessions conducted on two consecutive days for a group of 42 healthy adults. The reliability of measures was assessed by means of the test of no difference between two measurements (paired t-test), and relative reliability was based on the interclass correlation coefficient. The best results were obtained for Poincaré plots, detrended fluctuation analysis, and entropy measures; other indices performed rather poorly. These reliability measures were based on short-term measurements and two data sets only, so they have to be considered with care. In assessment of the utility of different methods other measures such as: sensitivity and specificity in distinguishing different conditions and robustness to noise have to be taken into account as well.

Although in principle, the above mentioned techniques have been shown to be good tools for characterization of various complex systems, no major breakthrough has yet been achieved by their application to biomedical signals including HRV analysis. No systematic study has been conducted by means of non-linear methods to investigate large patient populations. Therefore they were not recommended for diagnostic applications by the Task Force of the European Society of Cardiology and

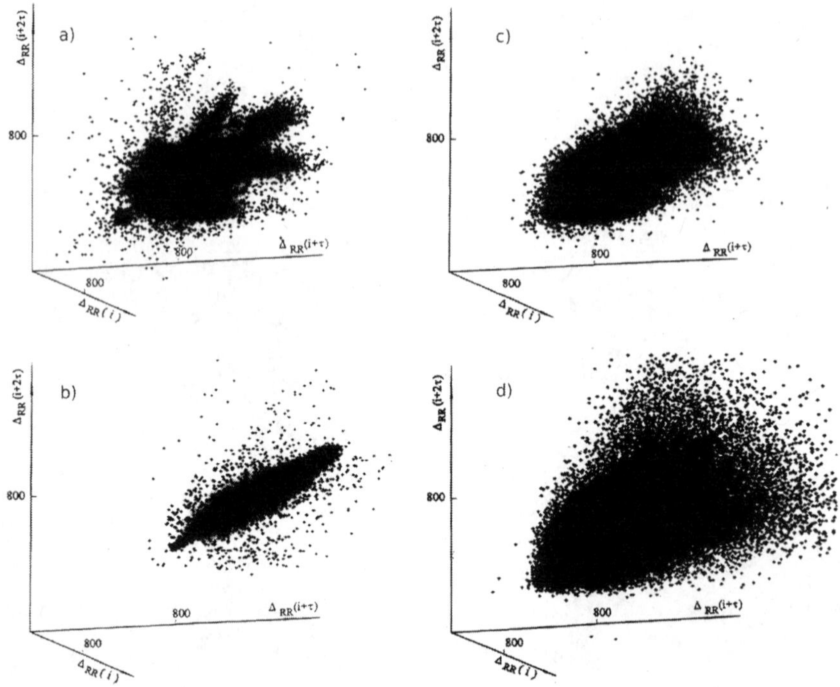

FIGURE 4.38: Three-dimensional images of RR interval return maps calculated from 24-h Holter ECG recordings. a) and b) data from the same subject before cardiac arrest and 1y after, respectively; plot b) is very similar to plots obtained for healthy subjects; c) example plot for a ventricular arrhythmia; d) example plot for atrial fibrillation. Adapted from [Żebrowski et al., 1994].

the North American Society of Electrophysiology [Malik, 1996]. Since 1996 considerable progress has been achieved in the field of applications of non-linear methods to HRV, but at present the standards are still lacking and none of these methods is in widespread clinical use. For validation of non-linear methods the studies involving application of both linear and non-linear approaches applied for the same data are needed.

4.2.3 Fetal ECG

The fetal electrocardiogram is the electrical activity of a fetus heart, which contains valuable information about the physiological state of a fetus and is an important tool in monitoring its well being. For instance pathological conditions like hypoxia or acidemia may be detected by means of fECG. The amplitude of fECG is in the range from 10 to 100 μV, whereas the maternal ECG (mECG) can be as high as:

0.5 mV–100 mV. When measured on maternal abdomen fECG is additionally disturbed by the muscular activity from abdomen and uterus, since the main frequency range of fetal and maternal ECG is in the frequency range 0.3–100 Hz, while for the abdominal muscle it is 10–400 Hz and for uterine contractions 0.01–0.6 Hz. Therefore the problem of extraction of fECG from other disturbing bioelectric signals is not a simple one. An example of mECG extraction from abdominal ECG is shown in Figure 4.39.

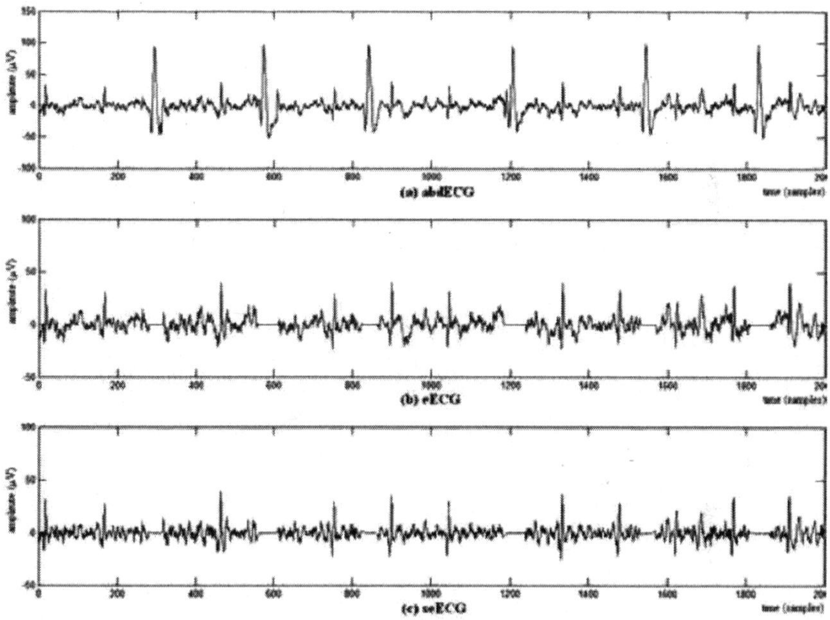

FIGURE 4.39: Example of abdominal ECG (abdECG) decomposition. Maternal QRS elimination and smoothing. (a) abdECG signal, (b) the signal after the maternal QRS elimination (eECG signal). (c) Smoothed eECG signal (seECG). From [Karvounis et al., 2007].

Several methods for extraction of fECG from maternal ECG (mECG) were devised including: adaptive filtering [Martinez et al., 1997], singular value decomposition [Kanjilal et al., 1997], fractals [Richter et al., 1998], neural networks [Camps et al., 2001], fuzzy logic [Al-Zaben and Al-Smadi, 2006], polynomial networks [Assaleh and Al-Nashash, 2005], combination of PCA and matched filtering [Martens et al., 2007], template construction [Taralunga et al., 2008]. Wavelet transform also found several applications to the fECG analysis, e.g., [Karvounis et al., 2007] and [Khamene and Negahdaripour, 2000] who used biorthogonal quadratic spline wavelet. Two-level wavelet transform (Daubechie wavelets) followed by low-pass filtering was applied by [Hassanpour and Parsaei, 2006]. [Jafari and Chambers, 2005] used for fECG extraction sequential source separation in the wavelet domain.

The problem of fECG extraction from the signal registered on the abdomen can be formulated as a blind source separation problem, since it involves the separation of the original source signals from a sensor array, without the knowledge on the transmission channels. BSS method involving singular value decomposition and ICA was applied to fECG by [De Lathauwer et al., 2000]. It was found that in the job of separation of fECG from mECG the independent component method performed better than PCA. The results of ICA application to eight channel data is shown in Figure 4.40.

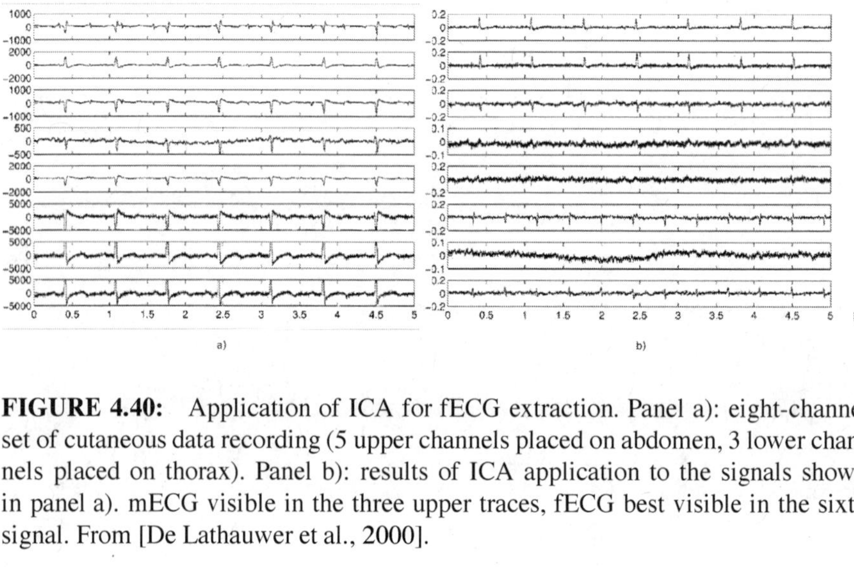

FIGURE 4.40: Application of ICA for fECG extraction. Panel a): eight-channel set of cutaneous data recording (5 upper channels placed on abdomen, 3 lower channels placed on thorax). Panel b): results of ICA application to the signals shown in panel a). mECG visible in the three upper traces, fECG best visible in the sixth signal. From [De Lathauwer et al., 2000].

In solving the problem of fECG – mECG separation, especially by ICA- based BSS approaches, some theoretical limitations appear, due to the requirement that the number of observations (registered signals) must be equal to or higher than the number of uncorrelated signal sources. However, in addition to heart signals, each electrode also picks up activity due to maternal myoelectric activity and other noises, which are the sources of partly uncorrelated noise different for each channel. Therefore BSS requires multielectrode arrays, usually including electrodes placed on thorax, which is not always practical in the clinical environment.

The aim of several works concerning fECG was the extraction of a fetal heart rate variability (fHRV) signal which can be used to find the synergetic control activity of the sympathetic and parasympathetic branches of autonomous nervous system

[Cerutti et al., 1986]. The method of fHRV estimation based on the minimization of a cost function that measures the differences between the discrete Fourier transform (DFT) of the fetal ECG waveform and the DFTs of its circularly shifted forms was proposed in [Sahin et al., 2010]. By using the linear phase shift property of the DFT, they showed that the minimization of this cost function is equivalent to finding the cosine waveform that matches best to the fECG power spectrum. The optimal cosine waveform was then used to estimate the fundamental period of fHRV.

The problem of separating the components of fHRV was approached as well by means of empirical mode decomposition by Ortiz et al. [Ortiz et al., 2005]. High frequency fHRV modes were obtained by application of EMD followed by the reconstruction of components above 0.3 Hz. The results showed differences in the power of frequency components for the episodes connected with breathing movements and body movements versus quiet epochs.

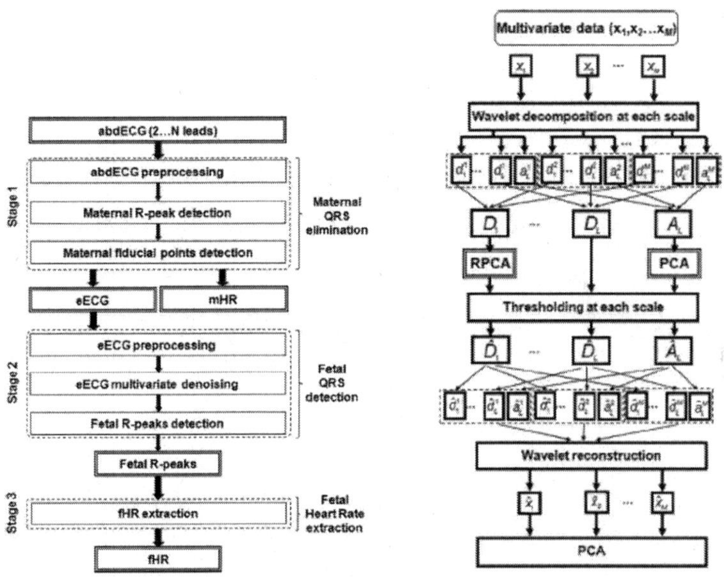

(a) (b)

FIGURE 4.41: (a) Flowchart of the methodology of fetal heat rate extraction. (b) Flowchart of the multivariate denoising technique. From [Karvounis et al., 2009].

A three stage method involving different signal analysis techniques for extracting fECG and determination of fHRV was proposed by [Karvounis et al., 2009]. The methodology can be applied to two or more channels. The flowchart of the method is shown in Figure 4.41 (a). The first stage involved mECG elimination. After filtering the signals in the 4–20 Hz band the parabolic fitting technique was used to detect maternal R-peaks. In order to determine maternal fiducial points a phase space thresholding method based on Poincaré maps was used. The axis in 3-D phase space were constructed from: signal, its first and second derivative. The obtained 3-D space plot is shown in Figure 4.42.

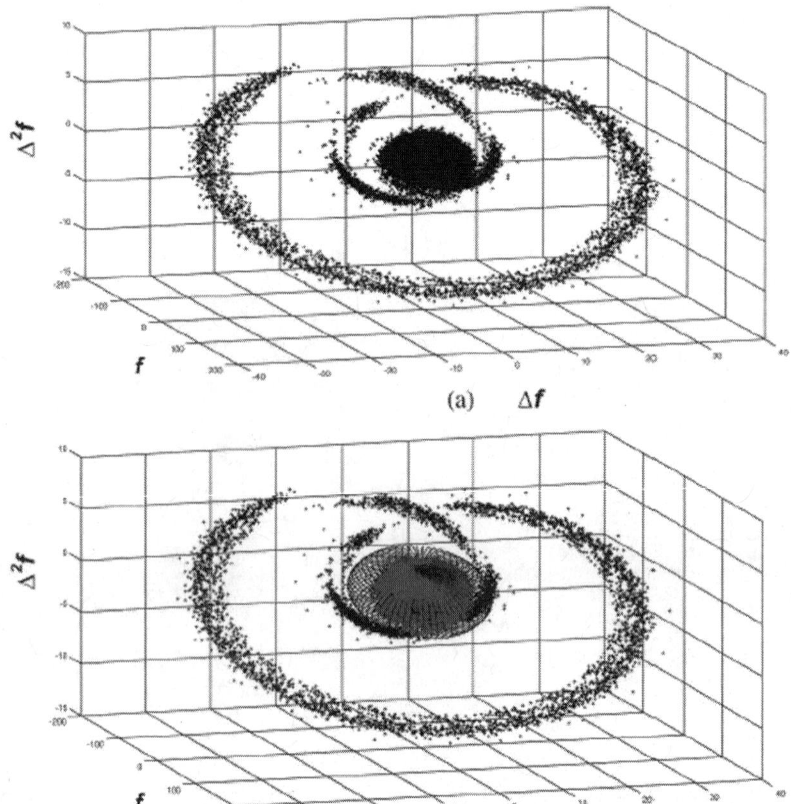

FIGURE 4.42: (a) Three-dimensional phase-space map and (b) ellipsoid used in the maternal QRS-points detection. From [Karvounis et al., 2009].

The thresholding method used for identification of QRS complex was based on observation that in the 3-D plot the points in the center correspond mainly to P and T waves and low frequency noise, while the points in the surrounding torus correspond to QRS and high frequency noise. As a threshold for QRS separation the 3-D ellipsoid was defined, which made it possible to determine the fiducial points of QRS and perform the subtraction of mECG from the original (not filtered) signal. In the second stage, the denoising was performed by means of the PCA application to wavelet coefficients (6-levels Daubechies) followed by reconstruction. The procedure of denoising is shown in Figure 4.41 (b). Then in the denoised signal R peaks of fECG not coincident with maternal R peaks were identified. In order to detect the superimposed peaks, a histogram of R-R intervals was constructed and the intervals of double length were considered in the procedure of finding the lost fetal R peaks. This information was used to construct the final fHRV series.

The above described method was tested on simulated signals with various signal to noise (SNR) ratios, generated according to [Sameni et al., 2007] and on experimental signals from the University of Nottingham database [Pieri et al., 2001]. Simulated data helped to improve the parameters of the ellipsoid used for thresholding, which resulted in better sensitivity values. The results for experimental time series also gave high values (around 90%) of selectivity. The described approach was compared with other methods: time-frequency based methodology [Karvounis et al., 2007], parabolic fitting [Zhang et al., 2006], template matching [Gibson et al., 1997], fast ICA algorithm [Hyvärinen, 1999]. The accuracy for real data showed the best results for the above described method and time-frequency method.

4.2.4 Magnetocardiogram and fetal magnetocardiogram

4.2.4.1 Magnetocardiogram

The magnetic field produced by a heart is around 50 pT, which is more than an order of magnitude bigger than the one generated by the brain (around 1 pT), which makes MCG easier to record than MEG. The first magnetocardiogram (MCG) was measured almost 50 years ago by Baule and Mc Fe by means of a gradiometer using a set of copper coils with several millions of windings. The era of magnetocardiology was opened by the introduction of SQUID. At first single sensors were used. At present commercially available devices include multiple MCG leads and the custom made systems have up to 128 sensors (the description of the magnetic field sensors may be found in Sect. 4.1.4).

One of the advantages of MCG is the lack of direct contact of the sensors with the skin which allows for fast screening of patients and is useful for people with burns. The same currents generate MCG and ECG signals and their time evolution is similar, but not identical. Tangential components of the fields are attenuated in the ECG. Contrary, the ideal magnetic lead is sensitive only to tangential components of electric sources and hence should be particularly responsive to abnormalities in activation, since normal activation sources are primarily radial. Another advantage of MCG is connected with the fact that magnetic permeability of the tissue is that of free space. This allows to record by means of MCG activity of the posterior side of the heart. In ECG measurement this activity is strongly attenuated by the resistivity of the lungs.

It was reported that MCG gives unique information on cardiac ischemia, arrhythmias, and fetal diagnosis [Yamada and Yamaguchi, 2005]. MCG was successfully applied in diagnosis of several pathological heart conditions. For example inspecting ST shifts in MCG during an exercise test [Cohen et al., 1983] demonstrated that injury currents resulting from ischemia, practically impossible to measure with ECG, may be detected by MCG. Usefulness of magnetic field maps to detect angina pectoris and myocardial infarction in cases when no ST complex elevation is present was reported in [Lim et al., 2009]. It was also reported that by means of MCG it is possible to distinguish the features characterizing coronary arterial disease and arrhythmia vulnerability [Stroink et al., 1999].

The MCG time evolution and an example of a measurement system are shown in Figure 4.43. Construction of maps representing magnetic field distribution measured over the thorax is one of the methods useful for clinical diagnosis. An example of such an approach may be the work by [Gireesan et al., 2010]. The MCG recorded by means of 36 sensors was filtered in the frequency band 0.01 to 120 Hz and the artifacts were removed by means of the wavelet transform. The MCG evolutions synchronized in respect to the R peak were averaged. In Figure 4.44 the averaged traces together with a magnetic field map are displayed. The MCG shows in several positions splitting of the R waves (not visible in ECG), which is a symptom of left bundle branch block. In the iso-magnetic field contour plot for the instant of R peak, the contributions of two equivalent current dipoles may be observed. In the

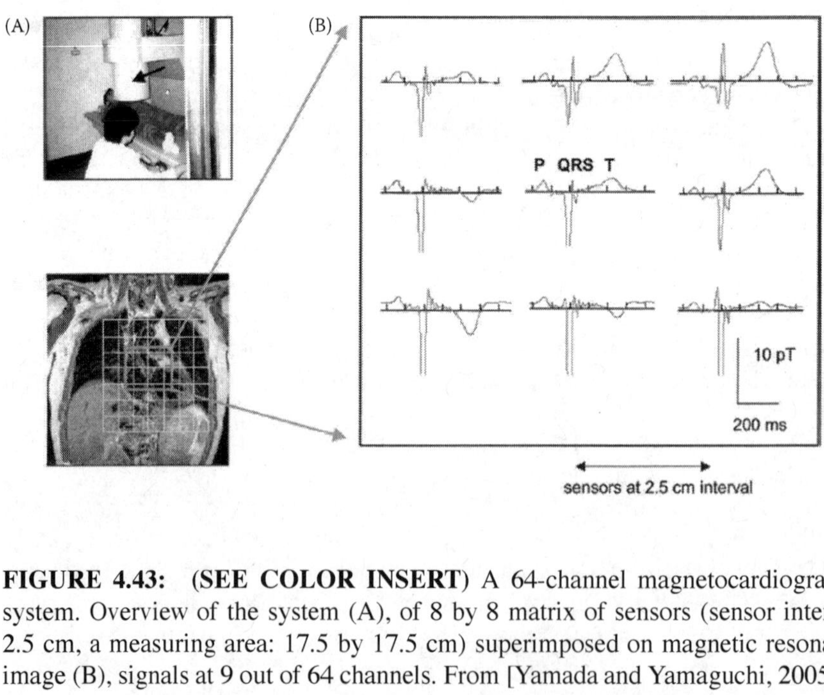

FIGURE 4.43: (SEE COLOR INSERT) A 64-channel magnetocardiographic system. Overview of the system (A), of 8 by 8 matrix of sensors (sensor interval: 2.5 cm, a measuring area: 17.5 by 17.5 cm) superimposed on magnetic resonance image (B), signals at 9 out of 64 channels. From [Yamada and Yamaguchi, 2005].

same study the fragmentation of the R peak of MCG and distinct contributions of more than one dipole in the contour maps was observed for several subjects, which didn't show distinct changes in ECG. It was postulated that the fragmentation of R peak of MCG was connected with altered ventricular depolarization. In the several investigated cases the MCG has revealed more distinct features connected with the disturbances in the heart action than ECG, which opens the possibility to improve the ability of early detection of myocardial infarction by means of MCG.

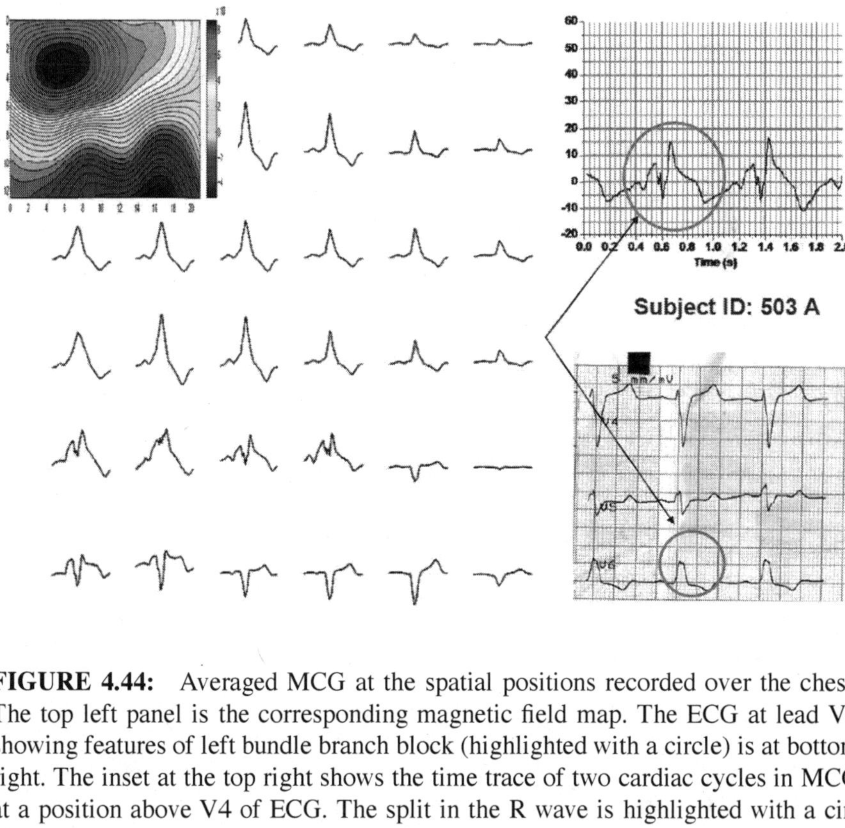

FIGURE 4.44: Averaged MCG at the spatial positions recorded over the chest. The top left panel is the corresponding magnetic field map. The ECG at lead V6 showing features of left bundle branch block (highlighted with a circle) is at bottom right. The inset at the top right shows the time trace of two cardiac cycles in MCG at a position above V4 of ECG. The split in the R wave is highlighted with a circle. From K. Gireesan, C. Sengottuvel, C. Parsakthi, P. Rajesh, M. Janawadkar, T. Radhakrishnan: Magnetocardiography study of cardiac anomalies, in Selma Supek, Ana Susac (Eds.): *Advances in Biomagnetism BIOMAG2010, IFMBE Proceedings* 28, 2010, pp. 431435, Figure: 1, www.springerlink.com/ifmbe ©International Federation for Medical and Biological Engineering 2010. (This figure is courtesy of IFMBE.)

The utility of the cardiac field mapping for the detection of coronary artery disease was studied in [Grapelyuk et al., 2010]. The comparison of the results with the ideal group mean maps was based on the three measures of similarity: Kullback-Leibler (KL) entropy (Sect. 3.4.2), normalized residual magnetic field strength, and deviations in the magnetic field map orientation. The mean values of these parameters during the depolarization and repolarization were used for the classification with the help of logistic regression. The features set based on KL entropy demonstrated the best classification results, namely sensitivity/specificity of 85/80% was reported.

The possibility of identification of inter-atrial conduction pathways by MCG was reported by [Jurkko et al., 2009]. The experiment involved the patients undergoing catheter ablation of paroxysmal atrial fibrillation. The intra-cardiac electroanatomic mapping was compared with pseudocurrent maps obtained by means of MCG recorded by 33 triple sensors, namely two planar gradiometers (x-y plane) and a magnetometer oriented along the z-axis (axis orientation as in Figure 4.27). Figure 4.45 shows averaged magnetic field density at each z-magnetometer, spatial distribution of B_z component, and pseudocurrent map. The pseudocurrent inversion is used to characterize the orientation of the magnetic field. The method is based on rotating the estimated planar gradients of B_z component by 90°:

$$b = \frac{\partial B_z}{\partial y}\bar{e}_x - \frac{\partial B_z}{\partial x}\bar{e}_y \tag{4.42}$$

where $\bar{e}_x$, $\bar{e}_y$ are the perpendicular unit vectors on the sensor array plane. The resulting arrow map provides a zero order approximation (pseudocurrent map) for the underlying electric current. The conclusions from the above studies were that by inspecting the pseudocurrent maps the current propagation may be determined and the damages in conducting pathways may be found. Overall findings imply that the non-invasive technique of pseudocurrent mapping may assist in the localization of the breakthroughs in conduction pathways and refining methods for the ablation treatment, if needed.

Biomagnetic methods are well suited to solve the problem of the localization of bioelectric sources in the body. Although magnetic fields are much less influenced by the body tissues than electric fields, MCG is to some extent modified by anisotropic properties of cardiac and torso tissues. The best localization results were obtained when activity starts in a small well defined area. For solution of inverse problems in magnetocardiography both the equivalent current dipole approach and effective magnetic dipole models have been used. It was reported that they may provide localization with accuracy of a few millimeters when realistic and patient tailored torso models are used [Pesola et al., 1999]. However, the approach is limited to localizing point-like focal sources such as those appearing, e.g., in focal arrhythmias or ventricular extrasystole [Fenici and Brisinda, 2006].

In general, multiple current sources are simultaneously activated during the cardiac cycle; thus the inverse solution based on current density distribution found several applications in magnetocardiography. For example in [Neonen et al., 2001] for equivalent current-density estimation in subjects with coronary artery disease, a patient-specific boundary-element torso model was used. Three different methods of regularization were confronted with the PET measurements. The results showed that

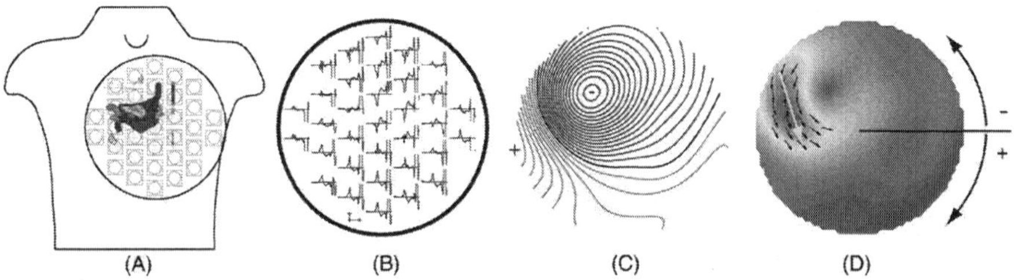

FIGURE 4.45: **(SEE COLOR INSERT)** Recording and analysis of atrial magnetic fields. (A) The sensor arrangement of a 33 triple sensor (99-channel) magnetometer. Superimposed on sensor array is the antero-posterior view of the left atrium by electroanatomic mapping. (B) Signal—averaged magnetic field density on each magnetometer channel over cardiac cycle. The onset and end of atrial signal are determined automatically using filtering technique. (C) Spatial distribution of the magnetic field B_z component over the middle part of atrial complex interpolated from the measurement using multipole expansion. The blue color indicates flux out of the chest (-) and red color flux into the chest (+). The step between two consecutive lines is 200 fT. (D) Pseudocurrent map derived by rotating magnetic field gradients by 90°. The red-yellow color indicates the area of the top 30% of strongest currents, and the large arrow indicates their mean direction. Zero angle direction is pointing from subject's right to left and positive clockwise. From [Jurkko et al., 2009].

non-invasive localization of ischemic areas in patients with coronary artery disease is possible when proper regularization is applied.

4.2.4.2 Fetal MCG

Magnetic field measurement is a promising technique for evaluation of fetal well-being. Fetal biomagnetic signals are unaffected by poor electrical conductivity of the vernix caseosa, a waxy substance which forms on the fetal skin at about 25 weeks' gestation and impedes the transmission of fetal bioelectric signals. Fetal magnetocardiography (fMCG) has the potential to provide beat-to-beat fetal heart rate in case of normal rhythm and its disturbances. Amplitude changes in fMCG and fetal heart rate acceleration make fetal movement assessment possible by means of a technique called actocardiography [Zhao and Wakai, 2002].

Various groups reported good quality fMCG recordings in which P, QRS, and T waves may be distinguished even without signal averaging. However, usually after bandpass filtering and maternal MCG removal, the averaged waveforms are computed after aligning fetal complexes, e.g., by means of autocorrelation. This kind of analysis was used for finding fetal cardiac repolarization abnormalities [Zhao et al., 2006].

For the determination of fetal heart rate variability (fHRV), methods similar to the ones described in the section concerning fECG may be used. In [Wilson et al., 2008] another approach was applied. The aim of the study was determination of the fetal heart rhythm; the shape of the complexes was not considered. After removal of maternal MCG four approaches for fetal QRS identification were tested: Hilbert transform, ICA, ICA followed by Hilbert transform, and filtering. The Hilbert transform (HT) yields an analytic signal whose amplitude is always positive independently of signal polarity (Sect. 2.4.1). The value called rate of change of HT amplitude (*RHA*) was defined as:

$$RHA(n) = \sqrt{(x_{n+1} - x_n)^2 + (y_{n+1} - y_n)^2} \quad (4.43)$$

where x_n and y_n are real and imaginary parts of the transformed signal at point n. The series $RHA(n)$ is always positive and signals from different channels may be easily averaged. The next step was application of FastICA. FastICA (Sect. 3.6.2.3) finds the columns of separating matrix by maximizing the absolute value of kurtosis, which works particularly well for QRS extraction due to the high value of kurtosis for QRS complexes. The third approach used in [Wilson et al., 2008] relied on application of HT to ICA components. The fourth method involved manual selection of QRS complexes in a signal filtered in the 1-100 Hz band. The output signals obtained by the above four methods are shown in Figure 4.46. The next step in analysis involved correction for missed beats which was performed in a way similar to the method described in the section on fECG. In the quantitative analysis of results the Hilbert method scored best, yielding the smallest number of total errors and efficiency of 99.6% in comparison to ICA (94.4%) and ICA-HT (96.4%).

Adaptive ICA was used in real-time fetal heart monitoring. The developed monitoring system [Waldert et al., 2007] consisted of a real-time access to fMCG data, an algorithm based on independent component analysis, and a graphical user interface. The authors reported that the algorithm extracts the current fetal and maternal heart signal from a noisy and artifact-contaminated data stream in real-time and is able to adapt automatically to continuously varying environmental parameters.

In spite of the numerous successful clinical applications based on the measurement of heart magnetic fields the advantages and disadvantages of ECG and MCG are still a matter of debate [Hren et al., 1998, Stroink, 2010]. Obviously the advantage of ECG is lower cost and easier support of the devices. The long years of clinical experience in ECG have some meaning as well. It seems that in future the decisive improvement may bring the fusion of both magnetic and electric data.

4.3 Electromyogram

Electromyogram (EMG) is a record of electrical muscle activity. In this section, only EMG of striated muscles will be discussed, the electrical activity of specific smooth muscles such as intestines—electroenterogram (EEnG) or stomach—electrogastrogram (EGG), will be discussed in the next section.

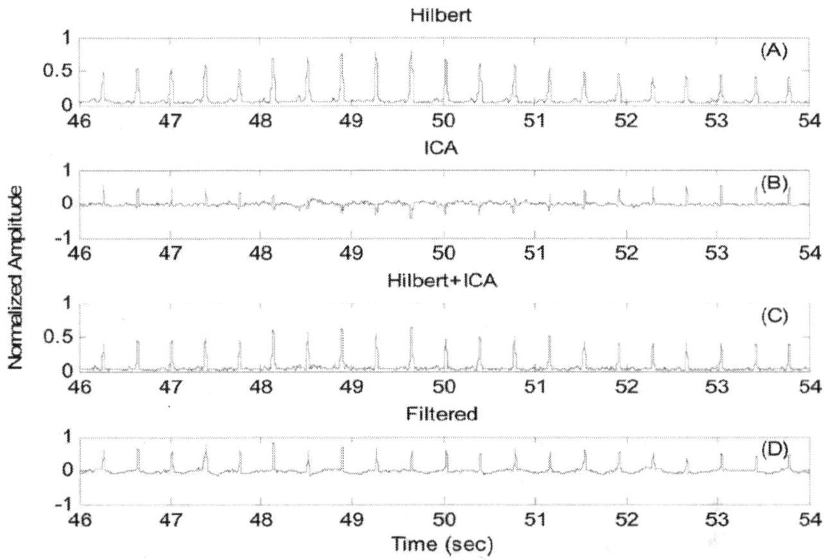

FIGURE 4.46: Plots of 8-s duration signal showing the output from the four post-processing methods. (A) Hilbert. (B) ICA. (C) Hilbert + ICA. (D) Filtered. In (B), the ICA component shows a momentary polarity inversion that results in missed beats. In (C), the Hilbert method applied to ICA components removes the polarity inversion. From [Wilson et al., 2008, Fig. 6].

4.3.1 Measurement techniques and physiological background

The EMG characteristics depend on the kind of applied electrodes. EMG may be recorded by means of intramuscular (needle or fine wire) or surface electrodes. The range of measured potential difference in case of intramuscular electrodes is 0.1–20 mV, and for surface electrodes 0.05–1 mV. In clinical practice EMG is used for routine diagnosis of neuromuscular disorders, based typically on single fiber (Figure 4.47 a) or concentric needle electrodes (Figure 4.47 b). Surface electrodes enable non-invasive EMG recording of the global muscle electrical activity. Earlier, surface electrodes were applied only in situations when the insertion of needle electrodes was difficult or not possible, e.g., for examination of children, long-term recordings, ergonomics, sports, or space medicine. Nowadays surface EMG electrode arrays (Figure 4.48) are available, with which more precise information on muscle activation can be obtained. This technique requires, however, advanced signal processing methods, which will be described later.

The registered EMG signal is a superposition of activity of single motor units (MUs). A single MU consists of a motoneuron and the group of muscle fibers innervated by its axon (Figure 4.49 a). Fibers belonging to a given MU are distributed throughout a muscle (Figure 4.49 b). Every motoneuron discharge evokes contrac-

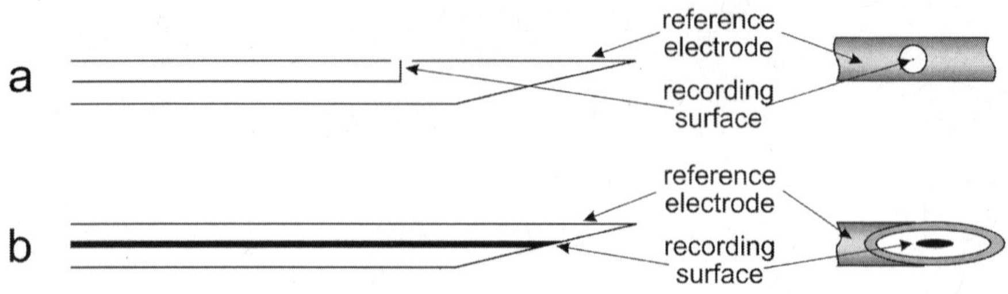

FIGURE 4.47: Types of EMG electrodes. a) Single wire electrode, b) needle electrode.

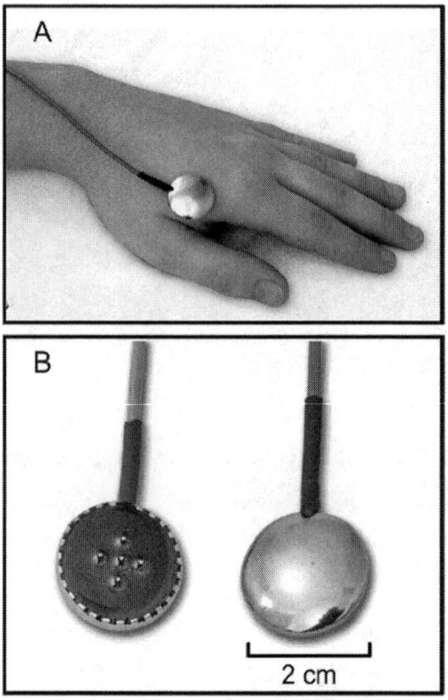

FIGURE 4.48: Surface electrodes array. From [Nawab et al., 2010].

tion of all its muscle fibers, preceded by their depolarization, which is detected as a waveform called motor unit action potential (MUAP). The spike component of MUAP (and its peak-to-peak amplitude) is determined by a few fibers closest to the electrode; the more remote fibers influence mainly MUAP duration. MUAP shape depends also on the temporal dispersion between single fiber potentials. The temporal dispersion is determined mainly by the distribution of motor endplates along muscle fibers, resulting in the differences in arrival times to the electrode. The shapes of consecutive MUAPs are not identical due to the variability of these arrival times caused by the small fluctuations in the synaptic transmission called jitter. In the healthy muscle, jitter is small (for biceps brachii around 20 μs) and for small force level the MUAPs generated by the given MU are easily identifiable by visual inspection.

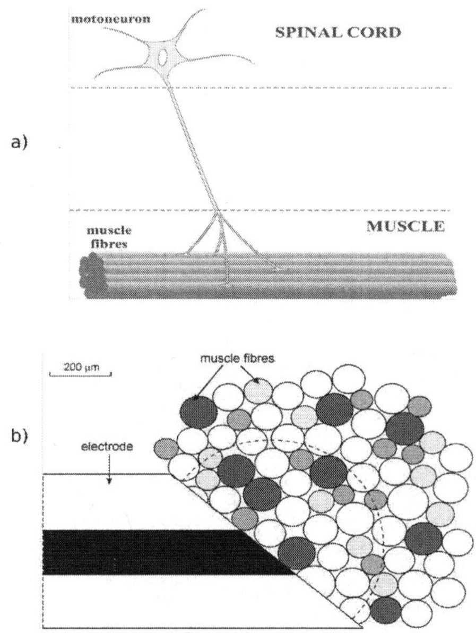

FIGURE 4.49: a) Motor unit, b) cross-section of a muscle and a needle electrode. Fibers belonging to the same MU are marked by the same shade.

A needle electrode typically detects the activity of several muscle fibers within its pick-up area, which belong to a few different MUs (Figure 4.49 b). Usually the shapes of MUAPs are different, since they depend on the geometrical arrangement of the fibers of given MU with respect to the electrode, thus at low force levels single MUAPs can be easily distinguished. During the steady muscle contraction MUs fire repetitively, generating MUAP trains with essentially constant mean firing rate. With the increasing force of muscle contraction, MU firing rate increases and additional MUs are recruited, so the probability of superposition of single MUAPs increases and the EMG shows a rise of amplitude and density (zero crossings of voltage). At strong muscle contractions a so-called interference pattern develops and EMG resembles stochastic signal. More detailed analysis of the relationship between EMG and muscle force can be found in [Piotrkiewicz, 1999].

In neuromuscular diseases, the shapes of MUAPs change (Figure 4.50), which makes their study useful for clinical diagnosis. Two main types of neuromuscular diseases are distinguished: neurogenic and myogenic. In neurogenic diseases the pathological process is related to the degeneration of the motoneuron; in myogenic ones the degeneration begins from the muscle fibers. In neurogenic processes some muscle fibers lose their innervation due to motoneuron death and are reinnervated by the surviving motoneurons; in effect the amplitude and duration of MUAP increases and the number of MUAPs in the pick-up area of electrode decreases. During the phase of acute reinnervation newly formed endplates are unstable and generate increased jitter. In myogenic disorders the amplitude and duration of MUAPs is reduced; they become polyphasic and contain satellite potentials. The typical examples of normal, neurogenic, and myogenic potentials are shown in Figure 4.50. More information about physiological and pathological processes in muscles and their influence on MUAP shapes may be found in [Buchthal, 1985].

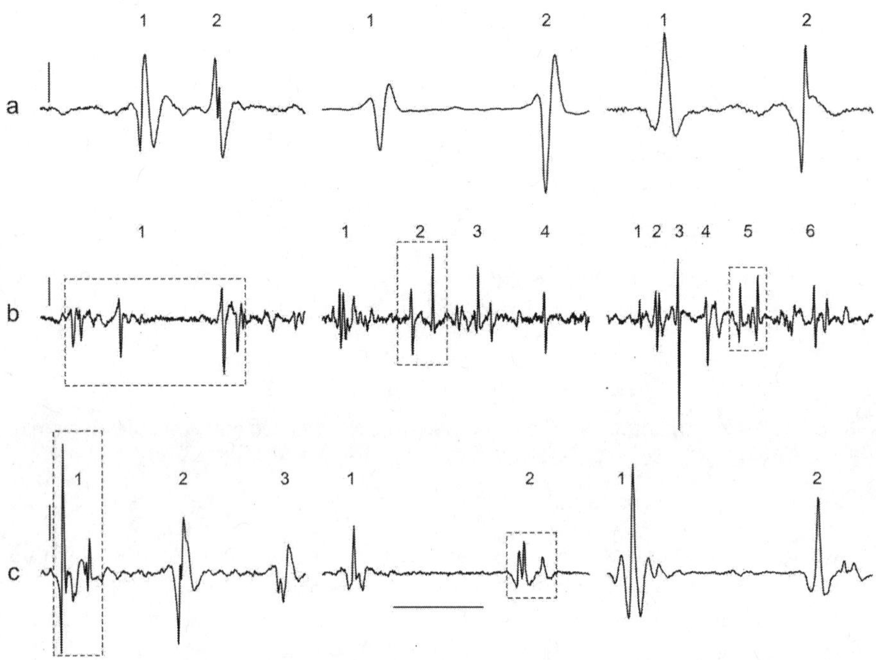

FIGURE 4.50: Examples of single MUAPs recorded during weak muscle contractions from normal and diseased muscles: a) control , b) Duchenne muscle dystrophy (myogenic disease), c) amyotrophic lateral sclerosis (neurogenic disease). For each type three 30-ms epochs from different subjects are shown, with a few single MUAPs marked by distinct numerals. Calibration bars: horizontal, 5 ms; vertical: 200 μV for a) and c), 100 μV for b). Note complex MUAPs in b) and c), indicated by broken line boxes. (By courtesy of M. Piotrkiewicz.)

4.3.2 Quantification of EMG features

The first effort to quantify EMG was connected with the manual evaluation of mean MUAP parameters from a representative number of MUs recorded during a weak effort [Buchthal, 1955, Buchthal, 1985]. The duration and the amplitude of MUAPs as well as the incidence of polyphasic potentials were evaluated from photographically recorded signals. The greater availability of computers in the 1970s and 1980s facilitated the development of computer-aided quantitative EMG processing methods. Automatic systems for EMG-based diagnosis including commercial systems evolved. These systems perform automatic identification of single MUAPs and extraction of their features, such as peak amplitude, rise time, duration, area, or number of phases (connected with baseline crossings). For interference patterns, usually the smaller fluctuations of amplitude are analyzed (so called turns-amplitude analysis). These parameters are illustrated in Figure 4.51. In order to increase the diagnostic yield the combination of the parameters was also introduced, e.g., the thickness– area divided by amplitude.

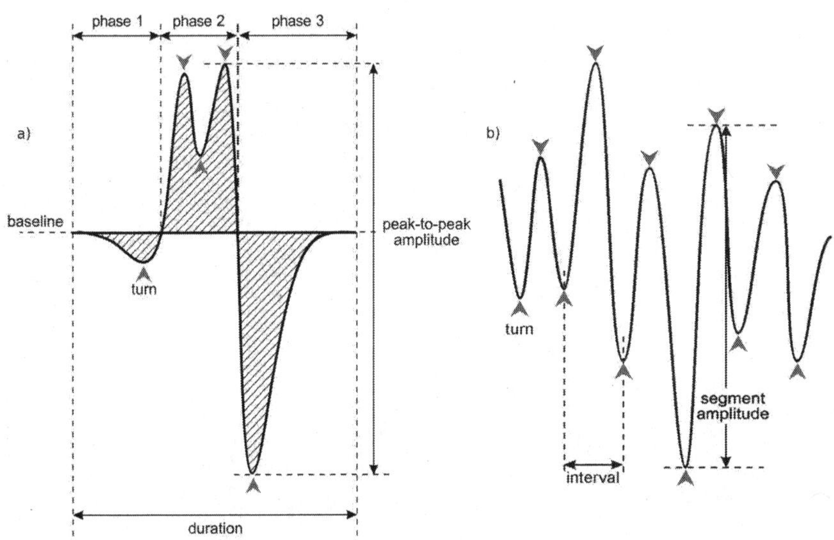

FIGURE 4.51: a) Parameters characterizing MUAP; b) parameters characterizing interference EMG. (By courtesy of M. Piotrkiewicz.)

Besides turns-amplitude analysis, the most popular automatic methods of interference pattern analysis include amplitude measurements, different spike counting methods, and power spectrum analyses. The methods of interference pattern analysis and their applications in diagnosis of myogenic disorders, in assessment of exerted force, muscle fatigue, and the condition of unused muscles and muscles in chronic pain syndromes are considered in [Fuglsang-Frederiksen, 2000].

In the system proposed by Pattichis and coworkers [Pattichis and Elia, 1999] for differentiation of myogenic and neurogenic signals from normal EMG, both time domain and frequency domain features of EMG were considered. The following time-domain parameters were automatically determined: 1) MUAP duration, 2) spike duration measured from the first to the last positive peak, 3) amplitude: difference between the minimum positive peak and maximum negative peak, 4) area: sum of the rectified MUAP integrated over its duration, 5) spike area: sum of the rectified MUAP integrated over spike duration, 6) phases, 7) turns. The frequency features were based on the spectra estimated by means of AR model of order 12. They included: 1) bandwidth, 2) quality factor—the ratio between the dominant peak frequency and bandwidth, 3) median frequency, 4) spectral moments describing the shape of the spectrum; additionally for classification AR coefficients were taken into account.

Univariate analysis was applied to find the coefficients which best separated the normal from pathological groups and multiple covariance analysis to select stepwise the best features and find the correlation between the parameters. For classification into three considered groups artificial neural networks (ANN) were applied. Three algorithms were used: back-propagation, the radial-basis function network, and self-organizing feature map. They are all available in MATLAB Neural Network Toolbox. The univariate and multivariate analysis indicated as the best classifying parameters: among time domain parameters—duration, and among AR spectral measures—median frequency. Median frequency and central frequency for the neurogenic group were higher, and for the myogenic group these parameters were lower than for the normal EMG. In classification by ANN the highest diagnostic yield was obtained for time domain features and next for frequency domain features; AR coefficients gave the poorest results.

4.3.3 Decomposition of needle EMG

The problem of automatic decomposition of an EMG signal into its constituent MUAP trains is important not only for medical diagnosis, but also for basic studies of the neuromuscular system. One of the first systems performing EMG decomposition was proposed by LeFever and de Luca [LeFever and De Luca, 1982]. The signal from the intramuscular needle electrode was sampled at 50 kHz and high-pass filtered in order to reduce the amplitude of slow rise-time MUAP waveforms recorded from fibers more distant from the electrode. The decomposition program was based on a template-matching algorithm and included a routine for continuous template updating, in which every consecutive MUAP could be classified as belonging to the existing template (then the template was updated), or used as the initial estimate of a new template, or discarded. The verification of the initial decomposition results was based on the a priori knowledge of firing statistics, i.e., inter-spike interval distribution within MUAP trains generated during isometric constant force contraction. The program might work automatically; however the reliability of the results was considerably higher if it was run interactively by an operator. To resolve parts of the EMG signal, presumably resulting from the overlapping MUAP waveforms, a two-step procedure was applied. The template fitted in the first step was subtracted from the signal. This produced a residue. In the second step, an attempt was made to fit another template to the residue.

In the relatively simple semi-automatic system proposed in [Mazurkiewicz and Piotrkiewicz, 2004] the first step of classification involved the criterion of similarity. Construction of a similarity vector was based on cross-correlation between the given MUAP and several templates indicated by an operator. In the next step the nearest neighbor classification algorithm was applied. The decomposition of complex potentials into superimposed MUAPs, based on consideration of the discharge rates of units, was performed interactively. The correct classification rate varied from 60% to 100% depending on the quality of the signal. The determined interspike intervals trains were inspected in the context of estimating afterhyperpolarization duration times of motoneurons. The results found application in the study of various neuromuscular disorders.

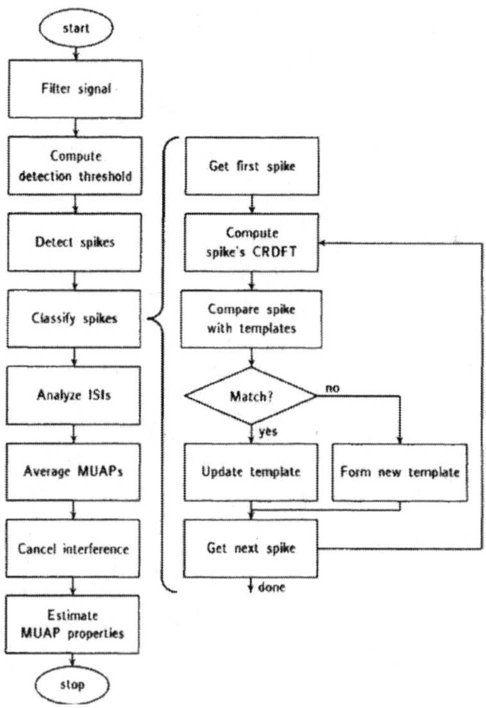

FIGURE 4.52: Flowchart of ADEMG program. From [McGill et al., 1985].

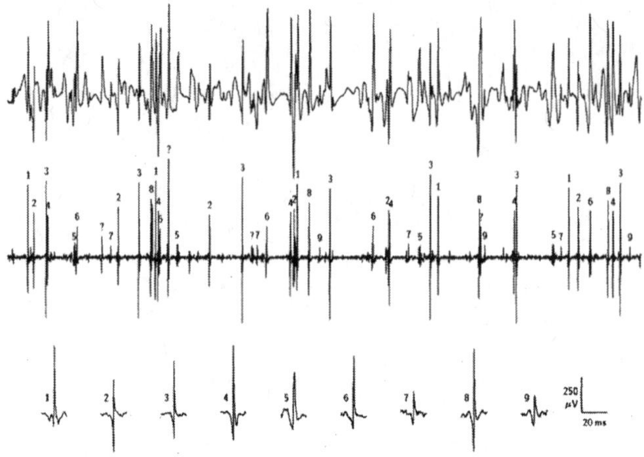

FIGURE 4.53: Illustration of MUAP identification by ADEMG. From the top: EMG signal, high-pass filtered signal, identified MUAPs shapes. From [McGill et al., 1985].

A program called ADEMG for extracting MUAPs from the EMG interference pattern for clinical diagnosis purposes was designed by McGill and coworkers [McGill et al., 1985]. The signal processing consisted of four steps: 1) EMG signal was digitally filtered to transform sharp rising edges into narrow spikes, 2) the spikes which exceeded a certain detection threshold were classified by a one-pass template matching method that automatically recognized and formed templates, 3) each tentatively identified train of MUAPs was verified by examining its firing rate, 4) the MUAPs corresponding to the verified spike trains were averaged from the raw signal using the identified spikes as triggers.

The conversion of MUAPs to spike trains was motivated by the fact that the spikes are better distinguishable than the MUAPs themselves, due to suppression of low-frequency noise, and can be reliably detected by a simple threshold crossing detector. ADMG in the last step of its operation adjusted the simple averages to cancel out possible interference from other MUAPs. Figure 4.52 shows the flowchart of the ADEMG program and Figure4.53 illustrates the extraction of single MUAPs.

In the system designed by Stashuk [Stashuk, 1999] MUAPs were identified by a similar procedure as in [McGill et al., 1985]. MUAP detection was followed by clustering and supervised classification procedures in which the shape and firing pattern of motor units were considered. The fraction of correctly detected MUAPS was reported as 89% and the maximum false detections rate was 2.5%.

The ADMG system was further developed into an interactive EMG decomposition program EMGLAB [McGill et al., 2005], which can handle single- or multi-channel signals recorded by needle or fine-wire electrodes during low and moderate levels of muscle contraction. EMGLAB is a MATLAB program for viewing EMG signals, decomposing them into MUAP trains, and averaging MUAP waveforms. It provides a graphical interface for displaying and editing results and algorithms for template matching and resolving superpositions. The software may be found at: `http://www.emglab.net`. At this location also MTLEMG [Florestal et al., 2006], a MATLAB function for multichannel decomposition of EMG, including a genetic algorithm for resolving superpositions may be found. EMG simulator available at `http://www.emglab.net` is a package for simulating normal and pathological EMG signals designed by [Hamilton-Wright and Stashuk, 2005]. The package contains executable code for PCs and Macintosh and a user interface written in MATLAB.

An automatic algorithm for decomposition of multichannel EMG recorded by wire electrodes (including signals from widely separated recording sites) into MUAPs was proposed by Florestal and coworkers [Florestal et al., 2009]. The program uses the multichannel information, looking for the MUAPs with the largest single component first, and then applies matches obtained in one channel to guide the search in other channels. Each identification is confirmed or refuted on the grounds of information from other channels. The program identified 75% of 176 MUAPs trains with accuracy of 95%. It is available at no cost at: `http://emglab.stanford.edu`.

As was mentioned above, both time and frequency features are useful for EMG quantitative analysis [Pattichis and Elia, 1999]. Wavelet transform, which provides description of signals in time and frequency domain, was applied for analysis of EMG for normal, myogenic and neurogenic groups [Pattichis and Pattichis, 1999]. Four kinds of wavelets were tested including: Daubechies4, Daubechies20, Chui (linear spline), and Battle-Lemarie (cubic spline) wavelets [Daubechies, 1992]. The scalogram was inspected for different wavelet types and in respect of characteristics of investigated groups of patients. Scalograms obtained by means of Daubechies4 wavelets had the shortest time spread in each band and they detected sharp spikes with good accuracy. Daubechies20 wavelets had lower time resolution, however they provided higher frequency resolution. The linear-spline scalograms showed a large energy concentration in the lowest energy bands; on the contrary scalograms for cubic spline wavelet were spread in the upper frequency band.

As might have been expected, scalograms for the neurogenic group had larger time spreads and those for the myogenic group had shorter time-domain spreads in respect to the normal group. Also shifts in frequency similar to these found in [Pattichis and Elia, 1999] were reported in pathological groups. The reduction of information was provided, since MUAP signals were described by a small number of wavelet coefficients located around the main peak. High frequency coefficients were well localized in time and captured the MUAP spike changes, whereas low frequency coefficients described the time span and average behavior of the signal. Wavelet coefficients characterizing MUAPs were used as the input parameters in classification procedures performed by ANN. The best diagnostic yield was obtained for Daubechies4 wavelets. Although the 16 wavelet coefficients captured most of the energy of the MUAP signals the results of classification were worse than those provided by the time domain parameters (listed above) defined in [Pattichis and Elia, 1999] (Sect. 4.3.2).

The EMG decomposition based on wavelet transform was proposed in [Zennaro et al., 2003] for analysis of long-term multichannel EMG. In the first step of analysis so-called active segments containing MUAPs were selected by thresholding based on an estimated signal to noise ratio. The Daubechies wavelets [Daubechies, 1992] were used and coefficients from lower frequency bands were applied for further steps of analysis. The high frequency coefficients were considered as more contaminated by noise and so called time-offset, which depended on the position of MUAP in the analysis window. The next steps of analysis included supervised classification and clustering based on the selected wavelet coefficients. The provisions were made to account for the possible changes of MUAP shape during the experimental session, by introducing adaptation of MUAPs templates. The achieved accuracy of classification was reported as 70%.

The problem inherent in application of discrete WT to EMG decomposition is connected with the fact that the description of the signal shape in terms of wavelet coefficients depends on the position of MUAP in the data window. This problem is usually alleviated by alignment of the main peaks of the analyzed potentials. Another approach which might be used for MUAP identification, which is free of restrictions imposed by the presence of the measurement window, is matching pursuit. To our knowledge such an attempt has not yet been undertaken.

4.3.4 Surface EMG

Because of its noninvasiveness, surface EMG (sEMG) has always attracted the attention of investigators seeking new methods for EMG analysis. One of the first applications of sEMG was in studying muscle fatigue. With increasing fatigue, the sEMG spectrum shifts toward lower frequencies. It was suggested that the spectral shift was caused by the decrease of conduction velocity of action potential along the muscle fibers [Lindstrom et al., 1970, Sadoyama and Miyano, 1981]. However, according to [Linssen et al., 1993] despite the definite and strong influence of the motor fiber conduction velocity on the power spectrum, the frequency shift cannot be explained by a change in propagation velocity alone. Nevertheless, reported changes of propagation velocity in some muscle diseases [Zwarts et al., 2000] indicate the usefulness of sEMG spectral analysis for medical diagnosis.

From sEMG spatial information concerning the MUAPs activity may also be gained. The analysis of surface EMG maps may provide indirect information on the spatial recruitment of motor units within a muscle [Falla and Farina, 2008].

The analysis of sEMG signals used for control of prostheses requires fast and efficient signal processing techniques. In this case the application of WT appeared quite successful [Englehart et al., 1999]. The sEMG signals were recorded from biceps and triceps during elbow and forearm movements. The aim of the study was the best distinction of the four classes of EMG patterns. Time-frequency analysis was performed by means of short-time Fourier transform (STFT), wavelets, and wavelet packets (WP). Different kinds of wavelets were considered. The best results in case of WT were obtained for Coiflet-4 and for WP for Symmlet-5. The task of the reduction of dimensionality of feature sets was performed by Euclidean distance class separability (CS) criterion and principal component analysis (PCA). For classification linear discriminant analysis (LDA) and multilayer perceptron (MLP) were applied. The best performance was obtained for WP/PCA/LDA combination, yielding a classification error of 6%. This result was an improvement in respect to procedures based on time features of MUAPs, which for the same data gave an average error of 9%.

WT was also successfully applied for diagnosis and follow up of Parkinson disease [De Michele et al., 2003]. The sEMG for ballistic movement was recorded from major pectoralis and posterior deltoid muscles. The Morlet complex wavelet was used and cross-correlation was computed between continuous wavelet transforms $W_f(a,\tau)$ and $W_g(a,\tau)$ of functions $f(t)$ and $g(t)$ describing signals from the considered muscles. The wavelet cross-scalogram is given by:

$$W_{fg}(a,\tau) = W_f^*(a,\tau)W_g(a,\tau) \tag{4.44}$$

The cross-correlation spectrum is defined as:

$$|W_{fg}(a,\tau)|^2 = |Re(W_{fg}(a,\tau))|^2 + |Im(W_{fg}(a,\tau))|^2 \tag{4.45}$$

The integration of the local wavelet cross-correlation spectrum over τ gives the global wavelet cross-spectrum. In distinction to FT here it is possible to select a proper integration time interval. In [De Michele et al., 2003] the threshold was chosen as 5% above maximum wavelet power peak. The time-frequency distributions of cross-correlation power for the normal group was more concentrated in time and frequency than for the Parkinsonian group, where a large spread was observed, especially in time. In order to quantify these differences a parameter called global power was determined by integration of scalograms in frequency and time. This parameter differentiated the Parkinsonian group from normal subjects better than conventional medical measure. The global power index was proposed as a parameter useful in estimation of the disease progress.

4.3.4.1 Surface EMG decomposition

Recently, the sEMG technique has drawn a lot of attention. One of the reasons is the fear of infection by needle electrodes. The problem inherent to sEMG technique is the influence of volume conduction and low-pass filtering effect of the tissues, which causes an increase in MUAPs duration, smoothing of their shapes, and a decrease of the differences between them. For these reasons sEMG was initially treated as an interference signal, from which only global properties of motor units and their firing rates could be assessed.

Nevertheless, the application of surface EMG (sEMG) for detection of single MU activity by means of small bipolar electrodes and their arrays was proposed by Gydikov and coworkers in the early 1970s [Gydikov and Kosarov, 1972, Gydikov and Gatev, 1982, Gydikov et al., 1986, Kosarov, 1974]. Further studies by these authors provided information on the velocity of the activity propagation along muscle fibers and on MU structure, in particular its size and location within the muscle, endplate distribution, muscle fibers topography [Gydikov and Gantchev, 1989].

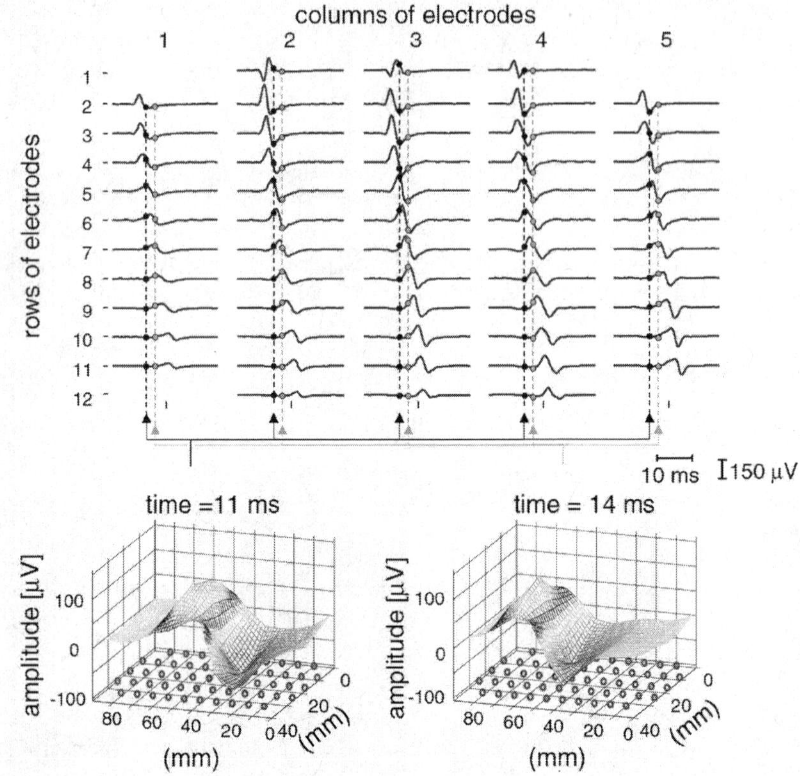

FIGURE 4.54: Motor unit action potential recorded with a grid of 61 electrodes (13 × 5 electrodes, inter-electrode distance 5 mm) from the biceps brachii muscle during an isometric contraction at 10% of the maximal force. The bipolar signals derived along the direction of the muscle fibers are shown in the top panel. The electrode grid was placed distal with respect to the innervation zone of the motor unit. The signals detected along the rows show similar action potential shapes with a delay corresponding to the propagation along the muscle fibers. The multi-channel action potential is a three-dimensional signal in time and space. The two-dimensional spatial representations for two time instants (11 ms and 14 ms after the generation of the action potential) are shown in the bottom panel. The circles on the plane representing the spatial coordinates mark the locations of the electrodes of the grid. From [Merletti et al., 2008].

The biggest challenge in sEMG processing—the extraction of individual MUAPs—became possible when the high-density grids, including sometimes hundreds of electrodes, became available. Figure 4.54 shows motor unit action potential recorded with a grid of 13x5 electrodes and illustrates tracking of the propagation of MUAP by multielectrode arrays. The process of the sEMG decomposition into

MUAPs is shown in Figure 4.55.

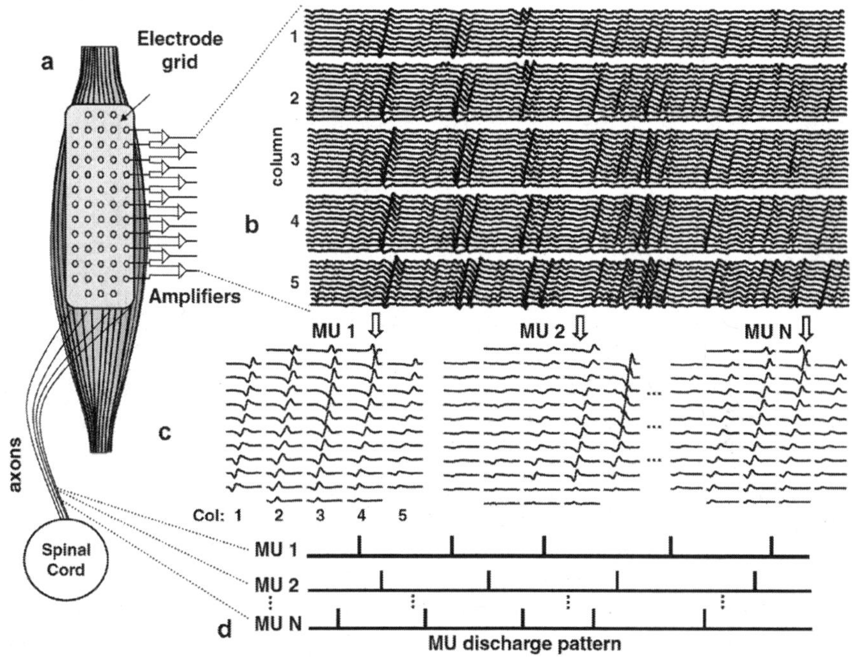

FIGURE 4.55: Representation of the process of decomposition of multi-channel surface EMG. (a) Surface EMG was recorded from the biceps brachii muscle with a 13× 5 electrode grid (corner electrodes are missing) with the columns parallel to the fiber direction. (b) Segment of 500 ms duration of bipolar EMG detected by each column of the grid. The action potentials propagation along the columns may be noticed. (c) Multi-channel action potentials for three motor units extracted from the interference signal with the decomposition algorithm described by [Holobar and Zazula, 2007]. (d) Estimated discharge patterns for the three motor units. From [Merletti et al., 2008].

The first step in analysis of sEMG is alleviation of the effect of volume conduction and suppression of signals from distant sources. This may be achieved by application of the Laplacian operator (Sect. 4.1.3). Spatial high-pass filtering effect may be achieved also by using bipolar montage [Kleine et al., 2007]. The example of spatially filtered signals is shown in Figure 4.56.

The procedure of MUAP identification is similar to these applied in case of signals recorded by the needle electrodes. It includes setting the threshold for candidate MUAPs, clustering, creating templates, template matching, decomposition. The serious problem is a small differentiation of the MUAP shapes recorded by sEMG in

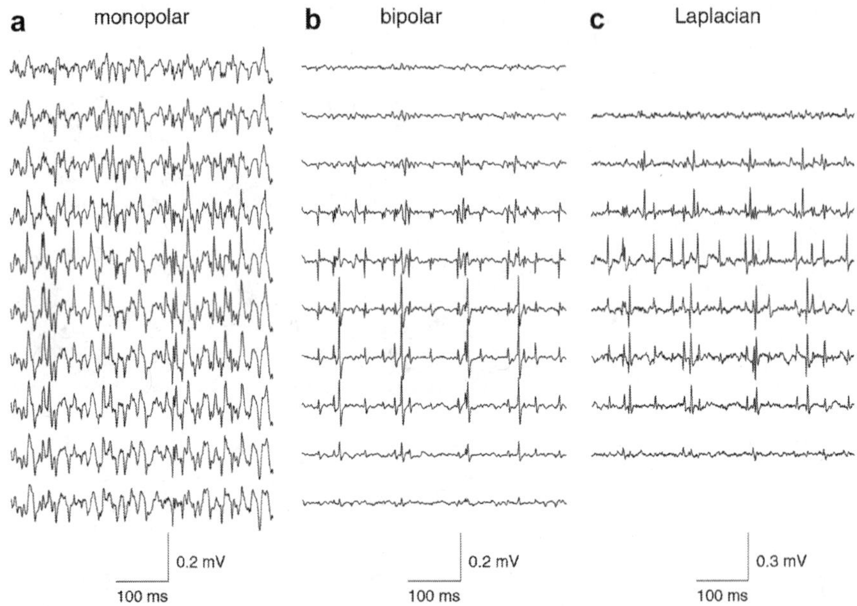

FIGURE 4.56: Example of spatially and temporally (high-pass at 15 Hz) filtered signals. Signal from the same 500 ms in monopolar (a), bipolar (b), and Laplacian (c) derivation. The upper trace corresponds to the medial, the lower to the lateral electrode positions. From [Kleine et al., 2007].

comparison to signals registered by needle electrode. This problem is now resolved by application of multielectrode grids, which provide the two-dimensional spatial information. The differences in location of single MUs with respect to the multielectrode and the information about the delay of the given MUAP connected with propagation of MUAPs along the muscle fibers, create so-called MU signatures, which are utilized in the algorithms of sEMG decomposition.

The semi-automatic techniques for extracting single MUs from sEMG including operator-computer interaction were proposed in 1999, e.g., [Disselhorst-Klug et al., 1999]. Later, the fully automatic system for extraction and classification of single MUAPs from sEMG was designed by Gazzoni et al. [Gazzoni et al., 2004]. The segmentation phase was based on the matched continuous wavelet transform, and was performed in two steps. In the first step candidate MUAPs were detected by multiscale matched filter. The second step was based on the properties of propagation of the action potentials along muscle fibers. Only MUAPs presenting clear propagation along the longitudinal fiber direction were selected. The clustering was performed by a set of neural networks, working in parallel, one for each available channel. They operated in an adaptive way updating the templates and creating new clusters for non-matching patterns. However, the system didn't offer the possibility of resolving the superimposed MUAPs.

In the system proposed by Kleine et al. [Kleine et al., 2007] a 10x13 grid with inter-electrode distance 5 mm was used. Bipolar and Laplacian reference systems were considered (Figure 4.56). The MUAPs were isolated from the background activity by peak detection. In the clustering procedure both the spatial (waveform and amplitude difference between channels) and the temporal (time-course of the potential in each channel) information were taken into account. Additional information was provided by the firing statistics. The clusters were inspected and adjusted interactively; that is, the operator could split or merge clusters. Then the MUAP templates were subtracted from continuous sEMG. In the decomposition procedure a single row (perpendicular to muscle fibers) of bipolar derivations was taken into account. The template matching procedure was repeated with different subsets of available data. It was confirmed that two bipolar montages, separated by a few millimeters in the direction parallel to the fibers will record almost the same waveform, delayed by a few milliseconds. On the other hand in the direction perpendicular to the muscle fibers the amplitude of bipolar MUAP is largely attenuated (see Figure 4.54 and Figure 4.55). In this case the muscle fibers of a particular MU cover only a small part of muscle fibers from another MU, so a single row of channels gives a reasonable separation of MUAPs.

The limitation of the method is the restriction to low force contractions, since at higher forces two or more MUAPs may have almost identical shapes and also the superimposed potentials become a problem. It was suggested that sEMG may be useful for quantification of neurogenic diseases, where the number of remaining MUs and MUAP interference are low.

In general, sEMG may be helpful in case of disturbances of the neuromuscular system leading to decrease of firing rates of MUs. An example may be the automatic classification of MUAPs in sEMG recorded from muscles paralyzed by spinal cord injury, where involuntary EMG activity occurs [Winslow et al., 2009]. However, it should be mentioned that changes in firing rates are not specific to any particular neuromuscular disorder.

The application of the blind source separation method for identification of motor unit potential trains in sEMG was proposed by Nakamura and coworkers [Nakamura et al., 2004]. The signals from eight electrodes placed perpendicularly to the muscle fibers were processed. Principal component analysis and independent component analysis were applied. The ICA approach performed better than PCA in separating groups of similar MUAP waveforms, however single MUAPs were not separated completely into independent components. In case of delays between potentials coming from the same motor unit ICA regards them as several different sources. The limitation of the method is connected also with the fact that the procedure does not take into account the firing rates of the motor units, which limits useful information available for classification. The BSS methods assume independence of the underlying sources which hampers their application for higher force level when the synchronization of the motor units is high. The advantages of the BSS are connected with the fact that the method does not rely on prior estimation of the shapes of MUAPs. It seems that the approach might be promising, for example, as a preprocessing step in the procedure of sEMG decomposition.

Blind source separation method was used by [Holobar et al., 2009] who measured sEMG from four muscles by means of 13x5 grid during low force contraction. The approach was based on the convolution kernel compensation (CKC) algorithm proposed by [Holobar and Zazula, 2007]. The signal was modeled as a convolutive mixture of sparse pulse trains, which carry information about the rising times of the detected symbols and the symbols themselves. The spatial and temporal statistics of symbols, i.e., convolution kernels, was combined with the information about their overlapping probability, in order to blindly reconstruct their pulse sequences. The model implied admixture of a white noise, however residual noise is dominated by the contributions of indecomposable MUAPs and does not meet the criterion of spatial and temporal independence. CKC assumed as well, that action potential of MU remains constant throughout a contraction. In [Holobar et al., 2009] the accuracy of the method reported as very good was assessed by the method proposed by [Farina et al., 2001] (described below). In case of one of the muscles the results give 98% of agreement in MUAPs identification with the intramuscular recordings. However, the method is limited to low force contractions (10% of maximal force) and biased toward the high-threshold MUs, which was admitted by the authors.

The performance of different methods of sEMG analysis is usually evaluated by comparison with the manual decomposition of a given signal. In view of the multitude of the approaches to the problem of EMG decomposition there emerged the need for a test tool to evaluate and compare different methods. Such a tool was proposed in [Farina et al., 2001]. The approach was based on the model for the generation of synthetic intra-muscular EMG signals. The library of 18 test signals was designed. The indexes estimating performance for segmentation and classification stages of decompositions and measure of association between model and detected classes of MUAPs were proposed. Additionally global indices were introduced, such as the difference between the number of model classes and the number of classes estimated by the algorithm, mean firing rate, and activation interval detection.

Yet another way of testing the decomposition algorithm was applied in the recent work of [Nawab et al., 2010]. It relied on reconstruction of the signal from the identified MUAPs and decomposition of this signal with some noise added. This procedure is illustrated in Figure 4.57. In the experiment a five-pin surface sensor was used (Figure 4.48) and the pair wise voltages between 5 pins were analyzed. The method of the decomposition was based on the algorithm proposed by [LeFever and De Luca, 1982]. The operation of the algorithm started by extracting as many MUAPs as possible from experimental sEMG action potential templates. Then it searched for signal regions where the extracted templates were in superposition with each other and with unidentified potentials. The algorithm required that the unidentified action potentials account for less than 25% of the signal energy. The assumption about inter-pulse intervals was only that they be less than 0.35 s. The algorithm allowed for changes in the action potential shape of each MUAP to evolve over the duration of the contraction. The authors considered 22 signals recorded from 5 muscles for contractions reaching 100% force levels. They reported the accuracy estimated by the reconstruct-and-test procedure (Figure 4.57) ranging from 77–97%. Upon analysis of decompositions performed on these signals it was found that 92% of over 3000 firings from

11 decomposed MUAPs of the signal from one sensor were locked in with firings of 11 corresponding MUAPs decomposed from the other sensors' signals.

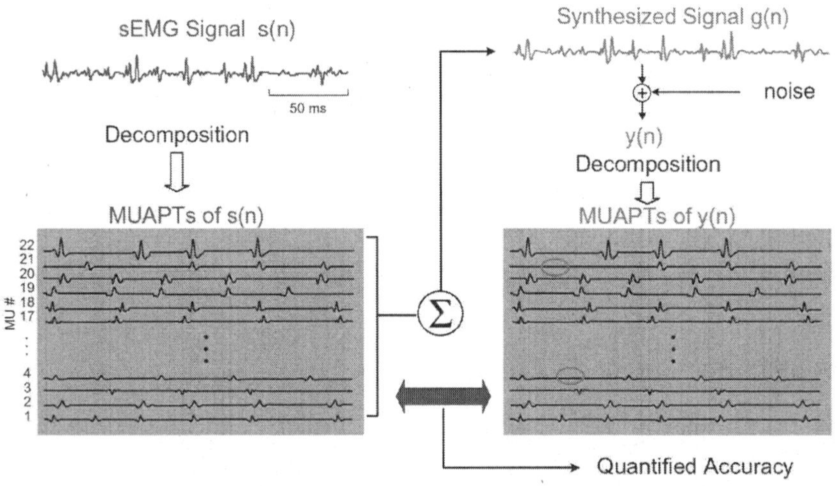

FIGURE 4.57: Illustration of "reconstruct-and-test" procedure for assessing the accuracy of the decomposition algorithm. An actual sEMG signal s(n) is decomposed to identify its MUAPTs (MUAP templates). Signal y(n) is synthesized by summing together the decomposed MUAPTs of s(n) and white Gaussian noise whose variance is set equal to that of the residual signal from the decomposition. The reconstituted y(n) signal is then decomposed and compared to the decomposed MUAPTs of s(n). The ellipses indicate discrepancies between the MUAPTs of y(n) and s(n). These are designated as errors. From [Nawab et al., 2010].

The method seems to be promising in respect to potential clinical applications, since the authors claimed that the system is able to detect morphology of the shapes of up to 40 concurrently active motor units without relocating the sensor and that the polyphasic MUAPs may be detected with the proposed technology.

The view on applicability of sEMG in clinical practice has changed over the years. In a technology assessment study issued by the American Association of Electrodiagnostic Medicine [Haig et al., 1996] it was concluded that there is no evidence to support the use of sEMG in the clinical diagnosis and management of nerve or muscle disease. However, the recent study based on the literature covering the period 1994–2006 [Meekins et al., 2008] stated that sEMG may be useful to detect the presence of some neuromuscular diseases and a study of fatigue, but there are insufficient data to support its utility for distinguishing between neuropathic and myopathic conditions. This statement creates a challenge for the scientist working in the field of sEMG to develop better techniques and signal processing methods.

The contemporary EMG diagnosis is largely based on the disease-specific differences in MUAP shapes which cannot be detected even by the most sophisticated methods of sEMG detection and processing due to the information loss caused by volume conduction. We can conclude therefore that at the moment surface EMG cannot replace needle EMG in diagnostic applications. However, sEMG is able to provide additional information, e.g., on MU size and topography or conduction velocity, which is complementary to the routine EMG diagnostic techniques [Merletti et al., 2008]. One of the most important advances in the sEMG field is the increasing possibility of extracting individual MUAPs from non-invasive recording. With the improvements in sensor design and further development of processing methods high resolution surface EMG may become in the near future a widespread tool for investigation of muscle function at the individual unit level.

4.4 Gastro-intestinal signals

An electrogastrogram (EGG) is a recording of electrical activity of stomach; an electroenterogram (EEnG) is a recording of both stomach and intestine electrical activity. EEnG in humans is recorded from three or more Ag-AgCl electrodes placed on the abdominal surface. Similarly, magnetogastrogram (MGG) is a record of the magnetic field generated by the electric field of the stomach; and magnetoenterogram (MEnG) corresponds to the magnetic field generated by the gastro-intestinal tract.

EGG and EEnG reflect the electrical activity of the smooth muscles of the gastro-intestinal tract, which drive the peristaltic movements. The patterns of electrical activity of EEnG consist of slow wave and superimposed action potentials generated by pacemakers—the interstitial cells of Cajal. The frequencies of slow waves change along the gastro-intestinal tract. They are respectively: in stomach—0.03–0.07 Hz, in large intestine 0.01–0.03 Hz, in ileum 0.07–0.13 Hz, in jejunum 0.13–0.18 Hz, in duodendum 0.18–0.25 Hz. In Figure 4.58 spectra presenting rhythms of different parts of the gastro-intestinal tract and their time fluctuations are shown.

The gastric and intestinal slow wave is present even during quiescent periods. When smooth muscle contraction occurs, spike potentials phase locked to it appear. Abnormality in the rhythmic electrical activity generation is called gastric dysrhythmia. In EGG the deviations from normal rhythm (3 cycles/minute) include bradygastria and tachygastria. Bradygastria involves a decreased rate of electrical activity in the stomach, below 2 cycles/minute for at least 1 minute. Tachygastria is connected with an increased rate of electrical activity above 4 cycles/minute for at least 1 minute. Dysrhythmic activities in the gastro-intestinal tract may lead to a variety of disorders such as gastroparesis, functional dyspepsia, and gastro-oseophageal reflux disease.

Although the first human EGG was recorded by Alvarez more than 80 years ago [Alvarez, 1922], the lack of standardized methodology in terms of electrode

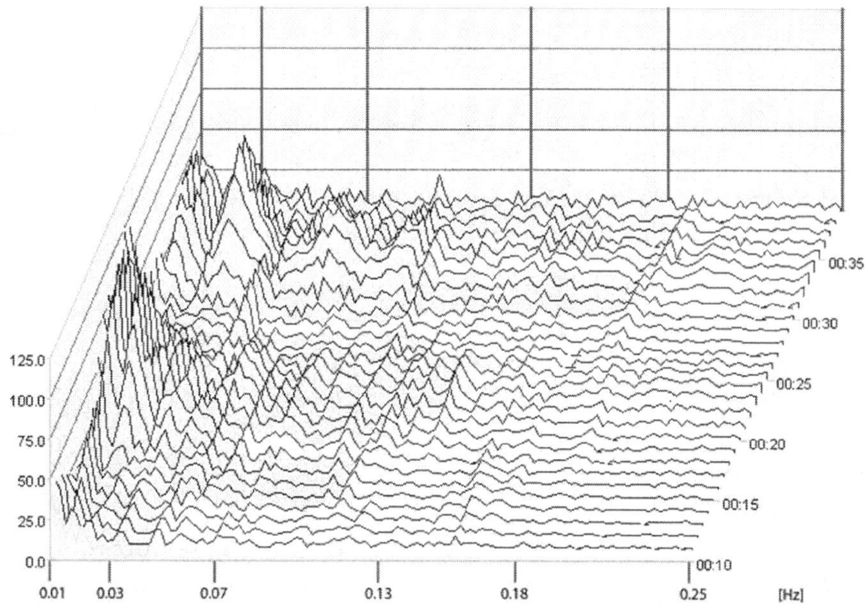

FIGURE 4.58: Power spectra showing rhythms of different parts of gastro-intestinal tract (marked on horizontal scale) and their evolution in time (oblique axis). On vertical axis, amplitude in μV.

positions, recording periods, test meals, methods of analysis, and normal reference diminishes the significance of EEnG in diagnosis of gastro-intestinal tract dysfunctions. Another obstacle is a low signal to noise ratio of EEnG, which is contaminated by respiration, ECG signal, movement artifacts, and other types of interference. The amelioration of the methods of EEnG processing together with standardization of procedures is a prerequisite for the broader application of this signal in medical diagnosis of gastro-intestinal disorders.

The most common method of EEnG evaluation is spectral analysis. From spectra the parameters characterizing the power in different frequency bands connected with activity of corresponding parts of the gastro-intestinal tract may be found. The dynamics of EEnG was usually estimated by means of STFT [van der Schee and Grashius, 1987, Pfister et al., 1988]. Discrete wavelet analysis was used in [Tokmakci, 2007] for evaluation of EGG from normal and diabetic gastroparetic subjects. In this study Daubechies db3 wavelets from the MathWorks MATLAB were used for three level decomposition. The detail coefficients were considered by the author as a noise and the analysis was conducted for three approximation coefficients: A1, A2, A3. Paired samples t-test reported (uncorrected) p-values below a 5% difference between the normal and pathological group for parameters: minimum values of A1, A2, A3, and for mean values of power spectral density obtained from these coefficients. However, it seems that a more appropriate statistical tool for evaluation of the significance of the parameters would be ANOVA.

The problem of improvement of signal to noise ratio in EEnG was approached in [Prats-Boluda et al., 2006] by application of discrete Laplacian, namely its approximation—the Hjorth transform (Sect. 4.1.3). The breath interference was removed by means of adaptive filter. From the periodogram it was found that the dominant frequency range of slow intestinal wave may be separated from respiration activity, which has a higher frequency range. The method of removing ECG influence and movement artifacts was tested on an animal model by comparison of internally recorded and surface EEnG in [Ye-Lin et al., 2010]. The applied approach involved combination of empirical mode decomposition and ICA. The increased correlation coefficient between internal and external recording indicated the improvement of signal to noise ratio.

The reduction of volume conduction influence may be achieved by recording of magnetic signals. The ability of a SQUID magnetometer to detect magnetic fields created by rhythmical activity of the bowels was demonstrated in [Richards et al., 1995]. Magnetic field was measured simultaneously with electrical activity recorded with electrodes surgically implanted in rabbit intestines. Transabdominal SQUID recording was highly correlated with the measurements obtained with electrodes during ischemic period and after the blood flow was restored. These positive findings have encouraged development of clinically useful, noninvasive detection of intestinal magnetic fields.

SQUID magnetometer was applied in [Bradshaw et al., 2006] to study the magnetogastrogram in 20 normal human subjects. In addition to computation of the frequency and amplitude parameters of the gastric slow wave, MGG surface current density mapping allowed for visualization of the propagating slow wave and estimation of its propagation velocity.

The propagation of gastric slow wave was also considered in [Chen et al., 1999]. The EGG was recorded by four electrodes placed along the gastric axis. Power spectra and cross-covariances between channels were calculated. The significant differences between time-lags were found. However, the delays obtained for the different combination of the electrodes were not consistent. This problem might be resolved by the application of multivariate method (see Sect. 3.3.2.3 and 3.5).

It seems that in EEnG analysis all potentially useful methods of signal processing have not been explored and it is still an open field for testing methods and models, particularly methods of time-frequency analysis.

4.5 Acoustic signals

4.5.1 Phonocardiogram

Phonocardiogram (PCG) is a recording of the acoustic signal produced by the mechanical action of the heart. The main frequency range of normal PCG is in the range of 10-35 kHz; however in case of artificial heart valves frequencies up to 50 kHz may be recorded. Heart sounds are generated by the vibration of the heart valves during their opening and closure and by the vibration of the myocardium and the associated structures. The sound generated by a human heart during a cardiac cycle consists of two dominant components called first heart sound, S1, and the second heart sound, S2. The S1 sound corresponds to the QRS complex of the ECG and the second heart sound, S2, follows the systolic pause in the normal cardiac cycle. In S1 two components, M1 and T1, may be distinguished which are due respectively to the closure of the mitral and of the tricuspid valve. The components of S2 are due to the closure of the aortic valve (A2) and pulmonary valve (P2). Additionally noise-like sounds called murmurs caused by the turbulence of the blood flow may be produced. They appear mostly during abnormal action of a heart. Examples of normal and pathological PCG are shown in Figure 4.59.

The technique of listening to the heart sounds, called auscultation, has been used for diagnostic purposes since the 16th century. However, the human ear is not very well suited to recognize the several events of short duration occurring in very small intervals of time, especially since they occur in the low frequency range, where the sensitivity of a human ear is not very good. Therefore signal processing techniques are very helpful in PCG analysis.

One of the first steps in PCG analysis is its segmentation. Usually ECG signal is used for this purpose, however the segmentation may be based on the PCG signal itself taking into account its time domain features and identification of S1 and S2 [Ari et al., 2008]. The recognition of S1 and S2 without referring to ECG was reported by [Ning et al., 2009], who used wavelet transform. PCG signal was found to be a convenient reference for establishing heart rate in fMRI studies, since PCG, contrary to ECG, is not disturbed by the electromagnetic fields of the equipment [Becker et al., 2010].

The PCG signal is non-stationary and consists of short transients of changing frequency. Hence time-frequency methods are the appropriate tools in its analysis. The early works on PCG analysis were based mainly on spectrograms calculated by the Fourier transform. Later the wavelet analysis was introduced to PCG analysis. The usefulness of different kinds of wavelets (Daubechies, Symlet, biorthogonal) for PCG reconstruction was tested in [Debbal and Bereksi-Reguig, 2008]. It was found that the best results were obtained by means of Daubechies db7 wavelet. The authors reported that the error of reconstruction can be used as a discriminatory parameter in classifying the pathological severity of the PCG. Daubechies wavelets were also used in the study of [Ning et al., 2009] for finding the offset of the systolic murmur,

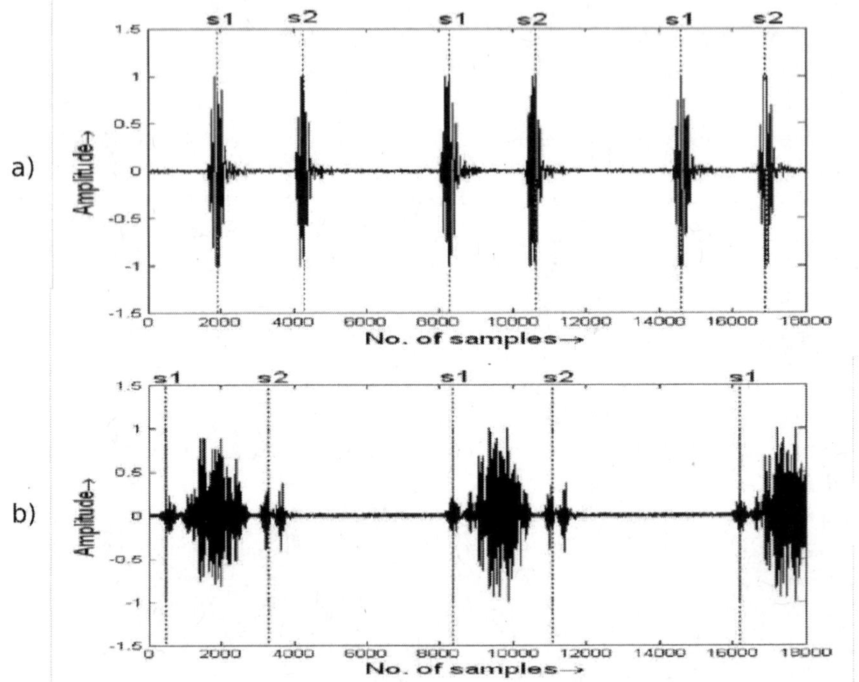

FIGURE 4.59: PCG signal a) for normal subject, b) for subject with pulmonary stenosis. Sampling frequency 8012 Hz. Adapted from [Ari et al., 2008].

which was then analyzed by means of the AR model of the order 2. The approach proved effective in delineating a set of clinically diagnosed systolic murmurs.

For evaluation of the condition of bioprosthethic valves, features of the power spectrum such as the location of maximum peak have been used to indicate whether the valve is normal or abnormal. For example, changes in the frequency content of PCGs from bioprosthetic valves were described by [Stein et al., 1981], who reported that the PCG of an abnormal valve has more high-frequency content than that of a normal valve, and in patients with mitral bioprotheses, implanted five years or longer, the higher frequencies have a greater proportion of sound energy compared to patients with implants of less than 1.5 years. These findings were based on the power spectra computed by means of Fourier transform. Joo and coworkers [Joo et al., 1983] reported that more accurate description of heart sounds in patients with a bioprosthetic heart valve may be obtained by a pole-zero model based on ARMA formalism.

In the study of performance of a heart valve implanted in the mitral position AR model was used by [Koymen et al., 1987]. The authors applied an approach similar to

FAD [Franaszczuk et al., 1989] (Sect. 2.3.2.2.4). Namely, they decomposed the impulse response function of the AR model into a sum of damped sinusoids. The very low level of residual signal after fitting the AR model suggested that the recorded PCG is essentially composed of delayed and overlapped repetitions of impulse response function of the AR model.

Matching pursuit approach is a high resolution method of time-frequency analysis (Sect. 2.4.2.2.7). It was applied by Zhang et al. [Zhang et al., 1998] for analysis-synthesis of PCG. The dyadic dictionary with certain limitations imposed on scale was used. The results showed that PCG can be synthesized from a limited number of the highest energy atoms and the method acts as a powerful de-noising filter. The separation of heart sounds from noise based on MP decomposition and fuzzy detection was also applied in [Tang et al., 2010]. In a practical experiment, heart sound signal was successfully separated from lung sounds and disturbances due to chest motion.

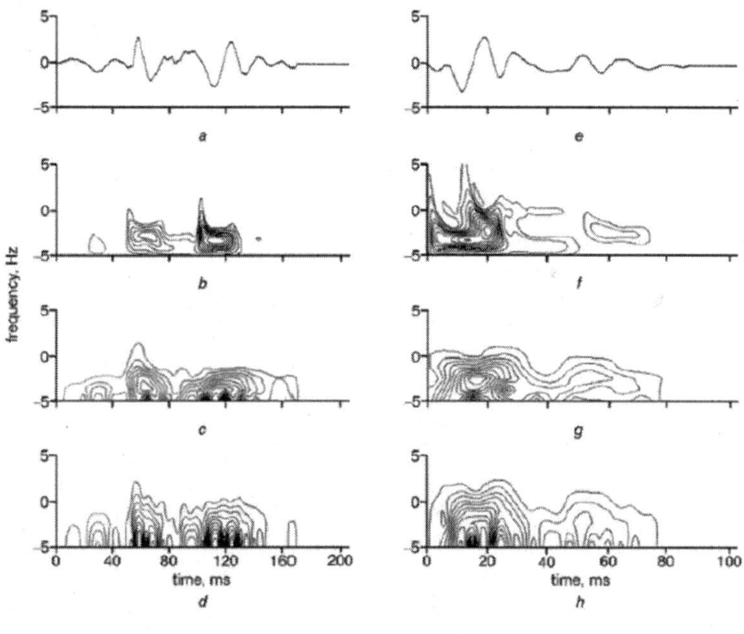

FIGURE 4.60: Comparison of time-frequency distributions obtained by means of MP and FT. (a) Averaged normal S1 of PCG and its contour plots using (b) MP-based time-frequency distribution, (c) spectrogram with 64-point STFT window length, and (d) spectrogram with 32-point STFT window length. (e) Averaged abnormal S1 of PCG and its contour plots using (f) MP-based time-frequency distribution, (g) spectrogram with 64-point STFT window length, and (h) spectrogram with 32-point STFT window length. From [Wang et al., 2001].

The usefulness of MP for PCG analysis was also confirmed by a study by Wang et al. [Wang et al., 2001], who reported that MP energy distribution provides better time-frequency representation of PCG than spectrogram. The authors reported good results of classification of mitral valve abnormality based on MP decomposition of first heart sound. Figure 4.60 shows spectrograms and MP time-frequency maps of first heart sound for normal and pathological case.

From the evidence gathered by the authors who applied different kinds of methods to PCG it follows that the time-frequency methods such as wavelets and MP, and also parametric methods based on the autoregressive approach are suitable for PCG analysis and the choice of the method depends on the problem to be solved.

4.5.2 Otoacoustic emissions

Otoacoustic emissions (OAEs) are weak acoustic signals generated by the inner ear in response to the stimulation and sometimes also spontaneously. OAE are measured by a sensitive microphone placed in the ear canal. Their typical frequencies range from 0.5 to 6 kHz [Probst et al., 1991]. Responses that are evoked by short acoustic stimuli are referred to as transiently evoked otoacoustic emissions (TEOAEs). They are usually measured in 20 ms window from stimulus onset. Long lasting OAEs recorded in 80 ms windows in response to broadband click stimulus are called synchronized spontaneous OAEs (SSOAEs). They are considered to represent spontaneous otoacoustic emissions (SOAEs). Distortion product otoacoustic emissions (DPOAEs) are the emissions which are generated by two tones f_1 and f_2, where f_1/f_2 is typically around $1.2f_1$. The emitted otoacoustic signal has the largest amplitude at $2f_1 - f_2$ frequency.

OAEs were first registered by Kemp [Kemp, 1978] and very quickly became an important tool in the diagnosis of hearing impairment. However, the mechanisms of their generation, which are closely related to the hearing process are still a matter of debate between two theories, namely traveling wave [Zweig and Shera, 1995, Shera and Guinan, 1999, Talmadge et al., 2000] and resonance theory [Sisto and Moleti, 1999, Bell, 2002, Bell and Fletcher, 2004]. OAEs are especially useful as a screening hearing test in neonates and small children. The review of OAE techniques and clinical applications may be found in [Robinette and Glattke, 2002].

During the recording procedure consecutive OAEs are stored in two buffers and the correlation between the averages from these two buffers serves as an index of the reproducibility and hence, the quality of the signal. Usually the discrimination between TEOAE responses from normal-hearing subjects and patients with sensorineural hearing losses is based on the reproducibility (or correlation) of the TEOAE responses obtained from two buffers, the signal to noise ratio (S/N) and overall TEOAE response level. Sometimes the frequency characteristics of the signal are considered; OAE measurement devices usually allow to calculate Fourier spectra.

More recently other TEOAE parameters have proven to be good signal descriptors and among these are the latencies of the TEOAE components. It was demonstrated in several papers that there are differences in latencies of OAEs between subjects with normal hearing and subjects exposed to noise, e.g., [Tognola et al., 1997, Sisto and

Moleti, 2002, Jedrzejczak et al., 2005]. OAE signal is a basic test of hearing impairment in small children and a lot of attention has been devoted to clinical significance and determination of latencies in neonates, e.g., [Tognola et al., 2005, Jedrzejczak et al., 2007].

In the research concerning the mechanisms of OAE generation the relation between frequency and latency of components is one of the important tests of the models, for example, the scale-invariance hypothesis [Talmadge et al., 2000] leaded to prediction of inverse relation between frequency and latency. However, this kind of relation was not confirmed experimentally and exponential dependence with different values of exponents was reported [Sisto and Moleti, 2002, Tognola et al., 1997, Jedrzejczak et al., 2004].

The non-stationary character of OAE and rising interest in frequency-latency dependencies promoted the application of time-frequency methods to this signal. Several methods have been devised to estimate the time-frequency distributions of OAEs, including short-time Fourier transform [Hatzopoulos et al., 2000], minimum variance spectral estimation [Zhang et al., 2008] methods based on the Wigner-Ville transform [Cheng, 1995], or Choi-Williams transform [Ozdamar et al., 1997]. However, the last two methods are biased by the presence of the cross-terms (Sect. 2.4.2).

One of the first applications of wavelets to the analysis of OAE concerned time-frequency decomposition of click evoked and synthesized emissions [Wit et al., 1994]. Both continuous [Tognola et al., 1997] and discrete [Sisto and Moleti, 2002] WT were used. For the purpose of improving pass/fail separation during transient evoked otoacoustic emission (TEOAE) hearing screening, the method which combined signal decomposition in scales by discrete WT, non-linear denoising, and scale-dependent time windowing was used [Januauskas et al., 2001]. WT was also applied for extraction of instantaneous frequencies of OAE [Delprat et al., 1992] and for construction of multiscale detector of TEOAE [Marozas et al., 2006]. In the above quoted contribution the detector performed adaptive splitting of the signal into different frequency bands using either wavelet or wavelet packet decomposition. The authors reported that the method performed significantly better than existing TEOAE detectors based on wave reproducibility or the modified variance ratio.

OAEs are a superposition of components characterized by specific frequencies and latencies. The dependence between their frequency and latency have been extensively studied in the context of verifying hypotheses concerning the mechanisms of OAE generation. To this avail the latency has to be determined. The problem of the identification of the OAE latencies was approached by discrete and continuous wavelet transform. The method based on the combination of spectral analysis and WT was proposed by Sisto and Moleti [Sisto and Moleti, 2002]. The limitation of WT in this respect is octave-band resolution, which influences the accuracy, especially for high frequencies. In order to surmount this difficulty the authors proposed a method relying on visual comparison between wavelet data and TEOAEs spectra, followed by identification of the wavelet contribution to a given spectral line. However, as was pointed out in [Moleti and Sisto, 2003] wavelet transform has a tendency to systematically underestimate the slope of the latency-frequency relation.

Substantial progress in the OAE field was achieved by the introduction to its analysis of matching pursuit [Jedrzejczak et al., 2004]. Matching pursuit decomposes signals into components of specific amplitudes, frequencies, latencies, and time spans, so these parameters are given explicitly. A comparison of different methods used for OAE analysis is shown in Figure 4.61. The simulated signal used for testing the methods was constructed from the so-called gammatones—the functions resembling the shape of click evoked OAE at single resonant frequency. The gammatone is expressed by a function, whose envelope rises as t^3 and decays exponentially with a constant Γ:

$$\gamma(t) = \gamma_0 t^3 e^{-2\pi\Gamma t} \sin(2\pi f t) \tag{4.46}$$

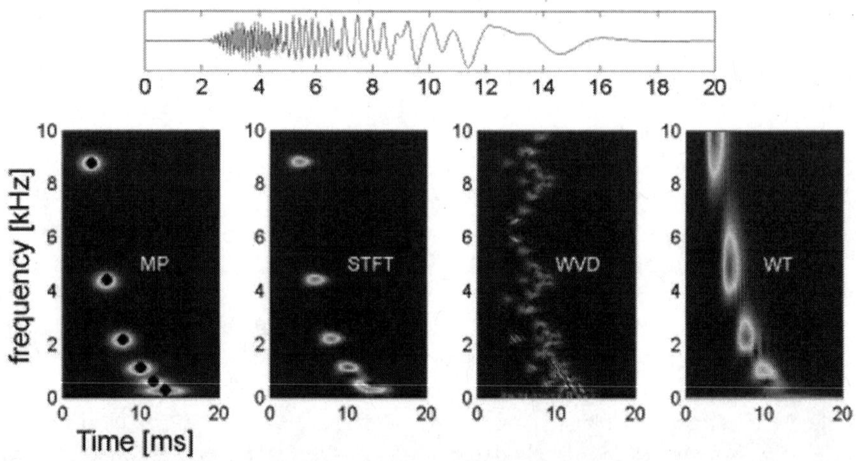

FIGURE 4.61: (SEE COLOR INSERT) Time-frequency energy distributions obtained by means of different methods (named on pictures) for simulated signal (shown above) consisting of six gamma tones of frequencies 280, 550, 1100, 2200, 4400, 8800 Hz.

The test signal was constructed from six gammatones of different frequencies (given in the legend to Figure 4.61). From Figure 4.61 it is easy to see that the components are best recognized by the MP method. Performance of a spectrogram is also quite good but two components of lowest frequencies are not distinguished. The additional advantage of MP is the parametric description of the components.

The MP method allowed for identification of the resonant modes of the OAE, which are characteristic for each ear [Jedrzejczak et al., 2004, Jedrzejczak et al., 2006]. It was found that in case of tonal stimulation the ear does not respond in the frequency of stimulus, but several resonant modes are excited (different in left and right ear). Those closest to the stimulation frequency have the highest amplitude (Figure 4.62). The explicit identification of the latency of components allowed for the determination of the frequency-latency dependence [Jedrzejczak et al., 2004] and found application for evaluation of the influence of an exposition of subjects to noise [Jedrzejczak et al., 2005].

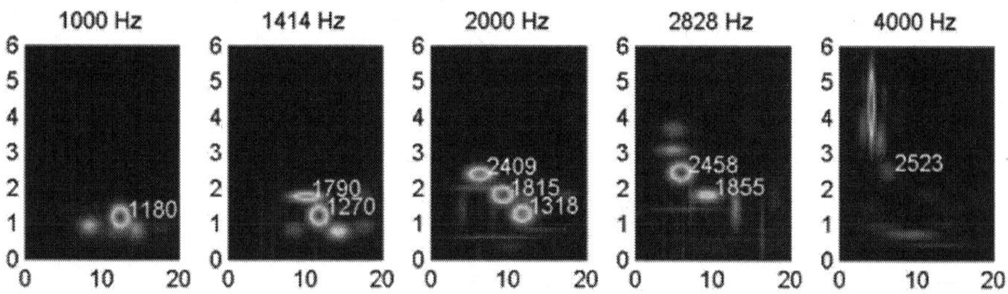

FIGURE 4.62: (SEE COLOR INSERT) Time-frequency energy distribution of OAE obtained for tonal stimulation of frequencies shown above the pictures. The same resonance modes are excited by tonal stimuli of different frequencies.

One of the parameters returned by MP is a time span of a component. It was observed that the histogram of time spans of TEOA components is bimodal [Jedrzejczak et al., 2004] and that long lasting components have a very narrow frequency band and do not obey exponential frequency-latency dependence. This kind of observation indicated that they are connected with SOAEs. The application of MP allowed for differentiation of TEOAE components into short and long lasting ones connected with SOAEs [Jedrzejczak et al., 2008] and explanation of paradoxically longer latencies found for pre-term neonates [Jedrzejczak et al., 2007].

Closer inspection of OAE components revealed that some of them, especially the long lasting ones, have an asymmetric shape. In order to better approximate the signal, asymmetric basic functions were introduced to the MP dictionary [Jedrzejczak et al., 2009]. These functions are composed of two parts; the first is based on Gabor, and the second on an exponential function. This kind of waveform can have different rise and fall times for the same frequency. Such a waveform can be described by the formula:

$$\Lambda(t;\mu,\sigma,\omega,\phi,T_f) = N \begin{Bmatrix} e^{-\frac{(t-\mu)^2}{2\sigma^2}} & \text{for } t \leq T_f \\ e^{-\alpha(t-\tau)} & \text{for } t > T_f \end{Bmatrix} \cos(\omega t + \phi) \tag{4.47}$$

where $\alpha = \frac{T_f-\mu}{\sigma^2}$ and $\tau = \frac{T_f+\mu}{2}$. The additional parameter $T_f > \mu$ determines the asymmetry of the atom. T_f describes the point at which the Gaussian envelope changes into an exponential decay function. The function thus obtained is continuous up to the first-order derivative.

Introduction of enriched dictionary encompassing asymmetric functions resulted in a better time resolution of time-frequency distributions of signal energy. In Figure

4.63 comparison of TOAE decomposition by means of enriched and Gabor dictionaries is shown. One can see that in some cases two Gabor atoms are needed to describe one component. The enriched dictionary provides a more sparse representation. Furthermore, the new approach gets rid of the "pre echo" effect, i.e., presence of false energy traces before the start of the signal visible on TF maps created from MP with Gabor dictionary. Analyses of simulated and real signals demonstrated that enriched dictionary provides correct estimation of latencies for components of short and long decay time.

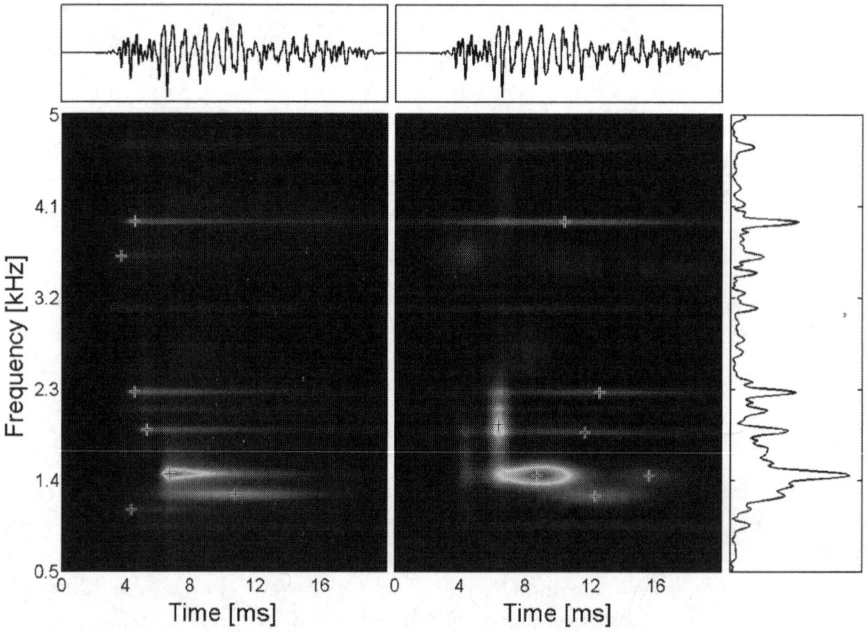

FIGURE 4.63: **(SEE COLOR INSERT)** Time-frequency distributions of OAE signal obtained by means of MP with dictionary containing asymmetric functions (left) and dictionary of Gabor atoms only (right). At the far right, Fourier spectrum of the signal is shown.

A hybrid matching pursuit algorithm that included Fourier spectral information was developed in [Notaro et al., 2007] with the aim of speeding-up computation times. Additionally the procedure was capable of identification of atoms whose latency-frequency relation was not compatible with the frequency-latency dependence. These atoms could be associated with several known phenomena, either intrinsic, such as intermodulation distortion, spontaneous emissions, and multiple internal reflections, or extrinsic, such as instrumental noise, linear ringing, and the acquisition window onset.

The inspection of the development of the techniques used for OAE analysis and the concrete results provided by them indicates that the matching pursuit is the most appropriate method for OAE analysis.

References

[Achermann et al., 1994] Achermann, P., Hartmann, R., Gunzinger, A., Guggenbhl, W., and Borbly, A. (1994). All night sleep and artificial stochastic control signals have similar correlation dimension. *Electroencephalogr Clin Neurophysiol*, 90:384–387.

[Achim et al., 1991] Achim, A., Richer, F., and Saint-Hilaire, J. M. (1991). Methodological considerations for the evaluation of spatio-temporal source models. *Electroencephalogr Clin Neurophysiol*, 79:227–240.

[Al-Angari and Sahakian, 2007] Al-Angari, H. M. and Sahakian, A. V. (2007). Use of sample entropy approach to study heart rate variability in obstructive sleep apnea syndrome. *IEEE Trans Biomed Eng*, 54:1900–1904.

[Al-Zaben and Al-Smadi, 2006] Al-Zaben, A. and Al-Smadi, A. (2006). Extraction of foetal ECG by combination of singular value decomposition and neuro-fuzzy inference system. *Phys Med Biol*, 51:137–143.

[Alvarez, 1922] Alvarez, W. (1922). The electrogastrogram and what it shows. *JAMA*, 78:1116–1119.

[Anderer et al., 1999] Anderer, P., Roberts, S., Schlogl, A., Gruber, G., Klosch, G., Herrmann, W., Rappelsberger, P., Filz, O., Barbanoj, M. J., Dorffner, G., and Saletu, B. (1999). Artifact processing in computerized analysis of sleep EEG - a review. *Neuropsychobiology*, 40:150–157.

[Andrzejak et al., 2003a] Andrzejak, R. G., Kraskov, A., Stogbauer, H., Mormann, F., and Kreuz, T. (2003a). Bivariate surrogate techniques: necessity, strengths, and caveats. *Phys Rev E Stat Nonlin Soft Matter Phys*, 68:066202.

[Andrzejak et al., 2003b] Andrzejak, R. G., Mormann, F., Kreuz, T., Rieke, C., Kraskov, A., Elger, C. E., and Lehnertz, K. (2003b). Testing the null hypothesis of the nonexistence of a preseizure state. *Phys Rev E*, 67(1):010901.

[Ari et al., 2008] Ari, S., Kumar, P., and Saha, G. (2008). A robust heart sound segmentation algorithm for commonly occurring heart valve diseases. *J Med Eng Technol*, 32:456–465.

[Armitage et al., 2002] Armitage, P., Berry, G., and Mathews, J.N.S. (2002). *Statistical Methods in Medical Research.* Blackwell Science, Oxford, 4th edition.

[Arnhold et al., 1999] Arnhold, J., Grassberger, P., Lehnertz, K., and Elger, C. (1999). A robust method for detecting interdependences: application to intracranially recorded EEG. *Physica D: Nonlinear Phenomena*, 134(4):419–430.

[Arnold et al., 1998] Arnold, M., Miltner, W. H., Witte, H., Bauer, R., and Braun, C. (1998). Adaptive AR modeling of nonstationary time series by means of Kalman filtering. *IEEE Trans Biomed Eng*, 45:553–562.

[Aschenbrenner-Scheibe et al., 2003] Aschenbrenner-Scheibe, R., Maiwald, T., Winterhalder, M., Voss, H. U., Timmer, J., and Schulze-Bonhage, A. (2003). How well can epileptic seizures be predicted? An evaluation of a nonlinear method. *Brain*, 126:2616–2626.

[Assaleh and Al-Nashash, 2005] Assaleh, K. and Al-Nashash, H. (2005). A novel technique for the extraction of fetal ECG using polynomial networks. *IEEE Trans Biomed Eng*, 52:1148–1152.

[Astolfi et al., 2005] Astolfi, L., Cincotti, F., Mattia, D., Babiloni, C., Carducci, F., Basilico, A., Rossini, P., Salinari, S., Ding, L., Ni, Y., He, B., and Babiloni, F. (2005). Assessing cortical functional connectivity by linear inverse estimation and directed transfer function: simulations and application to real data. *Clinical Neurophysiology*, 116:920–932.

[Astolfi et al., 2008] Astolfi, L., Cincotti, F., Mattia, D., De Vico Fallani, F., Tocci, A., Colosimo, A., Salinari, S., Marciani, M. G., Hesse, W., Witte, H., Ursino, M., Zavaglia, M., and Babiloni, F. (2008). Tracking the time-varying cortical connectivity patterns by adaptive multivariate estimators. *IEEE Trans Biomed Eng*, 55:902–913.

[Astolfi et al., 2006] Astolfi, L., Cincotti, F., Mattia, D., Marciani, M., Baccala, L., de Vico Fallani, F., Salinari, S., Ursino, M., Zavaglia, M., and Babiloni, F. (2006). Assessing cortical functional connectivity by partial directed coherence: simulations and application to real data. *IEEE Trans Biomed Eng*, 53(9):1802–12.

[Babiloni et al., 2009] Babiloni, C., Ferri, R., Binetti, G., Vecchio, F., Frisoni, G., Lanuzza, B., Miniussi, C., Nobili, F., Rodriguez, G., Rundo, F., Cassarino, A., Infarinato, F., Cassetta, E., Salinari, S., Eusebi, F., and Rossini, P. (2009). Directionality of EEG synchronization in Alzheimer's disease subjects. *Neurobiol Aging*, 30(1):93–102.

[Babiloni et al., 2000] Babiloni, F., Babiloni, C., Locche, L., Cincotti, F., Rossini, P. M., and Carducci, F. (2000). High-resolution electro-encephalogram: source estimates of Laplacian-transformed somatosensory-evoked potentials using a realistic subject head model constructed from magnetic resonance images. *Med Biol Eng Comput*, 38:512–519.

[Babloyantz and Destexhe, 1988] Babloyantz, A. and Destexhe, A. (1988). Is the normal heart a periodic oscillator? *Biol Cybern*, 58:203–211.

[Baccalá et al., 2004] Baccalá, L., Alvarenga, M., Sameshima, K., Jorge, C., and Castro, L. (2004). Graph theoretical characterization and tracking of the effective neural connectivity during episodes of mesial temporal epileptic seizure. *Journal of Integrative Neuroscience*, 3(4):379–395.

[Baccalá and Sameshima, 2001] Baccalá, L. and Sameshima, K. (2001). Partial directed coherence: a new concept in neural structure determination. *Biol Cybern*, 84:463–474.

[Bailon et al., 2006] Bailon, R., Sornmo, L., and Laguna, P. (2006). ECG derived respiratory frequency estimation. In Clifford, G., Azuaje, F., and McSharry, P., editors, *Advanced Methods and Tools for ECG Data Analysis*, pages 215–244. Artech House Inc., Norwood, MA.

[Ball et al., 2008] Ball, T., Demandt, E., Mutschler, I., Neitzel, E., Mehring, C., Vogt, K., Aertsen, A., and Schulze-Bonhage, A. (2008). Movement related activity in the high gamma range of the human EEG. *Neuroimage*, 41:302–310.

[Bartnik et al., 1992] Bartnik, E. A., Blinowska, K. J., and Durka, P. J. (1992). Single evoked potential reconstruction by means of wavelet transform. *Biol Cybern*, 67:175–181.

[Baselli et al., 1988] Baselli, G., Cerutti, S., Livraghi, M., Meneghini, C., Pagani, M., and Rimoldi, O. (1988). Causal relationship between heart rate and arterial blood pressure variability signals. *Med Biol Eng Comput*, 26:374–378.

[Baum et al., 1970] Baum, L., Petrie, T., Soules, G., and Weiss, N. (1970). A maximization technique occurring in the statistical analysis of probabilistic functions of Markov chains. *The Annals of Mathematical Statistics*, 41(1):164–171.

[Becker et al., 2010] Becker, M., Frauenrath, T., Hezel, F., Krombach, G. A., Kremer, U., Koppers, B., Butenweg, C., Goemmel, A., Utting, J. F., Schulz-Menger, J., and Niendorf, T. (2010). Comparison of left ventricular function assessment using phonocardiogram- and electrocardiogram-triggered 2D SSFP CINE MR imaging at 1.5 T and 3.0 T. *Eur Radiol*, 20:1344–1355.

[Bell, 2002] Bell, A. (2002). Musical ratios in geometrical spacing of outer hair cells in cochlea: strings of an underwater piano? In

C. Stevens, C., Burnham, D., McPherson, G., Schubert, E., and Renwick, J., editors, *Proc of the Intern Conference on Music Perception and Cognition.*

[Bell and Fletcher, 2004] Bell, A. and Fletcher, N. H. (2004). The cochlear amplifier as a standing wave: "squirting" waves between rows of outer hair cells? *J Acoust Soc Am*, 116(2):1016–1024.

[Bénar et al., 2009] Bénar, C. G., Papadopoulo, T., Torrésani, B., and Clerc, M. (2009). Consensus matching pursuit for multi-trial EEG signals. *J Neurosci Methods*, 180:161–170.

[Bendat and Piersol, 1971] Bendat, J. and Piersol, A. (1971). *Random Data: Analysis and Measurement Procedures.* Wiley & Sons, New York.

[Benjamini and Hochberg, 1995] Benjamini, Y. and Hochberg, Y. (1995). Controlling the false discovery rate: a practical and powerful approach to multiple testing. *Journal of the Royal Statistical Society*, 57(1):289–300.

[Benjamini and Yekutieli, 2001] Benjamini, Y. and Yekutieli, D. (2001). The control of the false discovery rate in multiple testing under dependency. *Annals of Statistics*, 29(4):1165–1188.

[Bergey and Franaszczuk, 2001] Bergey, G. K. and Franaszczuk, P. J. (2001). Epileptic seizures are characterized by changing signal complexity. *Clin Neurophysiol*, 112:241–249.

[Bertini et al., 2007] Bertini, M., Ferrara, M., De Gennaro, L., Curcio, G., Moroni, F., Vecchio, F., De Gasperis, M., Rossini, P., and Babiloni, C. (2007). Directional information flows between brain hemispheres during presleep wake and early sleep stages. *Cereb Cortex*, 17(8):1970–1978.

[Bezerianos et al., 2003] Bezerianos, A., Tong, S., and Thakor, N. (2003). Time-dependent entropy estimation of EEG rhythm changes following brain ischemia. *Ann Biomed Eng*, 31:221–232.

[Bianchi et al., 2010] Bianchi, A. M., Mendez, M. O., and Cerutti, S. (2010). Processing of signals recorded through smart devices: sleep-quality assessment. *IEEE Trans Inf Technol Biomed*, 14:741–747.

[Blackman and Tukey, 1958] Blackman, R. and Tukey, J. (1958). *The Measurement of Power Spectra.* Dover Publications, Dover, New York.

[Blinowska, 2002] Blinowska, K. (2002). Non-linear analysis of physiological signals—highlights and pitfalls. In Klonowski, W., editor, *Attractors, Signals and Synergetics. Proceedings of EUROATTRACTOR 2000*, pages 341–350. Pabst Science Publishers, Lengerich, Berlin.

[Blinowska and Durka, 1994] Blinowska, K. and Durka, P. (1994). *The Application of Wavelet Transform and Matching Pursuit to the Time Varying EEG Signals*, volume 4, chapter Intelligent Engineering Systems through Artificial Neural Networks, pages 535–540. ASME Press.

[Blinowska and Franaszczuk, 1989] Blinowska, K. and Franaszczuk, P. (1989). A model of the generation of electrocortical rhythms. In *Brain Dynamics*, Springer Series in Brain Dynamics 2, pages 192–201. Springer Verlag.

[Blinowska et al., 2004a] Blinowska, K., Kus, R., and Kaminski, M. (2004a). Granger causality and information flow in multivariate processes. *Phys Rev E*, 70:050902.

[Blinowska et al., 2010a] Blinowska, K., Kus, R., Kaminski, M., and Janiszewska, J. (2010a). Transmission of brain activity during cognitive task. *Brain Topogr*, 23:205–213.

[Blinowska and Malinowski, 1991] Blinowska, K. and Malinowski, M. (1991). Non-linear and linear forecasting of the EEG time series. *Biol Cybern*, 66:159–165.

[Blinowska et al., 2009] Blinowska, K., Muller-Putz, G., Kaiser, V., Astolfi, L., Vanderperren, K., Van Huffel, S., and Lemieux, L. (2009). Multimodal imaging of human brain activity: rational, biophysical aspects and modes of integration. *Comput Intell Neurosci*, page 813607.

[Blinowska et al., 2004b] Blinowska, K. J., Durka, P. J., and Zygierewicz, J. (2004b). Time-frequency analysis of brain electrical activity–adaptive approximations. *Methods Inf Med*, 43:70–73.

[Blinowska et al., 2010b] Blinowska, K. J., Kaminski, M., Kaminski, J., and Brzezicka, A. (2010b). Information processing in brain and dynamic patterns of transmission during working memory task by the SDTF function. *Conf Proc IEEE Eng Med Biol Soc*, 2010:1722–1725.

[Box and Cox, 1964] Box, G. and Cox, D. (1964). An analysis of transformations. *Journal of the Royal Statistical Society, Series B*, 26(2):211–252.

[Bradshaw et al., 2006] Bradshaw, L. A., Irimia, A., Sims, J. A., Gallucci, M. R., Palmer, R. L., and Richards, W. O. (2006). Biomagnetic characterization of spatiotemporal parameters of the gastric slow wave. *Neurogastroenterol Motil*, 18(8):619–631.

[Brazier and Casby, 1956] Brazier, M. and Casby, J. U. (1956). An application of the M.I.T. digital electronic correlator to a problem in EEG. *Electroencephalography*, 8:32.

[Brown et al., 2004] Brown, E. N., Kass, R. E., and Mitra, P. P. (2004). Multiple neural spike train data analysis: state-of-the-art and future challenges. *Nat Neurosci*, 7:456–461.

[Bryant and Yarnold, 1994] Bryant, F.B. and Yarnold, P.R. (1994). *Reading and Understanding Multivariate Statistics*, chapter Principal components analysis and exploratory and confirmatory factor analysis. American Psychological Association Books.

[Brzezicka et al., 2011] Brzezicka, A., Kaminski, M., Kaminski, J., and Blinowska, K. (2011). Information transfer during a transitive reasoning task. *Brain Topogr*, 24:1–8.

[Buchthal, 1955] Buchthal, F. (1955). Studies on muscular action potentials in the normal and diseased muscle. *Dtsch Z Nervrnheilkd*, 173:448–454.

[Buchthal, 1985] Buchthal, F. (1985). Electromyography in the evaluation of muscle diseases. *Neurol Clin*, 3:573–598.

[Bullock et al., 1995] Bullock, T. H., McClune, M. C., Achimowicz, J. Z., Iragui-Madoz, V. J., Duckrow, R. B., and Spencer, S. S. (1995). EEG coherence has structure in the millimeter domain: subdural and hippocampal recordings from epileptic patients. *Electroencephalogr Clin Neurophysiol*, 95:161–177.

[Buzsaki and Draguhn, 2004] Buzsaki, G. and Draguhn, A. (2004). Neuronal oscillations in cortical networks. *Science*, 304:1926–1929.

[Camps et al., 2001] Camps, G., Martinez, M., and Soria, E. (2001). Fetal ECG extraction using an FIR neural network. In *Proc Computers in Cardiology*, volume 28, pages 249–252. IEEE Press.

[Cardoso, 1999] Cardoso, J.-F. (1999). High-order contrasts for independent component analysis. *Neural Computation*, 11(1):157–192.

[Carli et al., 1999] Carli, F. D., Nobili, L., Ferrillo, F., Gelcichand, P., and Ferrillo, F. (1999). A method for the automatic detection of arousals during sleep. *Sleep*, 22:561–572.

[Cerutti et al., 1986] Cerutti, S., Baselli, G., Civardi, S., Ferrazzi, E., Marconi, A. M., Pagani, M., and Pardi, G. (1986). Variability analysis of fetal heart rate signals as obtained from abdominal electrocardiographic recordings. *J Perinat Med*, 14:445–452.

[Cerutti et al., 1987] Cerutti, S., Baselli, G., Liberati, D., and Pavesi, G. (1987). Single sweep analysis of visual evoked potentials through a model of parametric identification. *Biol Cybern*, 56:111–120.

[Cerutti et al., 1995] Cerutti, S., Bianchi, A., and Mainardi, L. (1995). Spectral analysis of heart rate variability signal. In Malik, M. and Camm, A., editors, *Heart Rate Variability*. Futura Publishing, Armonk, NY.

[Chee and Seow, 2007] Chee, J. and Seow, S.-C. (2007). Electrocardiogram. In Acharya, U., Suri, J., Spaan, J., and Krishnan, S., editors, *Advances in Cardiac Signal Processing*. Springer- Verlag, Berlin.

[Chen et al., 1999] Chen, J. D., Zou, X., Lin, X., Ouyang, S., and Liang, J. (1999). Detection of gastric slow wave propagation from the cutaneous electrogastrogram. *Am J Physiol*, 277:G424–430.

[Chen et al., 2004] Chen, Y., Rangarajan, G., Feng, J., and Ding, M. (2004). Analyzing multiple nonlinear time series with extended Granger causality. *Phys Lett A*, 324:26–35.

[Cheng, 1995] Cheng, J. (1995). Time-frequency analysis of transient evoked otoacoustic emissions via smoothed pseudo Wigner distribution. *Scand Audiol*, 24:91–96.

[Christov et al., 2006] Christov, I., Gomez-Herrero, G., Krasteva, V., Jekova, I., Gotchev, A., and Egiazarian, K. (2006). Comparative study of morphological and time-frequency ECG descriptors for heartbeat classification. *Med Eng Phys*, 28:876–887.

[Clifford, 2006] Clifford, G. (2006). Linear filtering methods. In Clifford, G., Azuaje, F., and McSharry, P., editors, *Advanced Methods and Tools for ECG Data Analysis*, pages 135–170. Artech House Inc., Norwood, MA.

[Clifford and Tarassenko, 2004] Clifford, G. D. and Tarassenko, L. (2004). Segmenting cardiac-related data using sleep stages increases separation between normal subjects and apnoeic patients. *Physiol Meas*, 25:27–35.

[Cohen et al., 1983] Cohen, D., Savard, P., Rifkin, R. D., Lepeschkin, E., and Strauss, W. E. (1983). Magnetic measurement of S-T and T-Q segment shifts in humans. Part II: Exercise-induced S-T segment depression. *Circ Res*, 53:274–279.

[Coifman and Wickerhauser, 1992] Coifman, R. and Wickerhauser, M. (1992). Entropy-based algorithms for best basis selection. *IEEE Transactions on Information Theory*, 38(2):713–718.

[Cook et al., 2009] Cook, I. A., Hunter, A., Korb, A., Farahbod, H., and Leuchter, A. (2009). *Quantitative EEG Analysis Methods and Clinical Applications*, chapter EEG signals in psychiatry: biomarkers for depression management. Artech House.

[Corlan and De Ambroggi, 2000] Corlan, A. D. and De Ambroggi, L. (2000). New quantitative methods of ventricular repolarization analysis in patients with left ventricular hypertrophy. *Ital Heart J*, 1(8):542–548.

[Costa et al., 2002] Costa, M., Goldberger, A., and Peng, C.-K. (2002). Multiscale entropy analysis of complex physiologic time series. *Physical Review Letters*, 89:068102.

[Croft et al., 2005] Croft, R. J., Chandler, J. S., Barry, R. J., Cooper, N. R., and Clarke, A. R. (2005). EOG correction: a comparison of four methods. *Psychophysiology*, 42:16–24.

[Crone et al., 1998] Crone, N. E., Miglioretti, D. L., Gordon, B., and Lesser, R. P. (1998). Functional mapping of human sensorimotor cortex with electrocorticographic spectral analysis. II. Event-related synchronization in the gamma band. *Brain*, 121 (Pt 12):2301–2315.

[Dalal et al., 2008] Dalal, S. S., Guggisberg, A. G., Edwards, E., Sekihara, K., Findlay, A. M., Canolty, R. T., Berger, M. S., Knight, R. T., Barbaro, N. M., Kirsch, H. E., and Nagarajan, S. S. (2008). Five-dimensional neuroimaging: localization of the time-frequency dynamics of cortical activity. *Neuroimage*, 40:1686–1700.

[Darvas et al., 2010] Darvas, F., Scherer, R., Ojemann, J. G., Rao, R. P., Miller, K. J., and Sorensen, L. B. (2010). High gamma mapping using EEG. *Neuroimage*, 49:930–938.

[Daubechies, 1992] Daubechies, I. (1992). *Ten Lectures on Wavelets*. SIAM.

[David et al., 2004] David, O., Cosmelli, D., and K.J., F. (2004). Evaluation of different measures of functional connectivity using a neural mass model. *Neuroimage*, 21:659–673.

[Davis, 1994] Davis, G. (1994). *Adaptive Nonlinear Approximations*. PhD thesis, New York University.

[Dawson, 1947] Dawson, G. D. (1947). Cerebral responses to electrical stimulation of peripheral nerve in man. *J Neurol Neurosurg Psychiatry*, 10:134–140.

[De Ambroggi et al., 1988] De Ambroggi, L., Bertoni, T., Breghi, M. L., Marconi, M., and Mosca, M. (1988). Diagnostic value of body surface potential mapping in old anterior non-Q myocardial infarction. *J Electrocardiol*, 21:321–329.

[De Lathauwer et al., 2000] De Lathauwer, L., De Moor, B., and Vandewalle, J. (2000). Fetal electrocardiogram extraction by blind source subspace separation. *IEEE Trans Biomed Eng*, 47:567–572.

[De Michele et al., 2003] De Michele, G., Sello, S., Carboncini, M. C., Rossi, B., and Strambi, S. K. (2003). Cross-correlation time-frequency analysis for multiple EMG signals in Parkinson's disease: a wavelet approach. *Med Eng Phys*, 25:361–369.

[De Vico Fallani et al., 2008] De Vico Fallani, F., Astolfi, L., Cincotti, F., Mattia, D., Marciani, M. G., Tocci, A., Salinari, S., Witte, H., Hesse, W., Gao, S., Colosimo, A., and Babiloni, F. (2008). Cortical network dynamics during foot movements. *Neuroinformatics*, 6:23–34.

[Debbal and Bereksi-Reguig, 2008] Debbal, S. M. and Bereksi-Reguig, F. (2008). Filtering and classification of phonocardiogram signals using wavelet transform. *J Med Eng Technol*, 32:53–65.

[Delorme and Makeig, 2004] Delorme, A. and Makeig, S. (2004). EEGLAB: an open source toolbox for analysis of single-trial EEG dynamics including independent component analysis. *J Neurosci Methods*, 134:9–21.

[Delorme et al., 2007] Delorme, A., Sejnowski, T., and Makeig, S. (2007). Enhanced detection of artifacts in EEG data using higher-order statistics and independent component analysis. *Neuroimage*, 34:1443–1449.

[Delprat et al., 1992] Delprat, N., Escudi, B., Guillemain, P., Kronland-Martinet, R., Tchamitchian, P., and Torrsani, B. (1992). Asymptotic wavelet and Gabor analysis: extraction of instantaneous frequencies. *IEEE Trans Inf Th*, 38:644–664.

[Dietch, 1932] Dietch, G. (1932). Fourier analyse von elektroenzephalogramen des menschen. *Pflugers Arch ges Physiol*, 230:106–112.

[Disselhorst-Klug et al., 1999] Disselhorst-Klug, C., Rau, G., Schmeer, A., and Silny, J. (1999). Non-invasive detection of the single motor unit action potential by averaging the spatial potential distribution triggered on a spatially filtered motor unit action potential. *J Electromyogr Kinesiol*, 9:67–72.

[Doyle, 1975] Doyle, D. (1975). Some comments on the use of Wiener filtering in the estimation of evoked potentials. *Electroenceph Clin Neurophysiol*, 38:533–534.

[Duncan et al., 2009] Duncan, C. C., Barry, R. J., Connolly, J. F., Fischer, C., Michie, P. T., Naatanen, R., Polich, J., Reinvang, I., and Van Petten, C. (2009). Event-related potentials in clinical research: guidelines for eliciting, recording, and quantifying mismatch negativity, P300, and N400. *Clin Neurophysiol*, 120:1883–1908.

[Durka, 2004] Durka, P. (2004). Adaptive time-frequency parametrization of epileptic EEG spikes. *Physical Review E*, 69:051914.

[Durka et al., 2002] Durka, P., Szelenberger, W., Blinowska, K., Androsiuk, W., and Myszka, M. (2002). Adaptive time-frequency parametrization in pharmaco EEG. *Journal of Neuroscience Methods*, 117:65–71.

[Durka, 2003] Durka, P. J. (2003). From wavelets to adaptive approximations: time-frequency parametrization of EEG. *Biomed Eng Online*, 2:1.

[Durka et al., 2001a] Durka, P. J., Ircha, D., and Blinowska, K. J. (2001a). Stochastic time-frequency dictionaries for matching pursuit. *IEEE Tran Signal Process*, 49(3):507–510.

[Durka et al., 2001b] Durka, P. J., Ircha, D., Neuper, C., and Pfurtscheller, G. (2001b). Time-frequency microstructure of event-related desynchronization and synchronization. *Med Biol Eng Comput*, 39(3):315–321.

[Durka et al., 2003] Durka, P. J., Klekowicz, H., Blinowska, K. J., Szelenberger, W., and Niemcewicz, S. (2003). A simple system for detection of EEG artifacts in polysomnographic recordings. *IEEE Trans Biomed Eng*, 50:526–528.

[Durka et al., 2005a] Durka, P. J., Malinowska, U., Szelenberger, W., Wakarow, A., and Blinowska, K. (2005a). High resolution parametric description of slow wave sleep. *Journal of Neuroscience Methods*, 147(1):15–21.

[Durka et al., 2005b] Durka, P. J., Matysiak, A., Montes, E. M., Sosa, P. V., and Blinowska, K. J. (2005b). Multichannel matching pursuit and EEG inverse solutions. *Journal of Neuroscience Methods*, 148(1):49–59.

[Durka et al., 2004] Durka, P. J., Zygierewicz, J., Klekowicz, H., Ginter, J., and Blinowska, K. J. (2004). On the statistical significance of event-related EEG desynchronization and synchronization in the time-frequency plane. *IEEE Trans Biomed Eng*, 51:1167–1175.

[Echeverria et al., 2001] Echeverria, J. C., Crowe, J. A., Woolfson, M. S., and Hayes-Gill, B. R. (2001). Application of empirical mode decomposition to heart rate variability analysis. *Med Biol Eng Comput*, 39:471–479.

[Eckmann et al., 1987] Eckmann, J.-P., Kamphorst, S. O., and Ruelle, D. (1987). Recurrence plots of dynamical systems. *Europhysics Letters*, 4(9):973–977.

[Efron and Tibshirani, 1993] Efron, B. and Tibshirani, R. (1993). *An Introduction to the Bootstrap.* Chapman & Hall, New York.

[Eichele et al., 2008] Eichele, T., Calhoun, V. D., Moosmann, M., Specht, K., Jongsma, M. L., Quiroga, R. Q., Nordby, H., and Hugdahl, K. (2008). Unmixing concurrent EEG-fMRI with parallel independent component analysis. *Int J Psychophysiol*, 67:222–234.

[Eichele et al., 2005] Eichele, T., Specht, K., Moosmann, M., Jongsma, M. L., Quiroga, R. Q., Nordby, H., and Hugdahl, K. (2005). Assessing the spatiotemporal evolution of neuronal activation with single-trial event-related potentials and functional MRI. *Proc Natl Acad Sci USA*, 102:17798–17803.

[Eichler, 2006] Eichler, M. (2006). On the evaluation of the information flow in multivariate systems by the Direct Transfer Function. *Biol Cybern*, 94:469–482.

[Elger and Lehnertz, 1998] Elger, C. E. and Lehnertz, K. (1998). Seizure prediction by non-linear time series analysis of brain electrical activity. *Eur J Neurosci*, 10:786–789.

[Engel and Singer, 2001] Engel, A. K. and Singer, W. (2001). Temporal binding and the neural correlates of sensory awareness. *Trends Cogn Sci (Regul Ed)*, 5:16–25.

[Englehart et al., 1999] Englehart, K., Hudgins, B., Parker, P. A., and Stevenson, M. (1999). Classification of the myoelectric signal using time-frequency based representations. *Med Eng Phys*, 21:431–438.

[Falla and Farina, 2008] Falla, D. and Farina, D. (2008). Non-uniform adaptation of motor unit discharge rates during sustained static contraction of the upper trapezius muscle. *Exp Brain Res*, 191:363–370.

[Fanselow et al., 2001] Fanselow, E., Sameshima, K., Baccala, L., and Nicolelis, M. (2001). Thalamic bursting in rats during different awake behavioral states. *Proc Nat Acad Sci USA*, 98(26):15330–15335.

[Farina et al., 2001] Farina, D., Colombo, R., Merletti, R., and Olsen, H. (2001). Evaluation of intra-muscular emg signal decomposition algorithms. *J Electromyogr Kinesiol*, 11:175–187.

[Fenici and Brisinda, 2006] Fenici, R. and Brisinda, D. (2006). Magnetocardiography provides non-invasive three-dimensional electroanatomical imaging of cardiac electrophysiology. *Int J Cardiovasc Imaging*, 22:595–597.

[Fereniec, 2008] Fereniec, M. (2008). *The Evaluation of Space Variability of the High-Resolution ECG during Atrial Heart Activity*. PhD thesis, IBIB PAN, Warsaw.

[Fereniec et al., 2008] Fereniec, M., Maniewski, R., Karpinski, G., Opolski, G., and Rix, H. (2008). High-resolution multichannel measurement and analysis of cardiac repolarization. *Biocybernetics and Biomedical Engineering*, 28(3):61–69.

[Fereniec and Karpinski, 2004] Fereniec, M. and Karpinski, G. (2004). Analysis of T-wave shape variability in hr ECG mapping. *Techn Health Care*, 12(2):125–127.

[Fischer et al., 2005] Fischer, M. J., Scheler, G., and Stefan, H. (2005). Utilization of magnetoencephalography results to obtain favourable outcomes in epilepsy surgery. *Brain*, 128:153–157.

[Florestal et al., 2006] Florestal, J. R., Mathieu, P. A., and Malanda, A. (2006). Automated decomposition of intramuscular electromyographic signals. *IEEE Trans Biomed Eng*, 53:832–839.

[Florestal et al., 2009] Florestal, J. R., Mathieu, P. A., and McGill, K. C. (2009). Automatic decomposition of multichannel intramuscular EMG signals. *J Electromyogr Kinesiol*, 19:1–9.

[Folland and Sitaram, 1997] Folland, G. and Sitaram, A. (1997). The uncertainty principle: A mathematical survey. *Journal of Fourier Analysis and Applications*, 3(3):207–238.

[Franaszczuk and Blinowska, 1985] Franaszczuk, P. and Blinowska, K. (1985). Linear model of brain electrical activity - EEG as a superposition of damped oscillatory modes. *Biol Cybern*, 53:19–25.

[Franaszczuk et al., 1985] Franaszczuk, P., Blinowska, K., and Kowalczyk, M. (1985). The application of parametric multichannel spectral estimates in the study of electrical brain activity. *Biol Cybern*, 51:239–247.

[Franaszczuk and Bergey, 1999] Franaszczuk, P. J. and Bergey, G. K. (1999). An autoregressive method for the measurement of synchronization of interictal and ictal EEG signals. *Biol Cybern*, 81:3–9.

[Franaszczuk et al., 1994] Franaszczuk, P. J., Bergey, G. K., and Kaminski, M. J. (1994). Analysis of mesial temporal seizure onset and propagation using the directed transfer function method. *Electroencephalogr Clin Neurophysiol*, 91:413–427.

[Franaszczuk et al., 1989] Franaszczuk, P. J., Mitraszewski, P., and Blinowska, K. (1989). FAD-parametric description of EEG time series. *Acta Physiol Pol*, 40:418–422.

[Freeman, 1975] Freeman, W. (1975). *Mass Action in the Nervous System.* Academic Press, New York.

[Friston et al., 2007] Friston, K., Ashburner, J., Kiebel, S., Nichols, T., and Penny, W., editors (2007). *Statistical Parametric Mapping: The Analysis of Functional Brain Images.* Academic Press.

[Fuglsang-Frederiksen, 2000] Fuglsang-Frederiksen, A. (2000). The utility of interference pattern analysis. *Muscle Nerve*, 23:18–36.

[Gazzoni et al., 2004] Gazzoni, M., Farina, D., and Merletti, R. (2004). A new method for the extraction and classification of single motor unit action potentials from surface EMG signals. *J Neurosci Methods*, 136:165–177.

[Ge et al., 2007] Ge, D., Srinivassan, N., and Krishnan, S. (2007). The application of autoregressive modeling in cardiac arrhythmia classification. In Acharya, U., Suri, J., Spaan, J., and Krishnan, S., editors, *Advances in Cardiac Signal Processing*. Springer- Verlag, Berlin.

[Gevins and Morgan, 1986] Gevins, A. S. and Morgan, N. H. (1986). Classifier-directed signal processing in brain research. *IEEE Trans Biomed Eng*, 33:1054–1068.

[Geweke, 1982] Geweke, J. (1982). Measurement of linear dependence and feedback between multiple time series. *J Am Stat Assoc*, 77(378):304–324.

[Giard et al., 1995] Giard, M. H., Lavikahen, J., Reinikainen, K., Perrin, F., Bertrand, O., Pernier, J., and Näätänen, R. (1995). Separate representation of stimulus frequency, intensity, and duration in auditory sensory memory: an event-related potential and dipole-model analysis. *J Cognitive Neuroscience*, 7(2):133–143.

[Gibson et al., 1997] Gibson, N. M., Woolfson, M. S., and Crowe, J. A. (1997). Detection of fetal electrocardiogram signals using matched filters with adaptive normalisation. *Med Biol Eng Comput*, 35:216–222.

[Ginter et al., 2001] Ginter, J., Blinowska, K. J., Kaminski, M., and Durka, P. J. (2001). Phase and amplitude analysis in time-frequency space–application to voluntary finger movement. *J Neurosci Methods*, 110:113–124.

[Ginter et al., 2005] Ginter, J., Blinowska, K. J., Kaminski, M., Durka, P. J., Pfurtscheller, G., and Neuper, C. (2005). Propagation of EEG activity in the beta and gamma band during movement imagery in humans. *Methods Inf Med*, 44:106–113.

[Gireesan et al., 2010] Gireesan, K., Sengottuvel, C., Parsakthi, C., Rajesh, P., Janawadkar, M., and Radhakrishnan, T. (2010). Magnetocardiography study of cardiac anomalies. In Supek, S. and Susac, A., editors, *Advances in Biomagnetism BIOMAG2010, IFMBE Proceedings*, volume 28, pages 431–435. Springer.

[Goldberger et al., 2000] Goldberger, A. L., Amaral, L. A., Glass, L., Hausdorff, J. M., Ivanov, P. C., Mark, R. G., Mietus, J. E., Moody, G. B., Peng, C. K., and Stanley, H. E. (2000). PhysioBank, PhysioToolkit, and PhysioNet: components of a new research resource for complex physiologic signals. *Circulation*, 101:E215–220.

[Gotman and Marciani, 1985] Gotman, J. and Marciani, M. G. (1985). Electroencephalographic spiking activity, drug levels, and seizure occurrence in epileptic patients. *Ann Neurol*, 17:597–603.

[Govindan et al., 2007] Govindan, R., Wilson, J., Eswaran, H., Lowery, C., and Preil, H. (2007). Revisiting sample entropy analysis. *Physica A: Statistical Mechanics and its Applications*, 376:158–164.

[Graimann et al., 2002] Graimann, B., Huggins, J. E., Levine, S. P., and Pfurtscheller, G. (2002). Visualization of significant ERD/ERS patterns in multichannel EEG and ECoG data. *Clin Neurophysiol*, 113:43–47.

[Granger, 1969] Granger, C. (1969). Investigating causal relations by econometric models and cross-spectral methods. *Econometrica*, 37:424–438.

[Grapelyuk et al., 2010] Grapelyuk, A., Schirdewan, A., Fischer, R., and Wessel, N. (2010). Cardiac magnetic field mapping quantified by Kullback-Leibler entropy detects patients with coronary artery disease. *Physiol Meas*, 31(10):1345–1354.

[Grassberger and Procaccia, 1983] Grassberger, P. and Procaccia, I. (1983). Measuring the strangeness of strange attractors. *Physica D*, 9:189–208.

[Gratton et al., 1983] Gratton, G., Coles, M. G., and Donchin, E. (1983). A new method for off-line removal of ocular artifact. *Electroencephalogr Clin Neurophysiol*, 55:468–484.

[Grau et al., 2007] Grau, C., Fuentemilla, L., and Marco-Pallars, J. (2007). Functional neural dynamics underlying auditory event-related n1 and n1 suppression response. *Neuroimage*, 36(6):522–531.

[Grave de Peralta Mendez and Gonzales Andino, 1999] Grave de Peralta Mendez, R. and Gonzales Andino, S. (1999). Distributed source models: standard solutions and new developments. In Uhl, C., editor, *Analysis of Neurophysiological Brain Functioning*, pages 176–291. Springer, Berlin.

[Groves and Thompson, 1970] Groves, P. and Thompson, R. (1970). Habituation: a dual-process theory. *Psychol Rev*, 77:419–450.

[Gunji et al., 2007] Gunji, A., Ishii, R., Chau, W., Kakigi, R., and Pantev, C. (2007). Rhythmic brain activities related to singing in humans. *Neuroimage*, 34:426–434.

[Gydikov and Gantchev, 1989] Gydikov, A. and Gantchev, N. (1989). Velocity of spreading of the excitation along the muscle fibres of human motor units measured by superficial electrodes. *Electromyogr Clin Neurophysiol*, 29:131–138.

[Gydikov and Gatev, 1982] Gydikov, A. and Gatev, P. (1982). Human single muscle fiber potentials at different radial distances from the fibers determined by a method of location. *Exp Neurol*, 76:25–34.

[Gydikov et al., 1986] Gydikov, A., Gerilovsky, L., Radicheva, N., and Trayanova, N. (1986). Influence of the muscle fibre end geometry on the extracellular potentials. *Biol Cybern*, 54:1–8.

[Gydikov and Kosarov, 1972] Gydikov, A. and Kosarov, D. (1972). Studies of the activity of alpha motoneurons in man by means of a new electromyographic method. In Somjem, G., editor, *Proceedings of Conference Neurophysiology Studies in Man*, pages 219–227.

[Haig et al., 1996] Haig, A. J., Gelblum, J. B., Rechtien, J. J., and Gitter, A. J. (1996). Technology assessment: the use of surface EMG in the diagnosis and treatment of nerve and muscle disorders. *Muscle Nerve*, 19:392–395.

[Hämäläinen et al., 1993] Hämäläinen, M., Hari, R., Ilmoniemi, R. J., Knuutila, J., and Lounasmaa, O. V. (1993). Magnetoencephalography—theory, instrumentation, and applications to noninvasive studies of the working human brain. *Rev Mod Phys*, 65(2):413–497.

[Hamilton-Wright and Stashuk, 2005] Hamilton-Wright, A. and Stashuk, D. W. (2005). Physiologically based simulation of clinical EMG signals. *IEEE Trans Biomed Eng*, 52:171–183.

[Hanninen et al., 2001] Hanninen, H., Takala, P., Makijarvi, M., Montonen, J., Korhonen, P., Oikarinen, L., Simelius, K., Nenonen, J., Katila, T., and Toivonen, L. (2001). Recording locations in multichannel magnetocardiography and body surface potential mapping sensitive for regional exercise-induced myocardial ischemia. *Basic Res Cardiol*, 96:405–414.

[Harris, 1978] Harris, F. (1978). On the use of windows for harmonic analysis with the Discrete Fourier transform. *Proc IEEE*, 66(1):51–83.

[Harrison et al., 2005] Harrison, M. A. F., Osorio, I., Frei, M. G., Asuri, S., and Lai, Y.-C. (2005). Correlation dimension and integral do not predict epileptic seizures. *Chaos: An Interdisciplinary Journal of Nonlinear Science*, 15(3):033106.

[Hassanpour and Parsaei, 2006] Hassanpour, H. and Parsaei, A. (2006). Fetal ECG extraction using wavelet transform. In *International Conference on Computational Intelligence for Modeling, Control and Automation, 2006 and International Conference on Intelligent Agents, Web Technologies and Internet Commerce*, page 179.

[Hatzopoulos et al., 2000] Hatzopoulos, S., Cheng, J., Grzanka, A., and Martini, A. (2000). Time-frequency analyses of TEOAE recordings from normals and SNHL patients. *Audiology*, 39:1–12.

[He et al., 2006] He, T., Clifford, G., and Tarassenko, L. (2006). Application of ICA in removing artifacts from the ECG. *Neural Comput and Applic*, 15(2):105–116.

[Hesse et al., 2003] Hesse, W., Moller, E., Arnold, M., and Schack, B. (2003). The use of time-variant EEG Granger causality for inspecting directed interdependencies of neural assemblies. *J Neurosci Methods*, 124:27–44.

[Hjorth, 1975] Hjorth, B. (1975). An on-line transformation of EEG scalp potentials into orthogonal source derivations. *Electroencephalogr Clin Neurophysiol*, 39:526–530.

[Hochberg and Tamhane, 1987] Hochberg, Y. and Tamhane, A. C. (1987). *Multiple Comparison Procedures*. Wiley, New York.

[Holmes et al., 1996] Holmes, A., Blair, R., Watson, J., and Ford, I. (1996). Non-parametric analysis of statistic images from functional mapping experiments. *Journal of Cerebral Blood Flow and Metabolism*, 16:7–22.

[Holobar et al., 2009] Holobar, A., Farina, D., Gazzoni, M., Merletti, R., and Zazula, D. (2009). Estimating motor unit discharge patterns from high-density surface electromyogram. *Clin Neurophysiol*, 120:551–562.

[Holobar and Zazula, 2007] Holobar, A. and Zazula, D. (2007a). Multichannel blind source separation using convolution kernel compensation. *IEEE Trans Signal Process*, 55:4487–4496.

[Hren et al., 1998] Hren, R., Stroink, G., and Horacek, B. M. (1998). Spatial resolution of body surface potential maps and magnetic field maps: a simulation study applied to the identification of ventricular pre-excitation sites. *Med Biol Eng Comput*, 36:145–157.

[Huang et al., 1998] Huang, N., Shen, Z., Long, S., Wu, M., Shih, H., Zheng, Q., Yen, N., Tung, C., and Liu, H. (1998). The empirical mode decomposition and the Hilbert spectrum for nonlinear and non-stationary time series analysis. *Proc R Soc Lond A*, 454:903–995.

[Hughes, 2006] Hughes, N. (2006). Probabilistic approaches to ECG segmentation and feature extraction. In Clifford, G., Azuaje, F., and McSharry, P., editors, *Advanced Methods and Tools for ECG Data Analysis*, pages 291–317. Artech House Inc., Norwood, MA.

[Huikuri et al., 1998] Huikuri, H. V., Makikallio, T. H., Airaksinen, K. E., Seppanen, T., Puukka, P., Raiha, I. J., and Sourander, L. B. (1998). Power-law relationship of heart rate variability as a predictor of mortality in the elderly. *Circulation*, 97:2031–2036.

[Hyde, 1999] Hyde, S. (1999). Likelihood based inference on the Box-Cox family of transformations: Sas and MATLAB programs. Technical report, Mathematical Sciences, Montana State University. Avaliable online at: http://www.math.montana.edu/hyde/papers/ms-paper99.pdf.

[Hyvärinen, 1999] Hyvärinen, A. (1999). Fast and robust fixed-point algorithms for independent component analysis. *IEEE Trans Neural Netw*, 10:626–634.

[Hyvärinen and Oja, 2000] Hyvärinen, A. and Oja, E. (2000). Independent component analysis: algorithms and applications. *Neural Networks*, 13(4-5):411–430.

[Iasemidis et al., 2003] Iasemidis, L. D., Shiau, D. S., Chaovalitwongse, W., Sackellares, J. C., Pardalos, P. M., Principe, J. C., Carney, P. R., Prasad, A., Veeramani, B., and Tsakalis, K. (2003). Adaptive epileptic seizure prediction system. *IEEE Trans Biomed Eng*, 50:616–627.

[Ibert et al., 2007] Ibert, C., Ancoli-Israel, S., Chesson, A., and Quan, S. (2007). *The AASM Manual for the Scoring of Sleep and Associated Events: Rules, Terminology and Technical Specification.* 1st edition. American Academy of Sleep Medicine, Westchester.

[Jafari and Chambers, 2005] Jafari, M. G. and Chambers, J. A. (2005). Fetal electrocardiogram extraction by sequential source separation in the wavelet domain. *IEEE Trans Biomed Eng*, 52:390–400.

[Jager, 2006] Jager, F. (2006). St analysis. In Clifford, G., Azuaje, F., and McSharry, P., editors, *Advanced Methods and Tools for ECG Data Analysis*, pages 269–290. Artech House Inc., Norwood, MA.

[Jager et al., 1998] Jager, F., Moody, G. B., and Mark, R. G. (1998). Detection of transient ST segment episodes during ambulatory ECG monitoring. *Comput Biomed Res*, 31:305–322.

[James and Gibson, 2003] James, C. J. and Gibson, O. J. (2003). Temporally constrained ICA: an application to artifact rejection in electromagnetic brain signal analysis. *IEEE Trans Biomed Eng*, 50:1108–1116.

[Januauskas et al., 2001] Januauskas, A., Marozas, V., Engdahl, B., Hoffman, H. J., Svensson, O., and Sornmo, L. (2001). Otoacoustic emissions and improved pass/fail separation using wavelet analysis and time windowing. *Med Biol Eng Comput*, 39:134–139.

[Jedrzejczak et al., 2008] Jedrzejczak, W. W., Blinowska, K. J., Kochanek, K., and Skarzynski, H. (2008). Synchronized spontaneous otoacoustic emissions analyzed in a time-frequency domain. *J Acoust Soc Am*, 124:3720–3729.

[Jedrzejczak et al., 2005] Jedrzejczak, W. W., Blinowska, K. J., and Konopka, W. (2005). Time-frequency analysis of transiently evoked otoacoustic emissions of subjects exposed to noise. *Hear Res*, 205:249–255.

[Jedrzejczak et al., 2006] Jedrzejczak, W. W., Blinowska, K. J., and Konopka, W. (2006). Resonant modes in transiently evoked otoacoustic emissions and asymmetries between left and right ear. *J Acoust Soc Am*, 119:2226–2231.

[Jedrzejczak et al., 2004] Jedrzejczak, W. W., Blinowska, K. J., Konopka, W., Grzanka, A., and Durka, P. J.

(2004). Identification of otoacoustic emissions components by means of adaptive approximations. *J Acoust Soc Am*, 115:2148–2158.

[Jedrzejczak et al., 2007] Jedrzejczak, W. W., Hatzopoulos, S., Martini, A., and Blinowska, K. J. (2007). Otoacoustic emissions latency difference between full-term and preterm neonates. *Hear Res*, 231:54–62.

[Jedrzejczak et al., 2009] Jedrzejczak, W. W., Kwaskiewicz, K., Blinowska, K. J., Kochanek, K., and Skarzynski, H. (2009). Use of the matching pursuit algorithm with a dictionary of asymmetric waveforms in the analysis of transient evoked otoacoustic emissions. *J Acoust Soc Am*, 126:3137–3146.

[Jerger et al., 2005] Jerger, K. K., Weinstein, S. L., Sauer, T., and Schiff, S. J. (2005). Multivariate linear discrimination of seizures. *Clin Neurophysiol*, 116:545–551.

[Jiang et al., 2007] Jiang, J. A., Chao, C. F., Chiu, M. J., Lee, R. G., Tseng, C. L., and Lin, R. (2007). An automatic analysis method for detecting and eliminating ECG artifacts in EEG. *Comput Biol Med*, 37:1660–1671.

[Jobert, 1992] Jobert, M. (1992). Pattern recognition by matched filtering; an analysis of sleep spindle and K-complex density under the influence of lormetazepam and zopiclone. *Electroencephalogr Clin Neurophysiol*, 26:100–107.

[John et al., 1978] John, E., Ruchkin, D., and Vidal, J. (1978). *Event Related Brain Potentials in Man*, chapter Measurment of event related potentials, pages 99–138. Academic Press.

[John and Thatcher, 1977] John, E. and Thatcher, R. (1977). *Neurometrics: Clinical Applications of Quantitative Electrophysiology*. Functional Neuroscience. Lawrence Erlbaum Associates.

[Jokiniemi et al., 2003] Jokiniemi, T., Simelius, K., Nenonen, J., Tierala, I., Toivonen, L., and Katilal, T.

(2003). Baseline reconstruction for localization of rapid ventricular tachycardia from body surface potential maps. *Physiol Meas*, 24(3):641–651.

[Joo et al., 1983] Joo, T. H., McClellan, J. H., Foale, R. A., Myers, G. S., and Lees, R. S. (1983). Pole-zero modeling and classification of phonocardiograms. *IEEE Trans Biomed Eng*, 30:110–118.

[Jouny et al., 2003] Jouny, C. C., Franaszczuk, P. J., and Bergey, G. K. (2003). Characterization of epileptic seizure dynamics using Gabor atom density. *Clin Neurophysiol*, 114:426–437.

[Jouny et al., 2005] Jouny, C. C., Franaszczuk, P. J., and Bergey, G. K. (2005). Signal complexity and synchrony of epileptic seizures: is there an identifiable preictal period? *Clin Neurophysiol*, 116:552–558.

[Joyce et al., 2004] Joyce, C. A., Gorodnitsky, I. F., and Kutas, M. (2004). Automatic removal of eye movement and blink artifacts from EEG data using blind component separation. *Psychophysiology*, 41:313–325.

[Jung et al., 2000] Jung, T., Makeig, S., Humphries, C., Lee, T., McKeown, M., Iragui, V., and Sejnowski, T. (2000). Removing electroencephalographic artifacts by blind source separation. *Psychophysiology*, 37(2):163–178.

[Jurkko et al., 2009] Jurkko, R., Mantynen, V., Tapanainen, J. M., Montonen, J., Vaananen, H., Parikka, H., and Toivonen, L. (2009). Non-invasive detection of conduction pathways to left atrium using magnetocardiography: validation by intracardiac electroanatomic mapping. *Europace*, 11:169–177.

[Kalitzin et al., 2002] Kalitzin, S., Parra, J., Velis, D. N., and Lopes da Silva, F. H. (2002). Enhancement of phase clustering in the EEG/MEG gamma frequency band anticipates transitions to paroxysmal epileptiform activity in epileptic patients with known visual sensitivity. *IEEE Trans Biomed Eng*, 49:1279–1286.

[Kalitzin et al., 2005] Kalitzin, S., Velis, D., Suffczynski, P., Parra, J., and da Silva, F. L. (2005). Electrical brain-stimulation paradigm for estimating the seizure onset site and the time to ictal transition in temporal lobe epilepsy. *Clin Neurophysiol*, 116:718–728.

[Kaminski and Blinowska, 1991] Kaminski, M. and Blinowska, K. (1991). A new method of the description of the information flow in brain structures. *Biol Cybern*, 65:203–210.

[Kaminski et al., 1997] Kaminski, M., Blinowska, K., and Szelenberger, W. (1997). Topographic analysis of coherence and propagation of EEG activity during sleep and wakefulness. *Electroencephalogr Clin Neurophysiol*, 102:216–227.

[Kaminski et al., 2001] Kaminski, M., Ding, M., Truccolo, W., and Bressler, S. (2001). Evaluating causal relations in neural systems: Granger causality, directed transfer function and statistical assessment of significance. *Biol Cybern*, 85:145–157.

[Kaminski et al., 2010] Kaminski, M., Szerling, P., and Blinowska, K. (2010). Comparison of methods for estimation of time-varying transmission in multichannel data. In *Proc 10th IEEE International Conference on Information Technology and Applications in Biomedicine*, Korfu.

[Kanjilal et al., 1997] Kanjilal, P. P., Palit, S., and Saha, G. (1997). Fetal ECG extraction from single-channel maternal ECG using singular value decomposition. *IEEE Trans Biomed Eng*, 44:51–59.

[Kantz and Schreiber, 2000] Kantz, H. and Schreiber, T. (2000). *Nonlinear Time Series Analysis*. Cambridge University press, Cambridge.

[Karvounis et al., 2009] Karvounis, E., Tsipouras, M., and Fotiadis, D. (2009). Detection of fetal heart rate through 3-d phase space analysis from multivariate abdominal recordings. *IEEE Trans Biomed Eng*, 56(5):1394–1406.

[Karvounis et al., 2007] Karvounis, E. C., Tsipouras, M. G., Fotiadis, D. I., and Naka, K. K. (2007). An automated methodology for fetal heart rate extraction from the abdominal electrocardiogram. *IEEE Trans Inf Technol Biomed*, 11:628–638.

[Katz et al., 1991] Katz, A., Marks, D. A., McCarthy, G., and Spencer, S. S. (1991). Does interictal spiking change prior to seizures? *Electroencephalogr Clin Neurophysiol*, 79(2):153–156.

[Kay, 1988] Kay, S. (1988). *Modern Spectral Estimation: Theory and Application.* Prentice-Hall, Englewood Cliffs, NJ.

[Kemp et al., 1985] Kemp, B., Jaspers, P., Franzen, J., and Janssen, A. (1985). An optimal monitor of the electroencephalographic sigma sleep state. *Biol Cybern*, 51:263–270.

[Kemp, 1978] Kemp, D. T. (1978). Stimulated acoustic emissions from within the human auditory system. *J Acoust Soc Am*, 64:1386–1391.

[Khaddoumi et al., 2006] Khaddoumi, B., Rix, H., Meste, O., Fereniec, M., and Maniewski, R. (2006). Body surface ECG signal shape dispersion. *IEEE Trans Biomed Eng*, 53(12):2491–2500.

[Khamene and Negahdaripour, 2000] Khamene, A. and Negahdaripour, S. (2000). A new method for the extraction of fetal ECG from the composite abdominal signal. *IEEE Trans Biomed Eng*, 47:507–516.

[Kleine et al., 2007] Kleine, B. U., van Dijk, J. P., Lapatki, B. G., Zwarts, M. J., and Stegeman, D. F. (2007). Using two-dimensional spatial information in decomposition of surface EMG signals. *J Electromyogr Kinesiol*, 17(5):535–548.

[Klekowicz et al., 2009] Klekowicz, H., Malinowska, U., Piotrowska, A. J., Wolynczyk-Gmaj, D., Niemcewicz, S., and Durka, P. J. (2009). On the robust parametric detection of EEG artifacts in polysomnographic recordings. *Neuroinformatics*, 7:147–160.

[Klimaszewska and Zebrowski, 2009] Klimaszewska, K. and Zebrowski, J. J. (2009). Detection of the type of intermittency using characteristic patterns in recurrence plots. *Phys Rev E Stat Nonlin Soft Matter Phys*, 80:026214.

[Klingenheben et al., 2005] Klingenheben, T., Ptaszynski, P., and Hohnloser, S. H. (2005). Quantitative assessment of microvolt T-wave alternans in patients with congestive heart failure. *J Cardiovasc Electrophysiol*, 16:620–624.

[Kocsis and Kaminski, 2006] Kocsis, B. and Kaminski, M. (2006). Dynamic changes in the direction of the theta rhythmic drive between supramammillary nucleus and the septohippocampal system. *Hippocampus*, 16:531–540.

[Kors et al., 1996] Kors, J. A., van Herpen, G., Wu, J., Zhang, Z., Prineas, R. J., and van Bemmel, J. H. (1996). Validation of a new computer program for Minnesota coding. *J Electrocardiol*, 29(1):83–88.

[Korzeniewska et al., 2008] Korzeniewska, A., Crainiceanu, C. M., Ku, R., Franaszczuk, P. J., and Crone, N. E. (2008). Dynamics of event-related causality in brain electrical activity. *Hum Brain Mapp*, 29:1170–1192.

[Korzeniewska et al., 1997] Korzeniewska, A., Kasicki, S., Kaminski, M., and Blinowska, K. J. (1997). Information flow between hippocampus and related structures during various types of rat's behavior. *J Neurosci Meth*, 73:49–60.

[Korzeniewska et al., 2003] Korzeniewska, A., Manczak, M., Kaminski, M., Blinowska, K. J., and Kasicki, S. (2003). Determination of information flow direction among brain structures by a modified directed transfer function (dDTF) method. *J Neurosci Methods*, 125:195–207.

[Kosarov, 1974] Kosarov, D. (1974). Vectorelectromiographic control on the position of surface electrodes in a relation to active motor units in the human muscles. *Acta Physiological et Pharmacologica Bulgarica*, 1:85–93.

[Koymen et al., 1987] Koymen, H., Altay, B. K., and Ider, Y. Z. (1987). A study of prosthetic heart valve sounds. *IEEE Trans Biomed Eng*, 34:853–863.

[Kreuz et al., 2004] Kreuz, T., Andrzejak, R. G., Mormann, F., Kraskov, A., Stögbauer, H., Elger, C. E., Lehnertz, K., and Grassberger, P. (2004). Measure profile surrogates: a method to validate the performance of epileptic seizure prediction algorithms. *Phys Rev E*, 69(6):061915.

[Krzyminiewski et al., 1999] Krzyminiewski, R., Panek, G., and Stepien, R. (1999). Correlation of results of coronarographic, SPECT examination and high-resolution vectorcardiography. *European Medical & Biological Engineering & Computing*, 37(2):514–515.

[Kubat et al., 1994] Kubat, M., Pfurtscheller, G., and Flotzinger, D. (1994). AI-based approach to automatic sleep classification. *Biol Cybern*, 70(5):443–448.

[Kubicki et al., 1989] Kubicki, S., Holler, L., Berg, I., Pastelak-Price, C., and Dorow, R. (1989). Sleep EEG evaluation: a comparison of results obtained by visual scoring and automatic analysis with the Oxford sleep stager. *Sleep*, 12:140–149.

[Kullback and Leibler, 1951] Kullback, S. and Leibler, R. (1951). On information and sufficiency. *Annals of Mathematical Statistics*, 22(1):79–86.

[Kus et al., 2006] Kus, R., Ginter, J. S., and Blinowska, K. J. (2006). Propagation of EEG activity during finger movement and its imagination. *Acta Neurobiol Exp (Wars)*, 66:195–206.

[Kus et al., 2004] Kus, R., Kaminski, M., and Blinowska, K. (2004). Determination of EEG activity propagation: pair-wise versus multichannel estimate. *IEEE Trans Biomed Eng*, 51:1501–1510.

[Lachaux et al., 2000] Lachaux, J.-P., Rodriguez, E., Martinerie, J., Adam, C., Hasboun, D., and Varela, F. J.

(2000). A quantitative study of gamma-band activity in human intracranial recordings triggered by visual simuli. *Eur J Neurosci*, 12:2608–2622.

[Lagerholm et al., 2000] Lagerholm, M., Peterson, C., Braccini, G., Edenbrandt, L., and Sornmo, L. (2000). Clustering ECG complexes using Hermite functions and self-organizing maps. *IEEE Trans Biomed Eng*, 47:838–848.

[Laguna et al., 1994] Laguna, P., Jane, R., and Caminal, P. (1994). Automatic detection of wave boundaries in multilead ECG signals: validation with the CSE database. *Comput Biomed Res*, 27:45–60.

[Laguna et al., 1999] Laguna, P., Moody, G. B., Garcia, J., Goldberger, A. L., Mark, R. G., and Goldberger, A. L. (1999). Analysis of the ST-T complex of the electrocardiogram using the Karhunen–Love transform: adaptive monitoring and alternans detection. *Med Biol Eng Comput*, 37:175–189.

[Lai et al., 2003] Lai, Y. C., Harrison, M. A., Frei, M. G., and Osorio, I. (2003). Inability of Lyapunov exponents to predict epileptic seizures. *Phys Rev Lett*, 91:068102.

[Lamothe and Stroink, 1991] Lamothe, R. and Stroink, G. (1991). Orthogonal expansions: their applicability to signal extraction in electrophysiological mapping data. *Medical and Biological Engineering and Computing*, 29(5):522–528.

[Lange et al., 1983] Lange, H. H., Lieb, J. P., Engel, J., and Crandall, P. H. (1983). Temporo-spatial patterns of pre-ictal spike activity in human temporal lobe epilepsy. *Electroencephalogr Clin Neurophysiol*, 56:543–555.

[Leanderson et al., 2003] Leanderson, S., Laguna, P., and Sornmo, L. (2003). Estimation of the respiratory frequency using spatial information in the VCG. *Med Eng Phys*, 25:501–507.

[Lee and Gevins, 1993] Lee, J. and Gevins, A. (1993). Method to reduce blur distortion from EEGs using a realistic head model. *IEEE Trans Biomed Eng*, 6:517–528.

[LeFever and De Luca, 1982] LeFever, R. S. and De Luca, C. J. (1982). A procedure for decomposing the myoelectric signal into its constituent action potentials–Part I: Technique, theory, and implementation. *IEEE Trans Biomed Eng*, 29:149–157.

[Lehmann and Skrandies, 1980] Lehmann, D. and Skrandies, W. (1980). Reference-free identification of components of checkerboard-evoked multichannel potential fields. *Electroencephalogr Clin Neurophysiol*, 48:609–621.

[LeVan et al., 2006] LeVan, P., Urrestarazu, E., and Gotman, J. (2006). A system for automatic artifact removal in ictal scalp EEG based on independent component analysis and Bayesian classification. *Clin Neurophysiol*, 117:912–927.

[Lewandowski et al., 2000] Lewandowski, P., Meste, O., Maniewski, R., Mroczka, T., Steinbach, K., and Rix, H. (2000). Risk evaluation of ventricular tachycardia using wavelet transform irregularity of the high-resolution electrocardiogram. *Med Biol Eng Comput*, 38:666–673.

[Li et al., 2008] Li, C., Tang, D., Zheng, D., Ding, G., Poon, C., and Wu, G. (2008). Comparison of nonlinear indices in analyses of heart rate variability. In *Proc 39th Annual International IEEE EMBES Conference*, pages 2145–2148.

[Lilliefors, 1967] Lilliefors, H. (1967). On the Kolmogorov-Smirnov test for normality with mean and variance unknown. *Journal of the American Statistical Association*, 62:399–402.

[Lim et al., 2009] Lim, H. K., Kwon, H., Chung, N., Ko, Y. G., Kim, J. M., Kim, I. S., and Park, Y. K. (2009). Usefulness of magnetocardiogram to detect unstable angina pectoris and non-ST elevation myocardial infarction. *Am J Cardiol*, 103:448–454.

[Lindstrom et al., 1970] Lindstrom, L., Magnusson, R., and Petersen, I. (1970). Muscular fatigue and action potential conduction velocity changes studied with frequency analysis of EMG signals. *Electromyography*, 10:341–356.

[Linssen et al., 1993] Linssen, W. H., Stegeman, D. F., Joosten, E. M., van't Hof, M. A., Binkhorst, R. A., and Notermans, S. L. (1993). Variability and interrelationships of surface EMG parameters during local muscle fatigue. *Muscle Nerve*, 16(8):849–856.

[Liston et al., 2006] Liston, A. D., De Munck, J. C., Hamandi, K., Laufs, H., Ossenblok, P., Duncan, J. S., and Lemieux, L. (2006). Analysis of EEG-fMRI data in focal epilepsy based on automated spike classification and Signal Space Projection. *Neuroimage*, 31:1015–1024.

[Liu et al., 2010] Liu, G., Wu, D., Zhao, G., Huang, B.-Y., Mei, Z.-Y., Y-W, G., and Wang, L. (2010). Use of refined sample entropy and heart rate variability to assess the effects of wearable respiratory biofeedback. In *3rd International Conference on Biomedical Engineering and Informatics (BMEI), 2010*, pages 1915–1919.

[Lombardi et al., 1987] Lombardi, F., Sandrone, G., Pernpruner, S., Sala, R., Garimoldi, M., Cerutti, S., Baselli, G., Pagani, M., and Malliani, A. (1987). Heart rate variability as an index of sympathovagal interaction after acute myocardial infarction. *Am J Cardiol*, 60:1239–1245.

[Lopes da Silva, 1996] Lopes da Silva, F. (1996). *Comprehensive Human Physiology*, chapter The generation of electric and magnetic signals of the brain by local networks. Springer-Verlag.

[Lopes da Silva et al., 2003a] Lopes da Silva, F., Blanes, W., Kalitzin, S., Parra, J., Suffczynski, P., and Velis, D. (2003a). Epilepsies as dynamical diseases of brain systems: basic models of the transition between normal and epileptic activity. *Epilepsia*, 44(s12):72–83.

[Lopes da Silva et al., 1989] Lopes da Silva, F., Pijn, J., and Boeijinga, P. (1989). Interdependence of EEG signals: linear vs. non-linear associations and the significance of time delays and phase shifts. *Brain Topogr*, 2:9–18.

[Lopes da Silva et al., 2003b] Lopes da Silva, F. H., Blanes, W., Kalitzin, S. N., Parra, J., Suffczynski, P., and Velis, D. N. (2003b). Dynamical diseases of brain systems: different routes to epileptic seizures. *IEEE Trans Biomed Eng*, 50:540–548.

[Lundin et al., 1994] Lundin, P., Eriksson, S., Strandberg, L., and Rehnqvist, N. (1994). Prognostic information from on-line vectorcardiography in acute myocardial infarction. *Am J Cardiol*, 74:1103–1108.

[Lutkepohl, 1993] Lutkepohl, H. (1993). *Introduction to Multiple Time Series Analysis*. Springer-Verlag, Berlin.

[Maestri et al., 2007] Maestri, R., Pinna, G. D., Porta, A., Balocchi, R., Sassi, R., Signorini, M. G., Dudziak, M., and Raczak, G. (2007). Assessing nonlinear properties of heart rate variability from short-term recordings: are these measurements reliable? *Physiol Meas*, 28:1067–1077.

[Mainardi et al., 1995] Mainardi, L. T., Bianchi, A. M., Baselli, G., and Cerutti, S. (1995). Pole-tracking algorithms for the extraction of time-variant heart rate variability spectral parameters. *IEEE Trans Biomed Eng*, 42:250–259.

[Makeig, 1993] Makeig, S. (1993). Auditory event-related dynamics of the EEG spectrum and effects of exposure to tones. *Electroencephalogr Clin Neurophysiol*, 86:283–293.

[Makeig, 2000] Makeig, S. (2000). EEGLAB: ICA toolbox for psychophysiological research. World Wide Web Publication, ¡www.sccn.ucsd.edu/eeglab/¿. WWW Site, Swartz Center for Computational Neuroscience, Institute of Neural Computation, University of San Diego California.

[Makeig et al., 1996] Makeig, S., Bell, A., Jung, T.-P., and Sejnowski, T. (1996). Independent component analysis of electroencephalographic data. In Touretzky, D., Mozer, M., and Hasselmo, M., editors, *Advances in Neural Information Processing Systems*, volume 8, pages 145–151. MIT Press, Cambridge, MA.

[Makeig et al., 2004] Makeig, S., Debener, S., Onton, J., and Delorme, A. (2004). Mining event-related brain dynamics. *Trends Cogn Sci (Regul Ed)*, 8:204–210.

[Makeig et al., 1997] Makeig, S., Jung, T. P., Bell, A. J., Ghahremani, D., and Sejnowski, T. J. (1997). Blind separation of auditory event-related brain responses into independent components. *Proc Natl Acad Sci USA*, 94:10979–10984.

[Malliani et al., 1991] Malliani, A., Pagani, A., Lombardi, F., and Cerutti, S. (1991). Cardiovascular neural regulation explored in the frequency domain. *Circulation*, 84:482–492.

[Malik, 1996] Malik, M. (1996). Heart rate variability: standards of measurement, physiological interpretation and clinical use. Task Force of the European Society of Cardiology and the North American Society of Pacing and Electrophysiology. *Circulation*, 93:1043–1065.

[Malik and Camm, 1995] Malik, M. and Camm, A. (1995). *Heart Rate Variability*. Futura Publishing, Armonk, NY.

[Malinowska et al., 2006] Malinowska, U., Durka, P., Blinowska, K., Szelenberger, W., and Wakarow, A. (2006). Micro- and macrostructure of sleep EEG. *IEEE BME Magazine*, 25:26–31.

[Malinowska et al., 2007] Malinowska, U., Durka, P., Zygierewicz, J., Szelenberger, W., and Wakarow, A. (2007). Explicit parameterization of sleep EEG transients. *Computers in Biology and Medicine*, 37:534–541.

[Malinowska et al., 2009] Malinowska, U., Klekowicz, H., Wakarow, A., Niemcewicz, S., and Durka, P. (2009). Fully parametric sleep staging compatible with the classic criteria. *Neuroinformatics*, 7:245–253.

[Mallat, 1989] Mallat, S. (1989). A theory for multiresolution signal decomposition: the wavelet representation. *IEEE Pattern Anal and Machine Intell*, 11(7):674–693.

[Mallat, 1999] Mallat, S. (1999). *A Wavelet Tour of Signal Processing*. Academic Press, London.

[Mallat and Zhang, 1993] Mallat, S. G. and Zhang, Z. (1993). Matching pursuits with time-frequency dictionaries. *IEEE Transactions on Signal Processing*, pages 3397–3415.

[Malmivuo and Plonsey, 1995] Malmivuo, J. and Plonsey, R. (1995). *Bioelectromagnetism – Principles and Applications of Bioelectric and Biomagnetic Fields.* Oxford University Press, New York.

[Mammone and Morabito, 2008] Mammone, N. and Morabito, F. C. (2008). Enhanced automatic artifact detection based on independent component analysis and Renyi's entropy. *Neural Netw*, 21:1029–1040.

[Marozas et al., 2006] Marozas, V., Janusauskas, A., Lukosevicius, A., and Sornmo, L. (2006). Multiscale detection of transient evoked otoacoustic emissions. *IEEE Trans Biomed Eng*, 53(8):1586–1593.

[Marple, 1987] Marple, S. L. J. (1987). *Digital Spectral Analysis with Applications.* Prentice-Hall Inc., Englewood Cliffs, NJ.

[Martens et al., 2007] Martens, S. M., Rabotti, C., Mischi, M., and Sluijter, R. J. (2007). A robust fetal ECG detection method for abdominal recordings. *Physiol Meas*, 28:373–388.

[Martin et al., 1972] Martin, W., Johnson, L., Viglione, S., et al. (1972). Pattern recognition of EEG-eog as a technique for all-night sleep stage scoring. *Electroencephalogr Clin Neurophysiol*, 32(4):17–427.

[Martinerie et al., 1998] Martinerie, J., Adam, C., Le Van Quyen, M., Baulac, M., Clemenceau, S., Renault, B., and

Varela, F. J. (1998). Epileptic seizures can be anticipated by non-linear analysis. *Nat Med*, 4:1173–1176.

[Martinez et al., 1997] Martinez, M., Soria, E., Calpe, J., Guerrero, J. F., and Magdalena, J. R. (1997). Application of the adaptive impulse correlated filter for recovering fetal electrocardiogram. In *Computers in Cardiology 1997*, pages 9–12. IEEE Comput Soc Press.

[Marwan et al., 2002] Marwan, N., Wessel, N., Meyerfeldt, U., Schirdewan, A., and Kurths, J. (2002). Recurrence-plot-based measures of complexity and their application to heart-rate-variability data. *Phys Rev E*, 66(2):026702.

[Matousek and Petersen, 1973] Matousek, M. and Petersen, I. (1973). *Automation of Clinical Electroencephalography*. Raven Press, New York.

[Matsuoka and Spencer, 1993] Matsuoka, L. and Spencer, S. (1993). Seizure localization using subdural grid electrodes. *Epilepsia*, 34(Suppl.6):8.

[Mazurkiewicz and Piotrkiewicz, 2004] Mazurkiewicz, L. and Piotrkiewicz, M. (2004). Computer system for identification and analysis of motor unit potential trains. *Biocybernetics and Biomedical Engineering*, 24(6):15–23.

[McKeown et al., 1998] McKeown, M., Humphries, C., Achermann, P., et al. (1998). A new method for detecting state changes in the EEG: exploratory application to sleep data. *J Sleep Res*, 7:48–56.

[McGill et al., 1985] McGill, K. C., Cummins, K. L., and Dorfman, L. J. (1985). Automatic decomposition of the clinical electromyogram. *IEEE Trans Biomed Eng*, 32(7):470–477.

[McGill et al., 2005] McGill, K. C., Lateva, Z. C., and Marateb, H. R. (2005). EMGLAB: an interactive EMG decomposition program. *J Neurosci Methods*, 149:121–133.

[McGillem and Aunon, 1977] McGillem, C. D. and Aunon, J. I. (1977). Measurements of signal components in single visually evoked brain potentials. *IEEE Trans Biomed Eng*, 24:232–241.

[McSharry et al., 2003] McSharry, P. E., Smith, L. A., and Tarassenko, L. (2003). Prediction of epileptic seizures: are nonlinear methods relevant? *Nat Med*, 9:241–242.

[Medvedev and Willoughby, 1999] Medvedev, J. and Willoughby, O. (1999). Autoregressive modeling of the EEG in systemic kainic acid-induced epileptogenesis. *Int J Neurosci*, 97:149–167.

[Meekins et al., 2008] Meekins, G. D., So, Y., and Quan, D. (2008). Electrodiagnostic medicine evidenced-based review: use of surface electromyography in the diagnosis and study of neuromuscular disorders. *Muscle Nerve*, 38:1219–1224.

[Merletti et al., 2008] Merletti, R., Holobar, A., and Farina, D. (2008). Analysis of motor units with high-density surface electromyography. *J Electromyogr Kinesiol*, 18:879–890.

[Meste et al., 2002] Meste, O., Blain, G., and Bermon, S. (2002). Analysis of the respiratory and cardiac systems coupling in pyramidal exercise using a time-varying model. In *Proc Computers in Cardiology*, volume 29, pages 429–432. IEEE Press.

[Minami et al., 1999] Minami, K., Nakajima, H., and Toyoshima, T. (1999). Real-time discrimination of ventricular tachyarrhythmia with Fourier-transform neural network. *IEEE Trans Biomed Eng*, 46:179–185.

[Mitraszewski et al., 1987] Mitraszewski, P., Blinowska, K., Franaszczuk, P., and Kowalczyk, M. (1987). A study of electrocortical rhythm generators. *Biol Cybern*, 56:255–260.

[Moleti and Sisto, 2003] Moleti, A. and Sisto, R. (2003). Objective estimates of cochlear tuning by otoacoustic emission analysis. *J Acoust Soc Am*, 113:423–429.

[Moller et al., 2001] Moller, E., Schack, B., Arnold, M., and Witte, H. (2001). Instantaneous multivariate EEG coherence analysis by means of adaptive high-dimensional autoregressive models. *J Neurosci Methods*, 105:143–158.

[Moody and Mark, 1990] Moody, G. and Mark, R. (1990). QRS morphology representation and noise estimation using the Karhunen-Loéve transform. In *Proc Computers in Cardiology*, pages 269–272. IEEE Press.

[Moosmann et al., 2008] Moosmann, M., Eichele, T., Nordby, H., Hugdahl, K., and Calhoun, V. D. (2008). Joint independent component analysis for simultaneous EEG-fMRI: principle and simulation. *Int J Psychophysiol*, 67:212–221.

[Moretti et al., 2003] Moretti, D. V., Babiloni, F., Carducci, F., Cincotti, F., Remondini, E., Rossini, P. M., Salinari, S., and Babiloni, C. (2003). Computerized processing of EEG-EOG-EMG artifacts for multi-centric studies in EEG oscillations and event-related potentials. *Int J Psychophysiol*, 47:199–216.

[Mormann et al., 2007] Mormann, F., Andrzejak, R. G., Elger, C. E., and Lehnertz, K. (2007). Seizure prediction: the long and winding road. *Brain*, 130(2):314–333.

[Mormann et al., 2005] Mormann, F., Kreuz, T., Rieke, C., Andrzejak, R., Kraskov, A., David, P., Elger, C., and Lehnertz, K. (2005). On the predictability of epileptic seizures. *Clin Neurophysiol*, 116(3):569–587.

[Mouraux and Iannetti, 2008] Mouraux, A. and Iannetti, G. D. (2008). Across-trial averaging of event-related EEG responses and beyond. *Magn Reson Imaging*, 26:1041–1054.

[Murray et al., 2008] Murray, M. M., Brunet, D., and Michel, C. M. (2008). Topographic ERP analyses: a step-by-step tutorial review. *Brain Topogr*, 20:249–264.

[Nakamura et al., 2004] Nakamura, H., Yoshida, M., Kotani, M., Akazawa, K., and Moritani, T. (2004). The application of independent component analysis to the multi-channel surface electromyographic signals for separation of motor unit action potential trains: part I-measuring techniques. *J Electromyogr Kinesiol*, 14:423–432.

[Nakamura et al., 1996] Nakamura, M., Sugi, T., Ikeda, A., Kakigi, R., and Shibasaki, H. (1996). Clinical application of automatic integrative interpretation of awake background EEG: quantitative interpretation, report making, and detection of artifacts and reduced vigilance level. *Electroencephalogr Clin Neurophysiol*, 98:103–112.

[Narayan, 2006] Narayan, S. (2006). The pathophysiology guided assessment of t-wave alternans. In Clifford, G., Azuaje, F., and McSharry, P., editors, *Advanced Methods and Tools for ECG Data Analysis*, pages 197–214. Artech House Inc., Norwood, MA.

[Nawab et al., 2010] Nawab, S. H., Chang, S. S., and De Luca, C. J. (2010). High-yield decomposition of surface EMG signals. *Clin Neurophysiol*, 121:1602–1615.

[Neonen et al., 2001] Neonen, J., Pesola, K., Hanninen, H., Lauerma, K., Takala, P., Makela, T., Makijarvi, M., Knuuti, J. Toivonen, L., and Katila, T. (2001). Current-density estimation of exercise induced ischemia in patients with multivessel coronary artery disease. *J Electrocardiography*, 34:37–42.

[Netoff et al., 2006] Netoff, T. I., Caroll, L., Pecora, L. M., and Schiff, S. J. (2006). Detecting coupling in the presence of noise and nonlinearity. In B. Scheleter, M. Winterhalder, J. T., editor, *Handbook of Time Series Analysis*. Wiley VCH.

[Newman, 2003] Newman, M. (2003). The structure and function of complex networks. *SIAM Rev*, 45:167–256.

[Nichols and Holmes, 2002] Nichols, T. E. and Holmes, A. P. (2002). Nonparametric permutation tests for functional neuroimaging: A primer with examples. *Human Brain Mapping*, 15:1–25.

[Niedermayer and Lopes da Silva, 2004] Niedermayer, E. and Lopes da Silva, F. H. (2004). *Electroencephalography, Basic Principles, Clinical Applications, and Re-*

lated Fields. 5th edition. Williams & Wilkins, Baltimore.

[Nieuwenhuis et al., 2005] Nieuwenhuis, S., Aston-Jones, G., and Cohen, J. D. (2005). Decision making, the P3, and the locus coeruleus-norepinephrine system. *Psychol Bull*, 131:510–532.

[Ning et al., 2009] Ning, J., Atanasov, N., and Ning, T. (2009). Quantitative analysis of heart sounds and systolic heart murmurs using wavelet transform and AR modeling. *Conf Proc IEEE Eng Med Biol Soc*, 2009:958–961.

[Noman et al., 2000] Noman, R., Pal, I., Steward, C., Walsleban, J., and Rapoport, D. (2000). Interobserver agreement among sleep scorers from different centers in a large dataset. *Sleep*, 23(7):901–908.

[Notaro et al., 2007] Notaro, G., Al-Maamury, A. M., Moleti, A., and Sisto, R. (2007). Wavelet and matching pursuit estimates of the transient-evoked otoacoustic emission latency. *J Acoust Soc Am*, 122:3576–3585.

[Nunez, 1981] Nunez, P. (1981). *Electric Fields of the Brain.* Oxford University Press, New York.

[Nunez and Pilgreen, 1991] Nunez, P. and Pilgreen, K. (1991). The spline-Laplacian in clinical neurophysiology: a method to improve EEG spatial resolution. *J Clin Neurophys*, 8:397–413.

[Nunez and Srinivasan, 2006] Nunez, P. and Srinivasan, R. (2006). *Electric Fields of the Brain: The Neurophysics of EEG.* 2nd edition. Oxford University Press, New York.

[Onton and Makeig, 2006] Onton, J. and Makeig, S. (2006). Information-based modeling of event-related brain dynamics. *Prog Brain Res*, 159:99–120.

[Orfanidis, 1988] Orfanidis, S. (1988). *Optimum Signal Processing*. Macmillan. Chapter 5.

[Ortiz et al., 2005] Ortiz, M., Bojorges, E., Aguilar, S., Echeveria, J., Gonzalez-Camarena, R., Carrasco, S.,

Gaitan, M., and Martinez, A. (2005). Analysis of high frequency fetal heart rate variability using empirical mode decomposition. *Computers in Cardiology*, 32:675–678.

[Osowski et al., 2006] Osowski, S., Hoai, L., and Markiewicz, T. (2006). Supervised learning methods for ECG classification/neural networks and svm approaches. In Clifford, G., Azuaje, F., and McSharry, P., editors, *Advanced Methods and Tools for ECG Data Analysis*, pages 319–337. Artech House Inc., Norwood, MA.

[Ozdamar et al., 1997] Ozdamar, O., Zhang, J., Kalayci, T., and Ulgen, Y. (1997). Time-frequency distribution of evoked otoacoustic emissions. *Br J Audiol*, 31:461–471.

[Pahlm and Sornmo, 1984] Pahlm, O. and Sornmo, L. (1984). Software QRS detection in ambulatory monitoring–a review. *Med Biol Eng Comput*, 22:289–297.

[Paiva and Rosa, 1991] Paiva, T. and Rosa, A. (1991). *K-complex Variability in Normal Subjects.* Physics Events and Dynamic Organization of Sleep. G. Terzano. Raven Press Ltd., New York.

[Pan and Tompkins, 1985] Pan, J. and Tompkins, W. J. (1985). A real-time QRS detection algorithm. *IEEE Trans Biomed Eng*, 32:230–236.

[Pantelopoulos and Bourbakis, 2010] Pantelopoulos, A. and Bourbakis, N. (2010). Efficient single-lead ECG beat classification using matching pursuit based features and artificial neural network. In *Proc of Inter Conference on Information Technology and Application in Biomedicine.*

[Pardey et al., 1996] Pardey, J., Roberts, S., Tarassenko, L., and Stradling, J. (1996). A new approach to the analysis of the human sleep-wakefulness continuum. *J Sleep Res*, 5:201–210.

[Park and Burrus, 1987] Park, T. and Burrus, C. (1987). *Digital Filter Design.* Wiley-Interscience.

[Parra et al., 2003] Parra, J., Kalitzin, S. N., Iriarte, J., Blanes, W., Velis, D. N., and Lopes da Silva,

F. H. (2003). Gamma-band phase clustering and photosensitivity: Is there an underlying mechanism common to photosensitive epilepsy and visual perception? *Brain*, 126:1164–1172.

[Pattichis and Elia, 1999] Pattichis, C. S. and Elia, A. G. (1999). Autoregressive and cepstral analyses of motor unit action potentials. *Med Eng Phys*, 21:405–419.

[Pattichis and Pattichis, 1999] Pattichis, C. S. and Pattichis, M. S. (1999). Time-scale analysis of motor unit action potentials. *IEEE Trans Biomed Eng*, 46:1320–1329.

[Paulini et al., 2007] Paulini, A., Fischer, M., Rampp, S., Scheler, G., Hopfengärtner, R., Kaltenhäuser, M., Dörfler, A., Buchfelder, M., and Stefan, H. (2007). Lobar localization information in epilepsy patients: MEG—a useful tool in routine presurgical diagnosis. *Epilepsy Research*, 76(2):124–130.

[Peng et al., 1995] Peng, C. K., Havlin, S., Stanley, H. E., Goldberger, A. L., and Goldberger, A. L. (1995). Quantification of scaling exponents and crossover phenomena in nonstationary heartbeat time series. *Chaos*, 5:82–87.

[Penzel et al., 2007] Penzel, T., Hirshkowitz, H., Harsh, J., Chervin, R., Butkov, N., Kryger, M., et al. (2007). Digital analysis and technical specifications. *J Clin Sleep Med*, 3(2):109–120.

[Pereda et al., 2005] Pereda, E., Quiroga, R. Q., and Bhattacharya, J. (2005). Nonlinear multivariate analysis of neurophysiological signals. *Progress in Neurobiology*, 77:1–37.

[Pereda et al., 2001] Pereda, E., Rial, R., Gamundi, A., and Gonzalez, J. (2001). Assessment of changing interdependences between human electroencephalograms using nonlinear methods. *Physica D*, 148:147–158.

[Perkiomaki et al., 2005] Perkiomaki, J. S., Makikallio, T. H., and Huikuri, H. V. (2005). Fractal and complexity measures of heart rate variability. *Clin Exp Hypertens*, 27:149–158.

[Pesola et al., 1999] Pesola, K., Nenonen, J., Fenici, R., Lotjonen, J., Makijarvi, M., Fenici, P., Korhonen, P., Lauerma, K., Valkonen, M., Toivonen, L., and Katila, T. (1999). Bioelectromagnetic localization of a pacing catheter in the heart. *Phys Med Biol*, 44:2565–2578.

[Pfister et al., 1988] Pfister, C. J., Hamilton, J. W., Nagel, N., Bass, P., Webster, J. G., and Tompkins, W. J. (1988). Use of spectral analysis in the detection of frequency differences in the electrogastrograms of normal and diabetic subjects. *IEEE Trans Biomed Eng*, 35:935–941.

[Pfurtscheller, 1991] Pfurtscheller, G. (1991). *Digital Biosignal Processing*, chapter Mapping procedures, pages 459–479. Elsevier.

[Pfurtscheller, 1999] Pfurtscheller, G. (1999). Quantification of ERD and ERS in the time domain. In Pfurtscheller, G. and Lopes da Silva, F. H., editors, *Event-related Desynchronization*, volume 6, pages 89–105. Elsevier.

[Pfurtscheller and Aranibar, 1979] Pfurtscheller, G. and Aranibar, A. (1979). Evaluation of event-related desynchronization (ERD) preceding and following voluntary self-paced movement. *Electroencephalogr Clin Neurophysiol*, 46:138–146.

[Pfurtscheller and Cooper, 1975] Pfurtscheller, G. and Cooper, R. (1975). Selective averaging of the intracerebral click evoked responses in man: an improved method of measuring latencies and amplitudes. *Electroencephalogr Clin Neurophysiol*, 38:187–190.

[Pfurtscheller and Lopes da Silva, 1999] Pfurtscheller, G. and Lopes da Silva, F. H. (1999). Event-related EEG/MEG synchronization and desynchronization: basic principles. *Clin Neurophysiol*, 110:1842–1857.

[Picton et al., 2000] Picton, T., Bentin, S., Berg, P., Donchin, E., Hillyard, S., Johnson, R., Miller, G., Ritter, W., Ruchkin, D., Rugg, M., and Taylor, M.

(2000). Guidelines for using human event-related potentials to study cognition: Recording standards and publication criteria. *Psychophysiology*, 37:127–152.

[Pieri et al., 2001] Pieri, J. F., Crowe, J. A., Hayes-Gill, B. R., Spencer, C. J., Bhogal, K., and James, D. K. (2001). Compact long-term recorder for the transabdominal foetal and maternal electrocardiogram. *Med Biol Eng Comput*, 39:118–125.

[Pijn et al., 1991] Pijn, J. P., Van Neerven, J., Noest, A., and Lopes da Silva, F. H. (1991). Chaos or noise in EEG signals; dependence on state and brain site. *Electroencephalogr Clin Neurophysiol*, 79:371–381.

[Pijn et al., 1997] Pijn, J. P., Velis, D. N., van der Heyden, M. J., DeGoede, J., van Veelen, C. W., and Lopes da Silva, F. H. (1997). Nonlinear dynamics of epileptic seizures on basis of intracranial EEG recordings. *Brain Topogr*, 9:249–270.

[Pincus, 1991] Pincus, S. M. (1991). Approximate entropy as a measure of system complexity. *Proc Natl Acad Sci USA*, 88:2297–2301.

[Pineas et al., 1982] Pineas, R., Crow, R., and Blackburn, H. (1982). *The Minnesota Code Manual of Electrocardiographic Findings*. John Wright-PSG, Inc., Littleton, MA.

[Pineda, 1995] Pineda, J. A. (1995). Are neurotransmitter systems of subcortical origin relevant to the electrogenesis of cortical ERPs? *Electroencephalogr Clin Neurophysiol Suppl*, 44:143–150.

[Pinkus and Singer, 1996] Pinkus, S. and Singer, B. (1996). Randomness and degress of irregularity. *Proc Natl Acad Sci USA*, 93:2083–2088.

[Pinsky, 2002] Pinsky, M. (2002). *Introduction to Fourier Analysis and Wavelets*. Brooks/Cole, isbn 0-534-37660-6 edition.

[Piotrkiewicz, 1999] Piotrkiewicz, M. (1999). Perspectives and drawbacks of surface electromyography. In

Kedzior, K. and Morecki, A., editors, *Lecture Notes of the 46th ICB Seminar Biomechanics of Musculoskeletal System. Medical Robotics*, pages 219–227.

[Poon, 2000] Poon, C. S. (2000). The chaos about heart rate chaos. *J Cardiovasc Electrophysiol*, 11:235–236.

[Porta et al., 1999] Porta, A., Baselli, G., Lombardi, F., Montano, N., Malliani, A., and Cerutti, S. (1999). Conditional entropy approach for the evaluation of the coupling strength. *Biol Cybern*, 81:119–129.

[Porta et al., 2000] Porta, A., Guzzetti, S., Montano, N., Pagani, M., Somers, V., Malliani, A., Baselli, G., and Cerutti, S. (2000). Information domain analysis of cardiovascular variability signals: evaluation of regularity, synchronization and co-ordination. *Med Biol Eng Comput*, 38:180–188.

[Prats-Boluda et al., 2006] Prats-Boluda, G., Martinez-de Juan, J. L., Garcia-Casado, J., Guardiola, J. L., and Ponce, J. L. (2006). Discrete Laplacian recordings of the electroenterogram from abdominal surface in humans. *Conf Proc IEEE Eng Med Biol Soc*, 1:6080–6083.

[Priestley, 1981] Priestley, M. (1981). *Spectral Analysis and Time Series*. Academic Press Inc., London.

[Probst et al., 1991] Probst, R., Lonsbury-Martin, B. L., and Martin, G. K. (1991). A review of otoacoustic emissions. *J Acoust Soc Am*, 89:2027–2067.

[Quian Quiroga and van Luijtelaar, 2002] Quian Quiroga, R. and van Luijtelaar, E. L. (2002). Habituation and sensitization in rat auditory evoked potentials: a single-trial analysis with wavelet denoising. *Int J Psychophysiol*, 43:141–153.

[Quian Quiroga et al., 2000] Quian Quiroga, R., Arnhold, J., and Grassberger, P. (2000). Learning driver-response relationships from synchronization patterns. *Phys Rev E Stat Phys Plasmas Fluids Relat Interdiscip Topics*, 61:5142–5148.

[Quian Quiroga et al., 2002] Quian Quiroga, R., Kraskov, R., Kreuz, T., and Grassberger, P. (2002). Performance of different synchronization measures in real data: a case study on electroencephalographic signals. *Phys Rev E*, 65:041903.

[Rabiner, 1989] Rabiner, L. (1989). A tutorial on Hidden Markov Models and selected applications in speech recognition. *Proc of the IEEE*, 77(2):257–286.

[Rampil, 2009] Rampil, I. (2009). EEG signal analysis in anesthesia. In Tong, S. and Thakor, N., editors, *Quantitative EEG Analysis Methods and Clinical Applications*. Artech House, Norwood, MA.

[Ray et al., 2008] Ray, S., Crone, N. E., Niebur, E., Franaszczuk, P. J., and Hsiao, S. S. (2008). Neural correlates of high-gamma oscillations (60-200 Hz) in macaque local field potentials and their potential implications in electrocorticography. *J Neurosci*, 28:11526–11536.

[Ray et al., 2003] Ray, S., Jouny, C. C., Crone, N. E., Boatman, D., Thakor, N. V., and Franaszczuk, P. J. (2003). Human ECoG analysis during speech perception using matching pursuit: a comparison between stochastic and dyadic dictionaries. *IEEE Trans Biomed Eng*, 50:1371–1373.

[Ray et al., 1986] Ray, S., Lee, W., Morgan, C., and Airth-Kindree, W. (1986). Computer sleep stage scoring an expert system approach. *International Journal of Bio-medical Computing*, 19(1):43–61.

[Rechtschaffen and Kales, 1968] Rechtschaffen, A. and Kales, A., editors (1968). *A manual of standardized terminology, techniques and scoring system for sleep stages in human subjects*. US Government Printing Office, Washington DC. Number 204 in National Institutes of Health Publications.

[Richards et al., 1995] Richards, W. O., Garrard, C. L., Allos, S. H., Bradshaw, L. A., Staton, D. J., and Wikswo,

J. P. (1995). Noninvasive diagnosis of mesenteric ischemia using a SQUID magnetometer. *Ann Surg*, 221:696–704.

[Richman and Moorman, 2000] Richman, J. S. and Moorman, J. R. (2000). Physiological time-series analysis using approximate entropy and sample entropy. *Am J Physiol Heart Circ Physiol*, 278:H2039–2049.

[Richter et al., 1998] Richter, M., Schreiber, T., and Kaplan, D. T. (1998). Fetal ECG extraction with nonlinear state-space projections. *IEEE Trans Biomed Eng*, 45:133–137.

[Rienzo et al., 1993] Rienzo, M. D., Mancia, G., Parati, G., Pedotti, A., and Zanchetti, A. (1993). *Blood Pressure and Heart Rate Variability, Computer Analysis, Methodology and Clinical Applications*. IOS Press, Amsterdam.

[Roberts and Tarassenko, 1992] Roberts, S. and Tarassenko, L. (1992). New method of automated sleep quantification. *Med Biol Eng Comput*, 30:509–517.

[Robinette and Glattke, 2002] Robinette, M. and Glattke, T. (2002). *Otoacoustic Emissions Clinical Applications*. Thieme, New York.

[Roche et al., 2003] Roche, F., Pichot, V., Sforza, E., Court-Fortune, I., Duverney, D., Costes, F., Garet, M., and Barthelemy, J. C. (2003). Predicting sleep apnoea syndrome from heart period: a time-frequency wavelet analysis. *Eur Respir J*, 22(6):937–942.

[Rogowski et al., 1981] Rogowski, Z., Gath, I., and Bental, E. (1981). On the prediction of epileptic seizures. *Bio Cybern*, 42:9–15.

[Rosa et al., 1999] Rosa, A., Parrino, L., and Terzano, M. (1999). Automatic detection of cyclic alternating pattern (CAP) sequences in sleep preliminary results. *Clin Neurophysiol*, 110:585–592.

[Rosik et al., 2001] Rosik, V., Tysler, M., Jurko, S., Raso, R., and Turzowa, M. (2001). Portable system for ECG mapping. *Measurement Science Review*, 1:27–30.

[Rulkov et al., 1995] Rulkov, N., Sushchik, M. M., Tsimring, L. S., and Abarbanel, H. D. I. (1995). Generalized synchronization of chaos in directionally coupled chaotic systems. *Phys Rev E*, 51:980– 994.

[Ruppert et al., 2003] Ruppert, D., Wand, M., and Carroll, R. (2003). *Semiparametric Regression*. Cambridge University Press, Cambridge.

[Saab and Gotman, 2005] Saab, M. E. and Gotman, J. (2005). A system to detect the onset of epileptic seizures in scalp EEG. *Clin Neurophysiol*, 116:427–442.

[Sadoyama and Miyano, 1981] Sadoyama, T. and Miyano, H. (1981). Frequency analysis of surface EMG to evaluation of muscle fatigue. *Eur J Appl Physiol Occup Physiol*, 47(3):239–246.

[Sahin et al., 2010] Sahin, I., Yilmazer, N., and Simaan, M. A. (2010). A method for subsample fetal heart rate estimation under noisy conditions. *IEEE Trans Biomed Eng*, 57:875–883.

[Salant et al., 1998] Salant, Y., Gath, I., and Henriksen, O. (1998). Prediction of epileptic seizures from two-channel EEG. *Medical and Biological Engineering and Computing*, 36:549–556.

[Sameni et al., 2007] Sameni, R., Clifford, G. D., Jutten, C., and Shamsollahi, M. B. (2007). Multichannel ECG and noise modeling: application to maternal and fetal ECG signals. *EURASIP J Adv Signal Process*, 2007. Article ID 043407.

[Sandwell, 1987] Sandwell, D. (1987). Biharmonic spline interpolation of geos-3 and seasat altimeter data. *Geoph Res Lett*, 2:139–142.

[Sanei and Chambers, 2009] Sanei, S. and Chambers, J. (2009). *EEG Signal Processing*. Wiley & Sons, Chichester.

[Sayers et al., 1974] Sayers, B., Beagley, H., and Henshall, W. (1974). The mechanism of auditory evoked EEG responses. *Nature*, 247:481–483.

[Schaltenbrand et al., 1993] Schaltenbrand, N., Lengelle, R., and Macher, J. P. (1993). Neural network model: application to automatic analysis of human sleep. *Comput Biomed Res*, 26:157–171.

[Schelter et al., 2009] Schelter, B., Timmer, J., and Eichler, M. (2009). Assessing the strength of directed influences among neural signals using renormalized partial directed coherence. *J Neurosci Methods*, 179:121–130.

[Schiff et al., 2005] Schiff, S. J., Sauer, T., Kumar, R., and Weinstein, S. L. (2005). Neuronal spatiotemporal pattern discrimination: the dynamical evolution of seizures. *Neuroimage*, 28:1043–1055.

[Schindler et al., 2007] Schindler, K., Leung, H., Elger, C. E., and Lehnertz, K. (2007). Assessing seizure dynamics by analysing the correlation structure of multichannel intracranial EEG. *Brain*, 130:65–77.

[Schlögl et al., 1999] Schlögl, A., Anderer, P., Barbanoj, M., Klösch, G., Gruber, G., Lorenzo, J., et al. (1999). Artefact processing of the sleep EEG in the siesta-project. *Proceedings EMBEC99*, pages 1644–1645.

[Schreiber, 2000] Schreiber, T. (2000). Measuring information transfer. *Phys Rev Lett*, 85:461–464.

[Semlitsch et al., 1986] Semlitsch, H. V., Anderer, P., Schuster, P., and Presslich, O. (1986). A solution for reliable and valid reduction of ocular artifacts, applied to the P300 ERP. *Psychophysiology*, 23:695–703.

[Senhadji et al., 1995] Senhadji, L., Carrault, G., Bellanger, J., and Passariello, G. (1995). Comparing wavelet transforms for recognizing cardiac patterns. *IEEE Eng Med Biol Mag*, 14:167–173.

[Serio, 1992] Serio, C. (1992). Discriminating low-dimensional chaos from randomness: A parametric time series modelling approach. *Il Nuovo Cimento B (1971-1996)*, 107:681–701.

[Shannon, 1948] Shannon, C. (1948). A mathematical theory of communication. *Bell System Technical Journal*, 27:379–423.

[Shannon, 1949] Shannon, C. E. (1949). Communication in the presence of noise. *Proc Institute of Radio Engineers*, 37(1):10–21. Reprint as classic paper in: Proc. IEEE, Vol. 86, No. 2, (Feb 1998).

[Shera and Guinan, 1999] Shera, C. A. and Guinan, J. J. (1999). Evoked otoacoustic emissions arise by two fundamentally different mechanisms: a taxonomy for mammalian OAEs. *J Acoust Soc Am*, 105:782–798.

[Shoeb et al., 2004] Shoeb, A., Edwards, H., Connolly, J., Bourgeois, B., Treves, S. T., and Guttag, J. (2004). Patient-specific seizure onset detection. *Epilepsy Behav*, 5:483–498.

[Sielużycki et al., 2009a] Sielużycki, C., König, R., Matysiak, A., Kuś, R., Ircha, D., and Durka, P. J. (2009a). Single-trial evoked brain responses modeled by multivariate matching pursuit. *IEEE Trans Biomed Eng*, 56:74–82.

[Sielużycki et al., 2009b] Sielużycki, C., Kuś, R., Matysiak, A., Durka, P. J., and König, R. (2009b). Multivariate matching pursuit in the analysis of single-trial latency of the auditory M100 acquired with MEG. *International Journal of Bioelectromagnetism*, 11(4):155–160.

[Simes, 1986] Simes, R. (1986). An improved Bonferroni procedure for multiple tests of significance. *Biometrika*, 73:751–4.

[Sinai et al., 2009] Sinai, A., Crone, N. E., Wied, H. M., Franaszczuk, P. J., Miglioretti, D., and Boatman-Reich, D. (2009). Intracranial mapping of auditory perception: event-related responses and electrocortical stimulation. *Clin Neurophysiol*, 120:140–149.

[Singer, 1993] Singer, W. (1993). Synchronization of cortical activity and its putative role in information processing and learning. *Annu Rev Physiol*, 55:349–374.

[Sisto and Moleti, 1999] Sisto, R. and Moleti, A. (1999). Modeling otoacoustic emissions by active nonlinear oscillators. *J Acoust Soc Am*, 106:1893–1906.

[Sisto and Moleti, 2002] Sisto, R. and Moleti, A. (2002). On the frequency dependence of the otoacoustic emission latency in hypoacoustic and normal ears. *J Acoust Soc Am*, 111:297–308.

[Skrandies, 1989] Skrandies, W. (1989). Data reduction of multichannel fields: global field power and principal component analysis. *Brain Topogr*, 2:73–80.

[Skrandies, 2005] Skrandies, W. (2005). Brain mapping of visual evoked activity–topographical and functional components. *Acta Neurol Taiwan*, 14:164–178.

[Smith and Karacan, 1971] Smith, J. and Karacan, I. (1971). EEG sleep stage scoring by an automatic hybrid system. *Electroencephalogr Clin Neurophysiol*, 31(3):231–237.

[Smylie et al., 1973] Smylie, D., Clarke, G., and T.J., U. (1973). Analysis of irregularities in the earth's rotation. *Comput Phys*, 13:391–430.

[Sokolov, 1960] Sokolov, E. (1960). Neuronal models and the orienting response. In Brazier, M., editor, *The Central Nervous System and Behavior III*. Macy Foundation.

[Souza Neto et al., 2004] Souza Neto, E. P., Custaud, M. A., Cejka, J. C., Abry, P., Frutoso, J., Gharib, C., and Flandrin, P. (2004). Assessment of cardiovascular autonomic control by the empirical mode decomposition. *Methods Inf Med*, 43:60–65.

[Spekreijse et al., 1976] Spekreijse, H., Esteves, O., and Reits, O. (1976). Visual evoked potentials and the physiological analysis of visual processes in man. In Desmedt, J., editor, *Visual Evoked Potentials in Man*, pages 3–15. Clarendon Press.

[Stam et al., 1999] Stam, C., Pijn, J., Suffczyski, P., and Lopes da Silva, F. (1999). Dynamics of the human alpha rhythm: evidence for non-linearity? *Clin Neurophys*, 110:1801–1813.

[Stashuk, 1999] Stashuk, D. W. (1999). Decomposition and quantitative analysis of clinical electromyographic signals. *Med Eng Phys*, 21:389–404.

[Stein et al., 1981] Stein, P. D., Sabbah, H. N., Lakier, J. B., Magilligan, D. J., and Goldstein, D. (1981). Frequency of the first heart sound in the assessment of stiffening of mitral bioprosthetic valves. *Circulation*, 63:200–203.

[Strang and Nguyen, 1996] Strang, G. and Nguyen, T. (1996). *Wavelets and Filter Banks.* Wellesley-Cambridge Press.

[Stroink, 2010] Stroink, G. (2010). Forty years of magnetocardiology. In Supek, S. and Susac, A., editors, *Advances in Biomagnetism BIOMAG2010, IFMBE Proceedings*, volume 28, pages 431–435. Springer.

[Stroink et al., 1999] Stroink, G., Meeder, R. J., Elliott, P., Lant, J., and Gardner, M. J. (1999). Arrhythmia vulnerability assessment using magnetic field maps and body surface potential maps. *Pacing Clin Electrophysiol*, 22:1718–1728.

[Stufflebeam et al., 2009] Stufflebeam, S. M., Tanaka, N., and Ahlfors, S. P. (2009). Clinical applications of magnetoencephalography. *Hum Brain Mapp*, 30:1813–1823.

[Suffczynski et al., 2008] Suffczynski, P., Kalitzin, S., da Silva, F. L., Parra, J., Velis, D., and Wendling, F. (2008). Active paradigms of seizure anticipation: computer model evidence for necessity of stimulation. *Phys Rev E*, 78(5):051917.

[Suffczynski et al., 2004] Suffczynski, P., Kalitzin, S., and Lopes Da Silva, F. H. (2004). Dynamics of non-convulsive epileptic phenomena modeled by a bistable neuronal network. *Neuroscience*, 126:467–484.

[Symms et al., 1999] Symms, M. R., Allen, P. J., Woermann, F. G., Polizzi, G., Krakow, K., Barker, G. J., Fish, D. R., and Duncan, J. S. (1999). Reproducible localization of interictal epileptiform discharges using EEG-triggered fMRI. *Phys Med Biol*, 44:N161–168.

[Taccardi, 1962] Taccardi, B. (1962). Distribution of heart potentials on dog's thoracic surface. *Circ Res*, 11(5):862–869.

[Taccardi, 1963] Taccardi, B. (1963). Distribution of heart potentials on the thoracic surface of normal human subjects. *Circ Res*, 12(4):341–352.

[Taddei et al., 1993] Taddei, A., Emdin, M., Varanini, M., Macerate, A., Pisani, P., Santarcangelp, E., and Marchesi, C. (1993). An approach to cardiorespiratory activity monitoring through principal component analysis. *Journal of Ambulatory Monitoring*, 5(2-3):167–173.

[Takahashi et al., 2007] Takahashi, D., Baccala, L., and Sameshima, K. (2007). Connectivity inference between neural structures via partial directed coherence. *Journal of Applied Statistics*, 34(10):1255–1269.

[Takens, 1981] Takens, F. (1981). Detecting strange attractors in turbulence. *Lecture Notes in Mathematics*, 898:366–381.

[Tallon-Baudry, 2003] Tallon-Baudry, C. (2003). Oscillatory synchrony and human visual cognition. *J Physiol Paris*, 97:355–363.

[Tallon-Baudry et al., 1996] Tallon-Baudry, C., Bertrand, O., Delpuech, C., and Pernier, J. (1996). Stimulus specificity of phase-locked and non-phase-locked 40Hz visual responses in human. *J Neurosci*, 16(13):4240–4249.

[Talmadge et al., 2000] Talmadge, C. L., Tubis, A., Long, G. R., and Tong, C. (2000). Modeling the combined effects of basilar membrane nonlinearity and roughness on stimulus frequency otoacoustic emission fine structure. *J Acoust Soc Am*, 108:2911–2932.

[Tang et al., 2010] Tang, H., Li, T., Park, Y., and Qiu, T. (2010). Separation of heart sound signal from noise in joint cycle frequency-time-frequency domains based on fuzzy detection. *IEEE Trans Biomed Eng*, 57(10):2438–2447.

[Taralunga et al., 2008] Taralunga, D. D., Wolf, W., Strungaru, R., and Ungureanu, G. (2008). Abdominal fetal ECG enhancement by event synchronous canceller. *Conf Proc IEEE Eng Med Biol Soc*, 2008:5402–5405.

[Ten Hoopen, 1975] Ten Hoopen, M. (1975). Variance in average response computation: regular versus irregular stimulation. In Remond, A., editor, *Handbook of Electroencephalography and Clinical Neurophysiology. Part A*, volume 8, pages 151–158. Elsevier, Amsterdam.

[Thakor et al., 2009] Thakor, N., Jia, X., and Geocardin, R. (2009). Monitoring neurological injury by qEEG. In Tong, S. and Thakor, N., editors, *Quantitative EEG Analysis Methods and Clinical Applications*. Artech House, Norwood, MA.

[Theiler et al., 1992] Theiler, J., Eubank, S., Longtin, A., Galdrikian, B., and Farmer, J. (1992). Testing for nonlinearity in time series: the method of surrogate data. *Physica D*, 58(1-4):77–94.

[Thompson and Spencer, 1966] Thompson, R. and Spencer, W. (1966). Habituation: a model phenomenon for the study of neuronal substrates of behavior. *Psychol Rev*, 73:16–43.

[Thomson, 1982] Thomson, D. (1982). Spectrum estimation and harmonic analysis. *Proceedings of the IEEE*, 70:1055–1096.

[Thurner et al., 1998] Thurner, S., Feurstein, M. C., and Teich, M. C. (1998). Multiresolution wavelet analysis of heartbeat intervals discriminates healthy patients from those with cardiac pathology. *Phys Rev Lett*, 80(7):1544–1547.

[Tognola et al., 1997] Tognola, G., Grandori, F., and Ravazzani, P. (1997). Time-frequency distributions of click-evoked otoacoustic emissions. *Hear Res*, 106:112–122.

[Tognola et al., 2005] Tognola, G., Parazzini, M., de Jager, P., Brienesse, P., Ravazzani, P., and Grandori, F. (2005). Cochlear maturation and otoacoustic emissions in preterm infants: a time-frequency approach. *Hear Res*, 199:71–80.

[Tokmakci, 2007] Tokmakci, M. (2007). Analysis of the electrogastrogram using discrete wavelet transform and statistical methods to detect gastric dysrhythmia. *J Med Syst*, 31:295–302.

[Toledo et al., 2003] Toledo, E., Gurevitz, O., Hod, H., Eldar, M., and Akselrod, S. (2003). Wavelet analysis of instantaneous heart rate: a study of autonomic control during thrombolysis. *Am J Physiol Regul Integr Comp Physiol*, 284(4):1079–1091.

[Tong et al., 2004] Tong, S., Hong, B., Vigderman, L., and Thakor, N. V. (2004). Subband EEG complexity after global hypoxic-ischemic brain injury. *Conf Proc IEEE Eng Med Biol Soc*, 1:562–565.

[Turin, 1960] Turin, G. (1960). An introduction to matched filters. *IRE Transactions on Information Theory 6*, 6(3):311– 329.

[Ulrych and Bishop, 1975] Ulrych, T. and Bishop, T. (1975). Maximum entropy spectral analysis and autoregressive decomposition. *Rev Geophys Space Phys*, 13:183–200.

[Urbano et al., 1998] Urbano, A., Babiloni, C., Onorati, P., Carducci, F., Ambrosini, A., Fattorini, L., and Babiloni, F. (1998). Responses of human primary sensorimotor and supplementary motor areas to internally triggered unilateral and simultaneous bilateral one-digit movements. A high-resolution EEG study. *Eur J Neurosci*, 10:765–770.

[Urbanowicz et al., 2007] Urbanowicz, K., Zebrowski, J., Baranowski, R., and Holyst, J. (2007). How random is your heart beat? *Physica A: Statistical Mechanics and Its Applications*, 384(2):439 – 447.

[van de Velde et al., 1999] van de Velde, M., Ghosh, I. R., and Cluitmans, P. J. (1999). Context related artefact detection in prolonged EEG recordings. *Comput Methods Programs Biomed*, 60:183–196.

[van de Velde et al., 1998] van de Velde, M., van Erp, G., and Cluitmans, P. J. (1998). Detection of muscle artefact in the normal human awake EEG. *Electroencephalogr Clin Neurophysiol*, 107:149–158.

[van der Schee and Grashius, 1987] van der Schee, E. and Grashius, J. (1987). Running spectral analysis as an aid in the representation and interpretation of electrogastrographic signals. *Med Biol Eng Comput*, 25:57–62.

[Viterbi, 1967] Viterbi, A. (1967). Error bounds for convolutional codes and an asymptotically optimum decoding algorithm. *IEEE Transactions on Information Theory*, 13(2):260–269.

[von Spreckelsen and Bromm, 1988] von Spreckelsen, M. and Bromm, B. (1988). Estimation of single-evoked cerebral potentials by means of parametric modeling and Kalman filtering. *IEEE Trans Biomed Eng*, 35:691–700.

[Waldert et al., 2007] Waldert, S., Bensch, M., Bogdan, M., Rosenstiel, W., Scholkopf, B., Lowery, C. L., Eswaran, H., and Preissl, H. (2007). Real-time fetal heart monitoring in biomagnetic measurements using adaptive real-time ICA. *IEEE Trans Biomed Eng*, 54:1867–1874.

[Walter, 1969] Walter, D. (1969). A posteriori Wiener filtering of average evoked response. *Electroenceph Clin Neurophysiol*, 27:61–70.

[Walter and Adey, 1963] Walter, D. and Adey, W. (1963). Spectral analysis of EEG recorded during learning in cat before and after subthalamic lesions. *Exp Neur*, 7:481.

[Wang and Azuaje, 2006] Wang, H. and Azuaje, F. (2006). An introduction to unsupervised learning for ECG classification. In Clifford, G., Azuaje, F., and McSharry, P., editors, *Advanced Methods and Tools for ECG Data Analysis*, pages 339–366. Artech House Inc., Norwood, MA.

[Wang et al., 2001] Wang, W., Guo, Z., Yang, J., Zhang, Y., Durand, L. G., and Loew, M. (2001). Analysis of the first heart sound using the matching pursuit method. *Med Biol Eng Comput*, 39(6):644–648.

[Weaver, 1968] Weaver, C. (1968). Digital filtering with application to electrocardiogram processing. *IEEE Trans on Audio and Electroacoustics*, 16(3):350–391.

[Welch, 1938] Welch, B. (1938). The significance of the difference between two means when the population variances are unequal. *Biometrika*, 29:350–362.

[Welch, 1947] Welch, B. (1947). The generalization of "Student's" problem when several different population variances are involved. *Biometrika*, 34(1–2):28–35.

[Welch et al., 1960] Welch, G., Bishop, G., and Kalman, R. E. (1960). A new approach to linear filtering and prediction problems. *Transaction of the ASME–Journal of Basic Engineering*, pages 35–45.

[Wendling et al., 2009] Wendling, F., Ansari-Asl, K., Bartolomei, F., and Senhadji, L. (2009). From EEG signals to brain connectivity: a model-based evaluation of interdependence measures. *J Neurosci Methods*, 183:9–18.

[Wendling et al., 2002] Wendling, F., Bartolomei, F., Bellanger, J. J., and Chauvel, P. (2002). Epileptic fast activity can be explained by a model of impaired GABAergic dendritic inhibition. *Eur J Neurosci*, 15:1499–1508.

[Wheless et al., 1999] Wheless, J. W., Willmore, L. J., Breier, J. I., Kataki, M., Smith, J. R., King, D. W., Meador, K. J., Park, Y. D., Loring, D. W., Clifton, G. L., Baumgartner, J., Thomas, A. B., Constantinou, J. E., and Papanicolaou, A. C. (1999). A comparison of magnetoencephalography, MRI, and V-EEG in patients evaluated for epilepsy surgery. *Epilepsia*, 40:931–941.

[Wiener, 1949] Wiener, N. (1949). *Extrapolation, Interpolation, and Smoothing of Stationary Time Series*. Wiley, New York.

[Wilson et al., 2008] Wilson, J. D., Govindan, R. B., Hatton, J. O., Lowery, C. L., and Preissl, H. (2008). Integrated approach for fetal QRS detection. *IEEE Trans Biomed Eng*, 55:2190–2197.

[Winslow et al., 2009] Winslow, J., Dididze, M., and Thomas, C. K. (2009). Automatic classification of motor unit potentials in surface EMG recorded from thenar muscles paralyzed by spinal cord injury. *J Neurosci Methods*, 185:165–177.

[Winterhalder et al., 2003] Winterhalder, M., Maiwald, T., Voss, H. U., Aschenbrenner-Scheibe, R., Timmer, J., and Schulze-Bonhage, A. (2003). The seizure prediction characteristic: a general framework to assess and compare seizure prediction methods. *Epilepsy Behav*, 4:318–325.

[Winterhalder et al., 2005] Winterhalder, M., Schelter, B., Hesse, W., Schwab, K., Leistritz, L., Klan, D., Bauer, R., Timmer, J., and Witte, H. (2005). Comparison of linear signal processing techniques to infer directed interactions in multivariate neural systems. *Signal Processing*, 85:2137–2160.

[Wit et al., 1994] Wit, H. P., van Dijk, P., and Avan, P. (1994). Wavelet analysis of real ear and synthesized click evoked otoacoustic emissions. *Hear Res*, 73:141–147.

[Woody, 1967] Woody, C. (1967). Characterization of an adaptive filter for the analysis of variable latency neuroelectric signals. *Medical and Biological Engineering and Computing*, 5:539–554.

[Wright et al., 2000] Wright, J., Bourke, P., and Chapman, C. (2000). Synchronous oscillations in the cerebral cortex and object coherence: simulation of basic electrophysiological findings. *Biol Cybern*, 83:341–353.

[Wright, 1989] Wright, J. J. (1989). *Brain Dynamics*, chapter Linearity and nonlinearity in electrocortical waves and their elementary statistical dynamics, pages 201–213. Springer Series in Brain Dynamics 2. Springer Verlag.

[Yamada and Yamaguchi, 2005] Yamada, S. and Yamaguchi, I. (2005). Magnetocardiograms in clinical medicine: unique information on cardiac ischemia, arrhythmias, and fetal diagnosis. *Intern Med*, 44:1–19.

[Ye-Lin et al., 2010] Ye-Lin, Y., Garcia-Casado, J., Prats-Boluda, G., and Martinez-de Juan, J. L. (2010). Combined method for reduction of high frequency interferences in surface electroenterogram (EEnG). *Ann Biomed Eng*, 38(7):2358–2370.

[Yu and Chen, 2007] Yu, S. and Chen, Y. (2007). Electrocardiogram beat classification based on wavelet transformation and probabilistic neural network. *Pattern Recognition Letters*, 28:1142–1150.

[Żebrowski et al., 1994] Żebrowski, J. J., Popawska, W., and Baranowski, R. (1994). Entropy, pattern entropy, and related methods for the analysis of data on the time intervals between heartbeats from 24-h electrocardiograms. *Phys Rev E*, 50(5):4187–4205.

[Zennaro et al., 2003] Zennaro, D., Wellig, P., Koch, V. M., Moschytz, G. S., and Laubli, T. (2003). A software package for the decomposition of long-term multichannel EMG signals using wavelet coefficients. *IEEE Trans Biomed Eng*, 50:58–69.

[Zhang et al. 2006] Zhang, Q., Manriquez, A. I., Medigue, C., Papelier, Y., and Sorine, M. (2006). An algorithm for robust and efficient location of T-wave ends in electrocardiograms. *IEEE Trans Biomed Eng* 53:2544–2552.

[Zhang et al., 1998] Zhang, X., Durand, L. G., Senhadji, L., Lee, H. C., and Coatrieux, J. L. (1998). Analysis-synthesis of the phonocardiogram based on the matching pursuit method. *IEEE Trans Biomed Eng*, 45(8):962–971.

[Zhang et al., 2008] Zhang, Z. G., Zhang, V. W., Chan, S. C., McPherson, B., and Hu, Y. (2008). Time-frequency analysis of click-evoked otoacoustic emissions by means of a minimum variance spectral estimation-based method. *Hear Res*, 243:18–27.

[Zhao et al., 2006] Zhao, H., Strasburger, J. F., Cuneo, B. F., and Wakai, R. T. (2006). Fetal cardiac repolarization abnormalities. *Am J Cardiol*, 98:491–496.

[Zhao and Wakai, 2002] Zhao, H. and Wakai, R. T. (2002). Simultaneity of foetal heart rate acceleration and foetal trunk movement determined by foetal magnetocardiogram actocardiography. *Phys Med Biol*, 47:839–846.

[Zoubir and Boashash, 1988] Zoubir, A. M. and Boashash, B. (1988). The bootstrap and its application in signal processing. *IEEE Sig Proc Mag*, 15:56–76.

[Zwarts et al., 2000] Zwarts, M. J., Drost, G., and Stegeman, D. F. (2000). Recent progress in the diagnostic use of surface EMG for neurological diseases. *J Electromyogr Kinesiol*, 10:287–291.

[Zweig and Shera, 1995] Zweig, G. and Shera, C. A. (1995). The origin of periodicity in the spectrum of evoked otoacoustic emissions. *J Acoust Soc Am*, 98:2018–2047.

[Zygierewicz et al., 1999] Zygierewicz, J., Blinowska, K., Durka, P., Szelenbeger, W., Niemcewicz, S., and Androsiuk, W. (1999). High resolution study of sleep spindles. *Clin Neurophysiol*, 110(12):2136–2147.

[Zygierewicz et al., 2005] Zygierewicz, J., Durka, P. J., Klekowicz, H., Franaszczuk, P. J., and Crone, N. E. (2005). Computationally efficient approaches to calculating significant ERD/ERS changes in the time-frequency plane. *J Neurosci Methods*, 145:267–276.

Index